I0072580

Micromachining Techniques and Applications

Micromachining Techniques and Applications

Edited by **Holly Dunham**

NY RESEARCH
P R E S S

New York

Published by NY Research Press,
23 West, 55th Street, Suite 816,
New York, NY 10019, USA
www.nyresearchpress.com

Micromachining Techniques and Applications
Edited by Holly Dunham

© 2015 NY Research Press

International Standard Book Number: 978-1-63238-326-6 (Hardback)

This book contains information obtained from authentic and highly regarded sources. Copyright for all individual chapters remain with the respective authors as indicated. A wide variety of references are listed. Permission and sources are indicated; for detailed attributions, please refer to the permissions page. Reasonable efforts have been made to publish reliable data and information, but the authors, editors and publisher cannot assume any responsibility for the validity of all materials or the consequences of their use.

The publisher's policy is to use permanent paper from mills that operate a sustainable forestry policy. Furthermore, the publisher ensures that the text paper and cover boards used have met acceptable environmental accreditation standards.

Trademark Notice: Registered trademark of products or corporate names are used only for explanation and identification without intent to infringe.

Printed in the United States of America.

Contents

Preface

The extensive techniques and applications of micromachining are elucidated in this profound book. Micromachining is used to fabricate 3D microstructures and it is the base of a technology called Micro-Electro-Mechanical-Systems (MEMS). Bulk micromachining and surface micromachining are the two most essential categories in this field. This book discusses the latest developments in micromachining technology. It offers reviews on diverse techniques and methods of micro/nano fabrications such as laser ablation, focused ion beams and various specialization techniques, compiled by eminent researchers and scientists around the world. It provides a comprehensive description of particular micromachining techniques, design, analytical studies, experimental paradigms and the final fabricated devices along with illustrative references regarding the various applications of this technique. Due to the multidisciplinary nature of this technology, this compilation will prove helpful to the scientists and researchers in the disciplines of engineering, materials sciences, physics, and chemistry.

This book has been the outcome of endless efforts put in by authors and researchers on various issues and topics within the field. The book is a comprehensive collection of significant researches that are addressed in a variety of chapters. It will surely enhance the knowledge of the field among readers across the globe.

It is indeed an immense pleasure to thank our researchers and authors for their efforts to submit their piece of writing before the deadlines. Finally in the end, I would like to thank my family and colleagues who have been a great source of inspiration and support.

Editor

Focused Ion Beam Based
Three-Dimensional Nano-Machining

Gunasekaran Venugopal[1,2],
Shrikant Saini[1] and Sang-Jae Kim[1,3]
[1]*Jeju National University, Department of Mechanical Engineering, Jeju,*
[2]*Karunya University, Department of Nanosciences and Technology, Tamil Nadu,*
[3]*Jeju National University, Department of Mechatronics Engineering, Jeju,*
[1,3]*South Korea*
[2]*India*

1. Introduction

In recent days, the micro/nano machining becomes an important process to fabricate micro/nano scale dimensional patterns or devices for many applications, especially in electrical and electronic devices. There are two kinds micro-machining in use. i) bulk micro-machining, ii) surface micro-maching. In the case of bulk micromaching, the structures can be made by etching inside a substrate selectively, however, in the case of surface micromachining; the patterns can be made on the top a desired substrate. FIB machining is considered as a one of famous bulk micro-machining processes. Many fabrication methods have been applied to fabricate the devices with smaller sizes (Kim, 1999; Latyshev, 1997). However, conventional until now the size of the smallest pattern was only 2×2 µm² was achieved with a lithography technique (Odagawa et al., 1998). Three dimensional as an alternative approach, focused-ion-beam (FIB) etching technique is the best choice for the micro/nano scale patterning. FIB 3-D etching technology is now emerged as an attractive tool for precision lithography. And it is a well recognized technique for making nanoscale stacked-junction devices, nano-ribbons and graphene based 3-D Single Electron Transistor (SET) devices.

FIB micro/nano machining is a direct etching process without the use of masking and process chemicals, and demonstrates sub-micrometer resolution. FIB etching equipments have shown potential for a variety of new applications, in the area of imaging and precision micromachining (Langford, 2001; Seliger, 1979). As a result, the FIB has recently become a popular candidate for fabricating high-quality micro-devices or high-precision microstructures (Melnagilis et al., 1998). For example, in a micro-electro-mechanical system (MEMS), this processing technique produces an ultra microscale structure from a simple sensor device, such as, the Josephson junction to micro-motors (Daniel et al., 1997). Also, the FIB processing enables precise cuts to be made with great flexibility for micro- and nano-technology. Also, the method of fabricating three-dimensional (3-D) micro- and nano-structures on thin films and single crystals by FIB etching have been developed in order to fabricate the 3-D sensor structures (Kim, 2008, 1999).

In this chapter, the focused ion beam (FIB) based three-dimensional nano-machining will be discussed in detail in which the nano-machining procedures are focused with fabricating nanoscale stacked junctions of layered-structured materials such as graphite, $Bi_2Sr_2Ca_{n-1}Cu_nO_{2n+4+x}$ (BSCCO) family superconductor (Bi-2212, Bi-2223, etc.,) and $YBa_2Cu_3O_7$ (YBCO) single crystals, etc. This work could show a potential future in further development of nano-quantum mechanical electron devices and their applications.

2. Classification of machining

Micromachining is the basic technology for fabrication of micro-components of size in the range of 1 to 500 micrometers. Their need arises from miniaturization of various devices in science and engineering, calling for ultra-precision manufacturing and micro-fabrication. Micromachining is used for fabricating micro-channels and micro-grooves in micro-fluidics applications, micro-filters, drug delivery systems, micro-needles, and micro-probes in biotechnology applications. Micro-machined components are crucial for practical advancement in Micro-electromechanical systems (MEMS), Micro-electronics (semiconductor devices and integrated circuit technology) and Nanotechnology. This kind of machining can be applicable for the bulk materials in which the unwanted portions of the materials can be removed while patterning.

In the bulk machining, the materials with the dimensions of more than in the range of micrometer or above centimetre scale are being used for the machining process. A best example for the bulk machining process is that the thread forming process on a screw or bolt, formation of metal components. Also this process can be applicable to produce 3D MEMS structures, which is now being treated as one of older techniques. This also uses anisotropic etching of single crystal silicon. For example, silicon cantilever beam for atomic force microscope (AFM).

Surface micro-machining is another new technique/process for producing MEMS structures. This uses etching techniques to pattern micro-scale structures from polycrystalline (poly) silicon, or metal alloys. Example: accelerometers, pressure sensors, micro gears and transmission, and micro mirrors etc. Micromachining has evolved greatly in the past few decades, to include various techniques, broadly classified into mask-based and tool-based, as depicted in the diagram below.

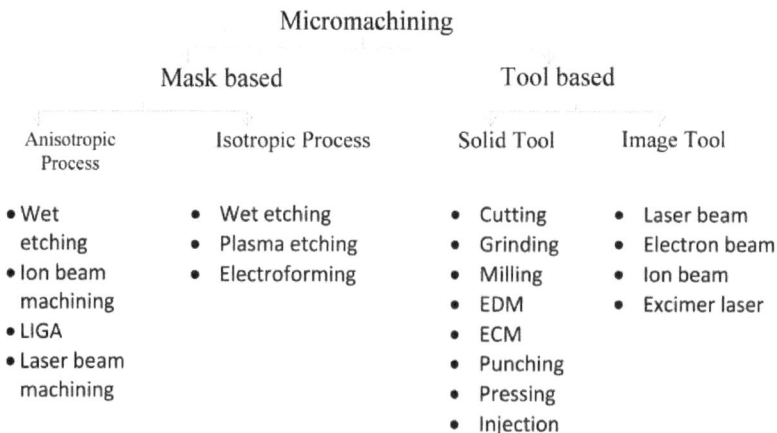

Micromachining

Mask based   Tool based

Anisotropic Process	Isotropic Process	Solid Tool	Image Tool
• Wet etching	• Wet etching	• Cutting	• Laser beam
• Ion beam machining	• Plasma etching	• Grinding	• Electron beam
• LIGA	• Electroforming	• Milling	• Ion beam
• Laser beam machining		• EDM	• Excimer laser
		• ECM	
		• Punching	
		• Pressing	
		• Injection	

While mask-based processes can generate 2-D/2.5-D features on substrates like semiconductor chips, tools-based processes have the distinct advantage of being able to adapt to metallic and non-metallic surfaces alike, and also generate 3-D features and/or free-form sculpted surfaces. However, the challenges of achieving accuracy, precision and resolution persist.

Internationally, the race to fabricate the smallest possible component has lead to realization of sizes ever below 10 μm, even though the peak industrial requirement has been recognized at 100s of μm. Thus, the present situation is particularly advantageous for the industry that develops/fabricates nano/micron scale components.

2.1 Various techniques of micromachining

Micromachining can be done by following various techniques.
a.   Photolithography
b.   Etching
c.   LIGA
d.   Laser Ablation
e.   Mechanical micromachining

Photolithography

This technique is being used in microelectronics fabrication and also used to pattern oxide/nitride/polysilicon films on silicon substrate. In this process, the basic steps involved are, photoresist development, etching, and resist removal. Photolithographic process can be described as follows:

The wafers are chemically cleaned to remove particulate matter, organic, ionic, and metallic impurities. High-speed centrifugal whirling of silicon wafers known as "Spin Coating" produces a thin uniform layer of photoresist (a light sensitive polymer) on the wafers. Photoresist is exposed to a set of lights through a mask often made of quartz. Wavelength of light ranges from 300-500 nm (UV) and X-rays (wavelengths 4-50 Angstroms). Two types of photoresist are used: (a) Positive: whatever shows, goes (b) Negative: whatever shows, stays. The photo resist characteristics after UV exposure are shown below in Fig. 1

UV Exposure Radiation

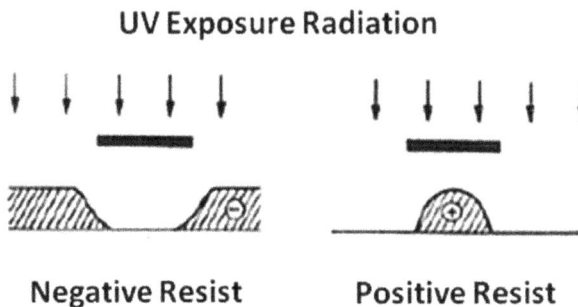

Negative Resist   **Positive Resist**

Fig. 1. Photoresist characteristics in UV exposure

Etching

Normally etching process can be classified in to two kinds. (a) Wet etching (b) Dry etching. The wet etching process involves transport of reactants to the surface, surface reaction and transport of products from surfaces. The key ingredients are the oxidizer (e.g. H_2O_2, HNO_3), the acid or base to dissolve the oxidized surface (e.g. H_2SO_4, NH_4OH) and dilutent media to transport the products through (e.g. H_2O). Dry etching process involves two kinds. (a) plasma based and (b) non plasma based.

LIGA

The LIGA is a German term which means LIthographie (Lithography) **G**alvanoformung (Electroforming) **A**bforming (Molding). The exact English meaning of LIGA is given in parenthesis. This process involves X-ray irradiation, resist development, electroforming and resist removal.

The detailed LIGA process description is discussed below:

- Deep X-ray lithography and mask technology
  - Deep X-ray (0.01 – 1nm wavelength) lithography can produce high aspect ratios (1 mm high and a lateral resolution of 0.2 μm).
  - X-rays break chemical bonds in the resist; exposed resist is dissolved using wet-etching process.
- Electroforming
  - The spaces generated by the removal of the irradiated plastic material are filled with metal (e.g. Ni) using electro-deposition process.
  - Precision grinding with diamond slurry-based metal plate used to remove substrate layer/metal layer.
- Resist Removal
  - PMMA resist exposed to X-ray and removed by exposure to oxygen plasma or through wet-etching.
- Plastic Molding
  - Metal mold from LIGA used for injection molding of MEMS.

LIGA Process Capability

- High aspect ratio structures: 10-50 μm with Max. height of 1-500 μm
- Surface roughness < 50 nm
- High accuracy < 1μm

Laser ablation

High-power laser pulses are used to evaporate matter from a target surface. In this process, a supersonic jet of particles (plume) is ejected normal to the target surface which condenses on substrate opposite to target. The ablation process takes place in a vacuum chamber - either in vacuum or in the presence of some background gas. The graphical process scheme is given below in Fig.2.

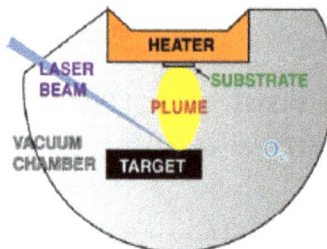

Fig. 2. Laser ablation experiment.

Mechanical micromachining

Lithography or etching methods are not capable of making true 3D structures e.g. free form surfaces and also limited in range of materials. Mechanical machining is capable of making free form surfaces in wide range of materials. Can we scale conventional/non-traditional machining processes down to the micron level? Yes! There are two approaches used to machine micron and sub-micron scale features.

1. Design ultra precision (nanometer positioning resolution) machine tools and cutting tools. For this, ultra precision diamond turning machines can be used.
2. Design miniature but precise machine tools
   Example: Micro-lathe, micro-mill, micro-EDM, etc

Mechanical micromachining process descriptions are given below:

* Can produce extremely smooth, precise, high resolution true 3D structures
* Expensive, non-parallel, but handles much larger substrates
* Precision cutting on lathes produces miniature screws, etc with 12 μm accuracy
* Relative tolerances are typically 1/10 to 1/1000 of feature
* Absolute tolerances are typically similar to those for conventional precision machining (Micrometer to sub-micrometer)

2.2 Focused-ion-beam (FIB) technique for nanofabrication

The focused ion beam based nanofabrication method can be followed for the fabricating the nanoscale devices on materials based on metal and non-metallic elements, particularly the layered structure materials like graphite, Bi-2212 and YBCO which are recently attracted the world scientific community due to their interesting electrical and electronic properties reported in recent reports (Venugopal, 2011; Kim, 2001).

Graphite is considered as a well known layered-structured material in which carbon sheets are arranged in a stacked-manner with interlayer distance of 0.34 nm. Each single graphite sheet is known as a graphene layer which is now becoming as one of hot topic in the world scientific community. In the recent reports (Venugopal, 2011a, 2011b, 2011c), the fabrication of submicron and below submicron stacked junctions were carved from the bulk graphite materials using FIB 3-D etching. The interesting results were obtained in those observations that the graphite stacked-junction with in-plane area A of 0.25 μm^2 showed nonlinear concave-like I–V characteristics even at 300 K; however the stack with A ≥ 1 μm^2 were shown an ohmic-like I–V characteristic at 300 K for both low and high-current biasing. It turned into nonlinear characteristics when the temperature goes down. These results may open road to develop further graphite based nonlinear electronic devices. Further researches are being carried out to find unexplored properties of graphite nano devices fabricated using FIB micro/nano machining technology.

The focused ion beam (FIB) machining to make micro-devices and microstructures has gained more and more attention recently (Tseng, 2004). FIB can be used as a direct milling method to make microstructures without involving complicated masks and pattern transfer processes. FIB machining has advantages of high feature resolution, and imposes no limitations on fabrication materials and geometry. Focused ion beams operate in the range of 10-200 keV. As the ions penetrate the material, they loose their energy and remove substrate atoms. FIB has proven to be an essential tool for highly localized implantation doping, mixing, micromachining, controlled damage as well as ion-induced deposition. The technological challenge to fabricate nanoholes using electron beam lithography and the minimal feature size accessible by these techniques is typically limited to tens of nanometers, thus novel procedures must be devised (Zhou, 2006).

The patterning of samples using the FIB (focused ion beam) technique is a very popular technique in the field of inspection of integrated circuits and electronic devices manufactured by the semi-conductor industry or research laboratories. This is the case mainly for prototyping devices. The FIB technique allowing us to engrave materials at very low dimensions is a complement of usual lithographic techniques such as optical lithography. The main difference is that FIB allows direct patterning and therefore does not require an intermediate sensitive media or process (resist, metal deposited film, etching process). FIB allows 3D patterning of target materials using a finely focused pencil of ions having speeds of several hundreds of km s^{-1} at impact. Concerning the nature of the ions most existing metals can be used in FIB technology as pure elements or in the form of alloys, although gallium (Ga$^+$ ions) is preferred in most cases.

Many device fabrication techniques based on electron beam lithography followed by reactive-ion etching (RIE), chemical methods, and evaporation using hard Si shadow masks, and including lithography-free fabrication, have been reported. The procedures, however, are complex and yield devices with dimensions of ~5 to 50 nm, which are restricted to simple geometries. RIE creates disordered edges, and the chemical methods produce irregular shapes with distributed flakes, which are not suitable for electronic-device application.

Practically, FIB patterning can be achieved either by local surface defect generation, by ion implantation or by local sputtering. These adjustments are obtained very easily by varying the locally deposited ion fluence with reference to the sensitivity of the target and to the selected FIB processing method (Gierak, 2009). The FIB milling involves two processes: 1) Sputtering, ions with high energy displace and remove atoms of substrate material, and the ions lose their energy as they go into the substrate; 2) Re-deposition, the displaced substrate atoms, that have gained energy from ions through energy transfer, go through similar process as ions, sputtering other atoms, taking their vacancy, or flying out.

A focused gallium ion beam having an energy typically around 30 keV is scanned over the sample surface to create a pattern through topographical modification, deposition or sputtering. A first consequence is that, mainly because of the high ion doses required (~10^{18} ions cm^{-2}) and of the limited beam particle intensity available in the probe, FIB etching-based processes remain relatively slow. We may recall that for most materials, the material removal rate for 30 keV gallium ions is around 1–10 atoms per incident ion, corresponding to a machining rate of around 0.1–1 μm^3 per nC of incident ions (Gierak, 2009). The second consequence is that for most applications the spatial extension of the phenomena induced by focused ion beam irradiation constitutes a major drawback.

In addition, there have been few reports of the fabrication of nano-structured materials, nano devices, and hierarchical nano-sized patterns with a 100 nm distance using a focused ion beam (FIB). Fabrication of graphene nanoribbons and graphene-based ultracapacitors were also reported recently. The above-discussed methods were followed by the two-dimensional (2D) fabrication methods and required extensive efforts to achieve precise control. Hence, a novel three-dimensional (3D) nanoscale approach to the fabrication of a stack of graphene layers via FIB etching is proposed, through which a thin graphite flake can be etched in the c-axis direction (stack height with a few tens of nanometers). Also the main purpose of describing graphite and other BSCCO based superconducting nanoscale devices is that these layered structured materials have shown an excellent device structures during fabrication and their electrical transport characteristics were interesting which will be useful to future works.

2.2.1 Nanoscale stack fabrication by focused-ion-beam

Using an FIB, perfect stacks can be fabricated more easily along the c-axis in thin films and single-crystal whiskers. FIB 3D etching has been recognized as a well-known method for fabricating high-precision ultra-small devices, in which etching is a direct milling process that does not involve the use of any masking and process chemicals and that demonstrates a submicrometer resolution. Thus, these our proposal is focused on the fabrication of a nanoscale stack from the layered structured materials like thin graphite flake and BSCCO, via FIB 3D etching. The detailed schematic of fabrication process is shown in Fig. 3.

The 3D etching technique is followed by tilting the substrate stage up to 90° automatically for etching thin graphite flake. We have freedom to tilt the substrate stage up to 60° and rotate up to 360°. To achieve our goal, we used sample stage that itself inclined by 60° with respect to the direction of the ion beam (fig 3a). The lateral dimensions of the sample were 0.5×0.5 µm². The in-plane area was defined by tilting the sample stage by 30° anticlockwise with respect to the ion beam and milling along the *ab*-plane.

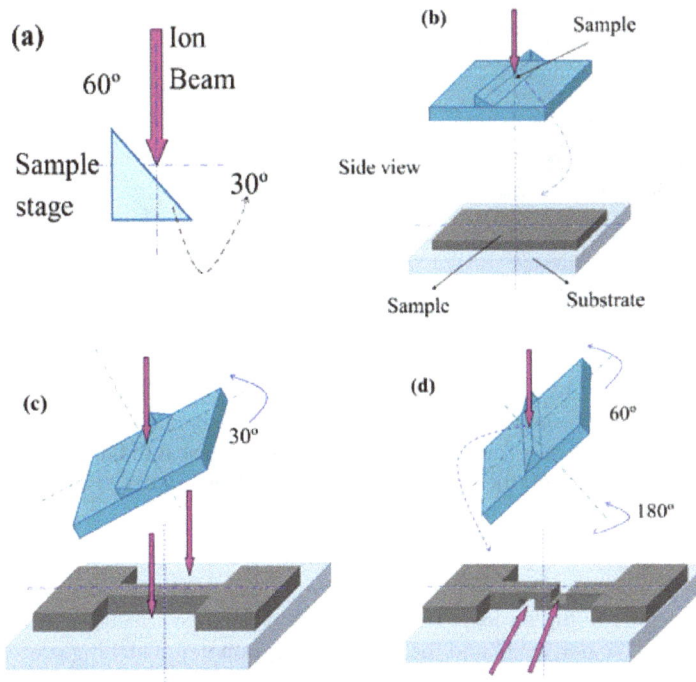

Fig. 3. FIB 3-D fabrication process (a) Scheme of the inclined plane has an angle of 60° with ion beam (where we mount sample). (b) The initial orientation of sample and sample stage. (c) Sample stage titled by 30° anticlockwise with respect to ion beam and milling along ab-plane. (d) The sample stage rotated by an angle of 180° and also tilted by 60° anticlockwise with respect to ion beam and milled along the c-axis.

The in-plane etching process is shown in Fig. 3(a)–(c). The out of plane or the c-axis plane was fabricated by rotating the sample stage by an angle of 180°, then tilting by 60° anticlockwise with respect to the ion beam, and milling along the c-axis direction. The schematic diagram of the fabrication process for the side-plane is shown in Fig. 3(d). The dimensions of the side-plane was W=0.5 μm, L=0.5 μm, and H=200 nm. The c-axis height length (H) of the stack was set as 200 nm. An FIB image of fabricated stack is shown in Fig. 4 in which the schematic of stack arrangement (graphene layers with interlayer distance 0.34 nm) was also shown in the inset (top right) in Fig. 4. The vertical red arrow indicates the current flow direction through the stack.

2.2.2 Transport characteristics of nanoscale graphite stacks

The electrical transport characteristics (including ρ-T and I-V) can be performed for the fabricated stack using closed-cycle refrigerator systems (CKW-21, Sumitomo) at various temperatures from 25 to 300 K with the use of the Keithley 2182A nanovoltmeters and AC & DC current source (6221). The I-V characteristics of the fabricated stack are shown in Fig.4.

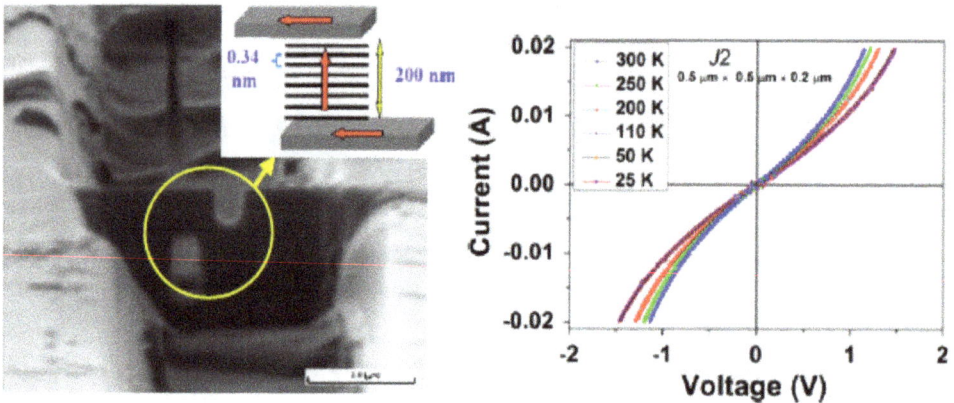

Fig. 4. FIB image of the nanoscale stack fabricated on a thin graphite flake along the c-axis height of 200 nm (image scale bar is 2 μm). Inset shows the schematic diagram of stack arrangement along the c-axis. (Venugopal et al, 2011). The vertical red arrow indicates the current flow direction through the nanoscale stack. I-V characteristics at various temperatures of the fabricated nanostack are also shown (right).

The FIB ion damage effect can be avoided if the device is fabricated at a 3D angle, in which the top layer of ab-plane will act as a masking layer and the ion beam is exactly perpendicular to the milling surface. The expected ion damage effect was simulated using the TRIM software (Ziegler, 1996) and the fabrication parameter of etching process for the 30 keV Ga⁺ ions was optimized. It was found from the simulation results that the depth of ion implantation is consistent with 10 nm. Majority (>95%) of the Ga⁺ ions are expected to be implanted within 10 nm of the side walls of stack surface, with a much smaller fraction, eventually stopping at as deep as 10 nm into the surface. Therefore, the proportion of the fabricated stack affected by ion beam damage is not very large, and it does not affect the quality of graphite devices in the c-axis direction.

By varying in-plane area (A) and stack height (H), several stacked-junctions with the dimensions of W = 1 μm, L = 1 μm, and H = 0.1 μm (denoted as J4) and W = 2 μm, L = 1 μm, and H = 0.3 μm (denoted as J5) were fabricated. The electrical transport characteristics were performed for these stacks and compared their results. The current-voltage (I-V) characteristics of the nanostack with in-plane area (A) of 0.25 μm² (J2) at various temperatures, are presented in Fig. 4. The stack showed a nonlinear concave-like I-V characteristics at all studied temperatures (25, 50, 110, 200, 250 and 300 K). At 300 K, the stack resistance was found as 75 Ω. The stack resistance found increases when the temperature goes down to 25 K.

The electrical characteristics of nanostack (J2) were analyzed and compared with bigger junctions J4 (1 × 1 × 0.1 μm³) and J5 (2 × 1 × 0.3 μm³). From the data analysis, it is clear that the stack with larger height and reduced in-plane effective area (A) has shown higher resistance than the stack with larger in-plane area (A). The I-V characteristics of junctions J4 and J5 at different temperatures are shown in Fig. 5 (a) and (b) respectively. A typical c-axis transport characteristics similar to junction J2 was observed. However the nonlinear I-V characteristics were not observed at 300 K, but ohmic like-linear behavior is observed. When the temperature goes down, this behavior is turned into curve-like nonlinear characteristics.

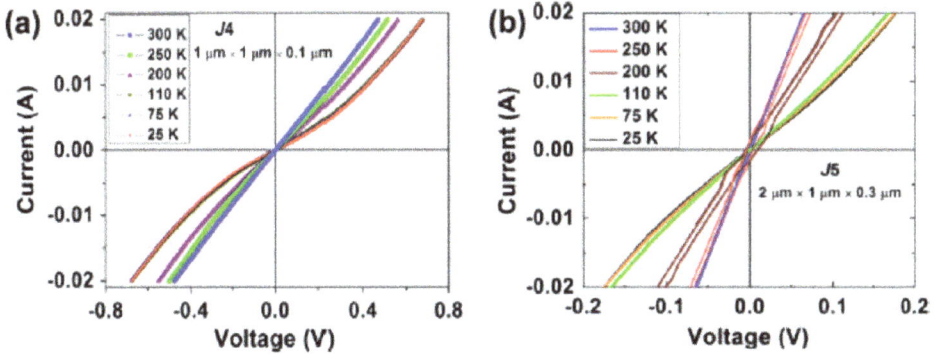

Fig. 5. (a) I–V characteristics of a bigger stacked-junction with A of 1 μm² (J4) at different temperature from 25 K to 300 K. (b) I–V characteristics of another bigger junction with A of 2 × 1 × 0.3 μm³ (J5) at different temperature from 25 K to 300 K. Both the junctions show ohmic like behavior at 300 K; however the same behavior turned into nonlinear characteristics when the temperature goes down (Venugopal et al, 2011).

There is a significant overlap of I-V curves for temperatures 110, 75 and 25 K. For graphite stacks with A ≥ 1 μm², there was no nonlinear I-V characteristics observed at 300 K even at high biasing. With a decrease of the stack size down to 0.25 μm², the junction shows clear nonlinear concave-like I-V characteristics for both 300 K and 25 K. Since the fabricated stack contains multiple elementary junctions along the c-axis, the nonlinear concave-like tunneling characteristics appeared from the I-V characteristics (Venugopal et al, 2011).

2.2.3 Temperature dependent resistivity of nanoscale graphite stack

Fig 6 represents the ρ–T characteristics of stacked-junction (J2). The junction J2 shows a semiconducting behavior for T > 65 K and metallic characteristics for T < 65 K. Above 65 K, thermal excitation of carriers plays a major role in semiconducting temperature dependence. However below 65 K, the interlayer hopping conduction combined with scattering of carriers by phonons can be responsible for the metallic-like temperature dependence. The ρ–T characteristics along the ab-plane transport are shown as inset in Fig. 6. A well understood metallic behavior was observed. This behavior is well agreed with earlier observations on c-axis characteristics of bulk graphite material (Matsubara, 1990).

An electron motion parallel to its plane is not affected by the stacking faults, however, but an electron motion in the c-axis direction is strongly impeded by the faults. The combined effects of impurity-assisted hopping, tunneling current, and the thermal excitation of the carriers on the plane of a stack play important roles in this temperature-dependent conduction mechanism in layered structured materials such as graphite.

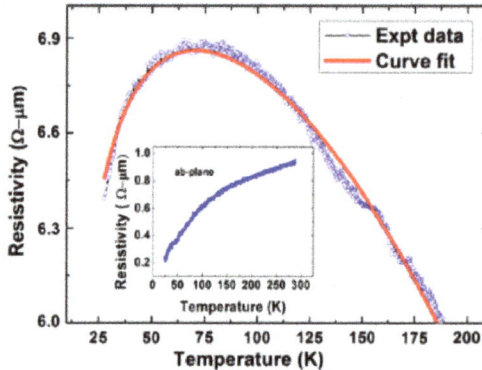

Fig. 6. The resistivity–temperature (ρ-T) characteristics of nanostack which shows a clear c-axis characteristics of graphite. A well agreed curve fitting to experimental data is also shown. A clear metallic behavior is observed for ab-plane transport of bare graphite flake which is shown as inset. (Venugopal, 2011)

2.3 FIB nano fabrication on superconducting devices

Considering Bi-family as a layered structure material, there are three compounds in the Bi-family high-temperature superconductors, differing in the type of planar CuO_2 layers; single-layered $Bi_2Sr_2CuO_{6+\delta}$ (Bi-2201) single crystal, double-layered $Bi_2Sr_2CaCu_2O_{8+\delta}$ (Bi-2212) single crystal, and triple-layered $Bi_2Sr_2Ca_2Cu_3O_{10+\delta}$ (Bi-2223) single crystal (Saini, 2010). This Bi-family material is a one of the famous emerging material for electron tunneling devices, such as intrinsic Josephson junctions (IJJ) in layered high-Tc superconductors. The spacing of consecutive copper-oxide double planes in the most anisotropic cuprate superconductors is greater than the coherence length in the out-of-plane c-direction. When a current flows along the c-direction in such a material, it therefore flows through a series array of "intrinsic" Josephson junctions (IJJs) (Kleiner, 1992). These junctions and junction arrays are showing promise for a wide variety of applications, including as voltage standards and sub-mm-wave oscillators (Wang, 2001). For sub-micron intrinsic junctions, there is an additional range of potential applications exploiting the Coulomb blockade effect, when the E_c is charging energy $E_c \geq E_J$, K_BT, where E_J is the Josephson energy & k_BT is thermal energy. These applications include electric-field sensors and quantum current standards (Bylander, 2005). In long arrays of junctions, E_c is enhanced by electron-electron interactions (Likharev, 1989, 1995) by a factor $[C/C_0]^{1/2}$, where C is the junction capacitance and C_0 is the stray capacitance to ground. The large ratio $C/C_0 \sim 10^6$ for intrinsic junctions makes them particularly suited to the applications involving Coulomb blockade effects. The features of the single Cooper-pair tunneling effect from the layered structure of Bi-family as well as for YBCO will also be discussed in detail.

Superconductivity is a phenomenon when the resistance of the material becomes zero and it expels all the magnetic field below a certain temperature usually at very low temperature. The phenomenon of superconductivity was discovered in 1911 by the Dutch physicist H. Kamerlingh Onnes. The quantum application of superconductivity was introduced in 1962. B. D. Josephson discovered a tunnel junction consists of two strips of superconductors separated by an insulator where the insulator is so thin that electrons can tunnel through it known as Josephson junction.

The schematic of different types of Josephson junctions are shown below in Fig.7. S stands for superconductor, S' for a superconductor above T_c, N for normal metal, Se for semiconductor, and I for an insulator.

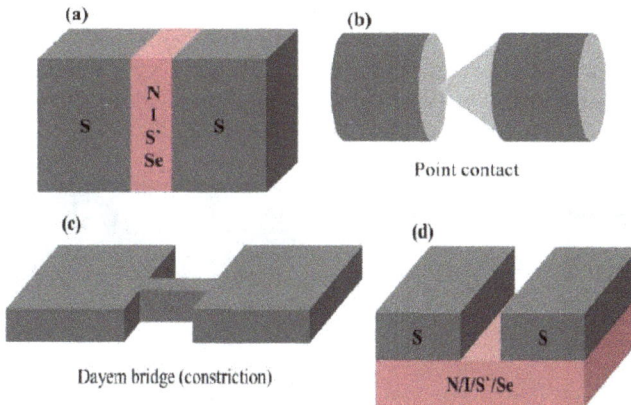

Fig. 7. The schematics of different types of superconducting devices.

The term high-temperature superconductor was first introduced in 1986 to designate the new family of cuprate-perovskite ceramic materials discovered by Johannes George Bednorz and Karl Alexander Müller [J. G. Bednorz, K. A. Mueller (1986) "Possible high T_C superconductivity in the Ba-La-Cu-O system", *Zeitschrift für Physik B* 64 (2) 189–193 doi:10.1007/BF01303701] for which they won the Nobel Prize in Physics in the following year. Their discovery of the first high-temperature superconductor, LaBaCuO, with a transition temperature of 30 K, generated great excitement. In 1988, BSCCO ($Bi_2Sr_2Ca_{n-1}Cu_nO_{2n+4+x}$, with $n=2$ being the most commonly studied compound, though $n=1$ and $n=3$ have also received significant attention) as a new class of superconductor was discovered by Maeda and coworkers [H. Maeda, Y. Tanaka, M. Fukutumi, and T. Asano (1988) "A New High-T_c Oxide Superconductor without a Rare Earth Element" *Jpn. J. Appl. Phys.* 27 (2) L209–L210. doi:10.1143/JJAP.27.L209.] at the National Research Institute for Metals in Japan, though at the time they were unable to determine its precise composition and structure. The discovery of these high temperature superconductors gave a path for the application of the superconductivity at higher temperature.

2.3.1 FIB nanomachining of Intrinsic Josephson Junctions (IJJs) on BSCCO and Y123/Pr123 multilayered thin films

Many fabrication methods based on high-resolution patterning have been applied to develop high-Tc superconducting devices. Very small structures are needed in the fabrication of tunneling devices, such as intrinsic Josephson junctions (IJJ) in layered high Tc superconductors $Bi_2Sr_2CaCu_2O_{8+\delta}$ (Bi-2212). Perfect stacks are more easily obtained in c-axis high-quality thin films than in a-axis films or single-crystal whiskers. However, the IJJ fabrication process using c-axis thin films and single crystals requires intricate processes and limits the junction size in mesa structures.

As per previous reports, the fabrication of IJJs by the focused ion beam (FIB) etching method using single-crystal whiskers as a base material requires some complicated processes, including turning over of the sample. As an alternative approach, in this chapter, a three-dimensional IJJ fabrication method is presented using c-axis thin films. The fabrication steps using c-axis single crystal are also simplified by the *in situ* process. Here, the 3D FIB etching methods using YBCO thin films and Bi-2212 single-crystal whiskers were described as examples with a successive decrease of their in-plane area, S, down to a submicron scale. Also, there was a possibility to identify the features of the single Cooper-pair tunneling effect from the layered structure of Bi-2212 with very narrow interval between layers.

FIB image of a submicron stack fabricated on Bi-2212 single crystal whiskers with in-plane area of 0.4 μm × 0.4 μm and schematic of the IJJs configuration are shown in Fig. 8, in which FIB fabrication procedures followed as described in section 2.1.2.

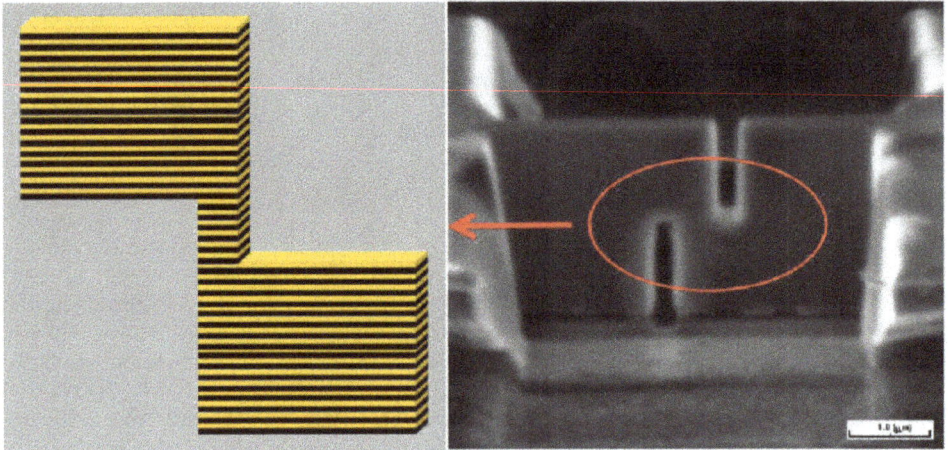

Fig. 8. FIB image of a submicron stack fabricated on Bi-2212 single crystal whisker. The red color circular part shown in stack contains many IJJs.

FIB image of submicron stack (scale bar of 1 μm) and schematic of the Josephson junctions configuration in the submicron stack fabricated on a-axis oriented $YBa_2Cu_3O_7/PrBa_2Cu_3O_7$ multi layered thin films are shown in Fig.9. The arrow indicates the direction of current to observe the effect of Josephson junctions. The axial direction of thin film is shown in the expended view.

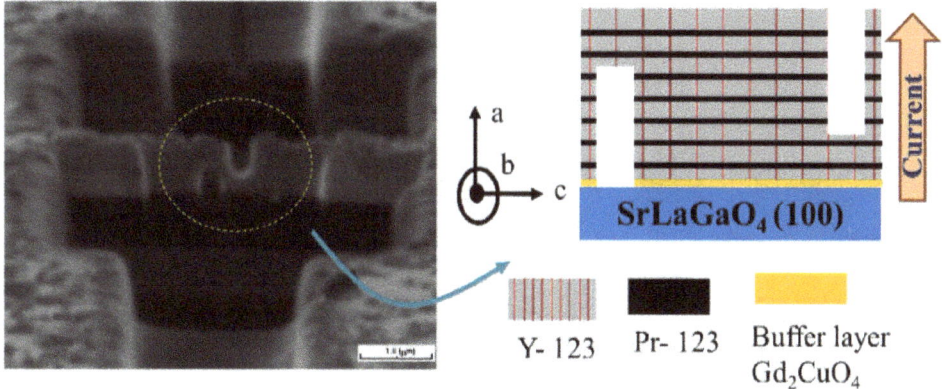

Fig. 9. FIB image of a submicron stack fabricated on a-axis oriented Y123/Pr123 multi layered thin films.

2.3.2 Electrical transport characteristics of Josephson junctions fabricated on multi layered thin films of Y123/Pr123

Fig. 10 represents R-T characteristics of the Josephson junctions fabricated on multi layered thin films of Y123/Pr123 which shows T_c about 83 K.

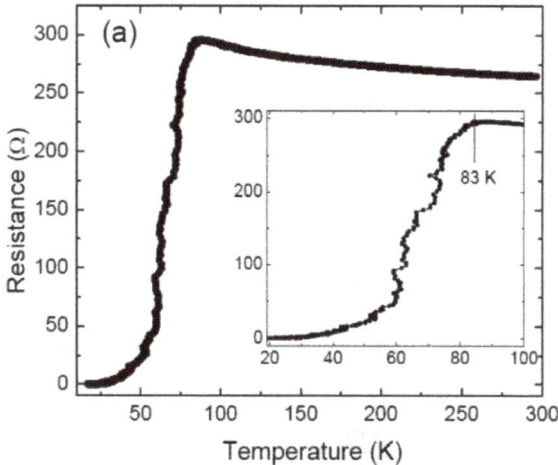

Fig. 10. R-T characteristics of the device show T_c about 83 K.

I-V characteristics of the same device were studied without microwave irradiation at different temperature of 10, 20, and 30 K, shown in Fig. 11. As temperature decreases, the critical current density of superconducting device is increases gradually.

The above discussed nanomachining/milling techniques followed by focused ion beam 3-D technique shall be applicable to other layered-structured materials rather than graphite flake, BSCCO, YBCO and multilayered thin films, etc,. This may have great potential in future nanodevice development and applications.

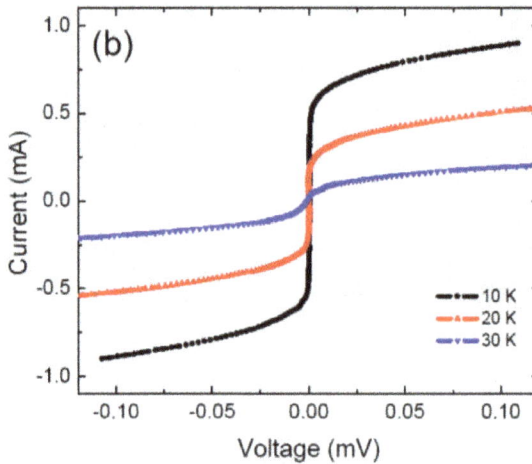

Fig. 11. *I-V* characteristics of the device without microwave irradiation at different temperature of 10, 20, and 30 K. The critical current density J_c about 2.2 X 10^5 A/cm^2 is measured at 20 K.

2.4 Future advances

In the future, micromachining is destined to improve upon its shortcomings, as the various micromachining processes become accurate, reliable, versatile and cost-effective. In India, BARC has established premier micromachining and nano-finishing facilities along with state-of-the art metrology systems. On the other hand, IIT Bombay has also taken a lead in establishing tool-based micromachining facilities. Even at South Korea, the technology towards nanomachining becomes popular nowadays and the active research is now under progress through which an interesting studies may be explored in near future.

FIB technology is still relatively young compared with other semiconductor fabrication processes. One of the major challenges for all of the microfabrication and nanofabrication technologies is to downscale the feature size while maintaining a high throughput. To increase the throughput and the ability to be used in production, the milling rate of the existing FIB milling systems has to be improved. A variable-diameter beam system should be developed to provide multi-resolution milling to cope with different accuracy or tolerance requirements. It is ideal that the beam diameter can be continuously changed in situ. (Tseng, 2004). This type of system has been available for many macro-scale fabrication processes. With this system, a larger beam can be used for roughing 'cut' (milling) to increase the milling rate in regions where only lower resolution is needed. The advantages to use a heavy-duty two-lens system with improved automation should be examined with the goal to develop a system for limited production usage first. Once the high-performance FIB system is used in production, it can be a vital candidate to become the mainstream tool for the future microtechnology and nanotechnology industry.

With an increasing awareness about the advantages of manufacturing micro-components indigenously instead of importing at high costs, the researchers and industrialists are in need of the knowledge of micromachining technology.

3. Conclusion

In conclusion, the focused ion beam based nanomachining have been discussed in detail for the layered structured materials, BSCCO superconducting devices, YBCO based thin film devices, and a-axis oriented Y123/Pr123 multi layered thin film devices. The development of focused ion beam technology based nanomachining is one amongst many examples on how research results may have found unexpected applications in totally different application areas. This is particularly true for the FIB technology development itself that has benefited from all the previously made advances in field emission physics, charged particle optics theory or modelling and in fundamental instrumentation or applied metrology. All these advances were very quickly and efficiently integrated into FIB instruments, so that in less than one decade FIB instruments have moved out from some specialist laboratories to enter almost every modern laboratory, research institute or processing environment. This is also true for the semiconductor industry that has been almost immediately applying FIB systems for device inspection failure analysis and reverse engineering with roaring success. The FIB processing methods which we discussed in this chapter, appear now to be well suited and very promising for several diverse nanotechnology applications, and may be of major interest for future applications to spin-electronics, nano-electronics, nano-optics or nanomagnetism.

4. Acknowledgment

This research was supported by National Research Foundation of Korea Grant under contract numbers 2009-0087091 and 2011-0015829 through the Human Resource Training Project for Regional Innovation. A part of this research was also supported by the 2012 Jeju Sea Grant College Program funded by Ministry of Land, Transport and Maritime Affairs, Republic of Korea

5. References

Bylander, J. (2005). Current Measurement by Real-time Counting of Single Electrons.. *Nature.*, Vol.434, pp. 361-364.

Daniel, J. H. (1997). Focused Ion Beams in Microsystem Fabrication. *Microelectron. Eng.*, Vol.35, No.1-4, pp. 431-434.

Gierak, J. (2009). Focused Ion Beam Technology and Ultimate Applications. *Semicond. Sci. Technol.*, Vol.24, pp. 043001-043022.

Kim, S. J. (2001). Fabrication and Characteristics of Submicron Tunnelling Junctions on High Tc Superconducting c-axis Thin Films and Single Crystals. *J. Appl. Phys.*, Vol.89, No.11, pp. 7675-7677.

Kim, S. J. (1999). Submicron Stacked-junction Fabrication from $Bi_2Sr_2CaCu_2O_{8+\delta}$ Whiskers by Focused-Ion-Beam Etching . *Appl. Phys. Lett.*, Vol.74, No.8, pp. 1156-1158.

Kim, S. J. (2008). Development of Focused Ion Beam Machining Systems for Fabricating Three-dimensional Structures. *Jpn. J. Appl. Phys.*, Vol.47, No.6, pp. 5120-5122.

Kim, S. J. (1999). 3D intrinsic Josephson junctions using c-axis thin films and single crystals. *Supercond. Sci. Technol.*, Vol.12, pp. 729-731.

Kleiner, R. (1992). Intrinsic Josephson Effects in BiSrCaCuO Single-Crystals. *Phys. Rev. Lett.*, Vol.68, pp. 2394-2396.

Latyshev, Y. I. (1997). Intrinsic Josephson Effects on Stacks Fabricated from High Quality BSCCO 2212 Single Crystal Whiskers. *Physica C.*, Vol.293, pp. 174-180.

Langford, R. M. (2001). Preparation of Site Specific Transmission Electron Microscopy Plan-view Specimens using a Focused Ion Beam System. *J. Vac. Sci. Technol. B.*, Vol.19, No.755. doi:10.1116/1.1371317.

Likharev, K. K. (1989). Single-electron Tunnel Junction Array: An Electrostatic Anolog of the Josephson Transmission Line. *IEEE Trans. Mag.*, Vol.25, pp. 1436. DOI: 10.1109/20.92566.

Likharev, K. K. (1995). Electron-electron Interaction in Linear Arays of Small Tunnel Junctions. *Appl. Phys. Lett.*, Vol.67, pp. 3037-3039.

Matsubara, K. (1990). Electrical Resistance in c-direction of Graphite. *Phys. Rev. B.*, Vol.41, pp. 969-974.

Melnagilis, J. (1998). A Review of Ion Projection Lithography. *J. Vac. Sci. Technol. B.*, Vol.16, No.3, pp. 927-957.

Odagawa, A. (1998). Characteristics of Intrinsic Josephson Junctions in Thin Stack on Bi-2223 Thin Films. *Jpn. J. Appl. Phys.*, Vol.37, No.1, pp. 486-491.

Saini, S. (2010). Characterization of Submicron Sized Josephson Junction Fabricated in a $Bi_2Sr_2Ca_2Cu_3O_{10+\delta}$ (Bi-2223) Single Crystal Whisker. *J. Supercond. Nov. Magn.* Vol.23, pp. 811-813.

Seliger, R. L. (1979). High-resolution, Ion-beam Processes for Microstructure Fabrication. *J. Vac. Sci. Technol. B.*, Vol.16, No.6, pp.1610-1612.

Tseng, A. A. (2004). Recent Developments in Micromilling using Focused Ion Beam Technology. *J. Micromech. Microengg.*, Vol.14, pp. R15-R34.

Venugopal, G. (2011a). Fabrication of Nanoscale Three-dimensional Graphite Stacked-junctions by Focused-ion-beam and Observation of Anomalous Transport Characteristics. *Carbon*, Vol.49, No.8, pp. 2766-2772.

Venugopal, G. (2011b). Temperature Dependence of Transport Anisotrophy of Planar-type Graphite Nano-structures Fabricated by Focused Ion Beam. *J. Nanosci. Nanotechnol.* Vol.11, No.1, pp. 296-300.

Venugopal, G. (2011c). Fabrication and Characteristics of Submicron Stacked-Junctions on Thin Graphite Flakes. *J. Nanosci. Nanotechnol.* Vol.11, No.2, pp. 1405-1408.

Wang, H. B. (2001). Terahertz Responses of Intrinsic Josephson Junctions in High Tc Superconductors. *Phys. Rev. Lett.*, Vol.87, pp. 107002-107005.

Ziegler, J. F; Biersack, J. P. & Littmark, U. (1996). *The Stopping and Range of Ions in Solids*, Pergamon, New York.

Zhou, J.; Yang, G. (2006). *Proceedings of the 7th ICFDM 2006 International Conference on Frontiers of Design and Manufacturing*, pp. 453-458, Guangzhou, China, June 19-22, 2006

2

Fundamentals of Laser Ablation of the Materials Used in Microfluidics

Tai-Chang Chen and Robert Bruce Darling
University of Washington,
USA

1. Introduction

Microfluidics falls into an intermediate range within the spectrum of applications for microfabrication techniques. The width and depth of most microfluidic channels fall in the range of 10-1000 μm, and this feature size is thus small for conventional machine tool microfabrication, but quite large for photolithographically defined etching processes of the type used within the microelectronics industry. In addition, most microfluidic channels occupy only ~10% or less of the surface area of a microfluidic device. Wet chemical or plasma etching processes to produce microfluidic devices therefore take considerable time to complete, based upon the comparatively deep depths that are required for the channels. A comparatively fast wet or dry etching rate of 1 μm/min would still require up to several hours per wafer to achieve these depths. The small surface areas that are etched within this time make conventional batch processing of wafers less attractive economically. In many cases, photolithographically defined microfluidic features with micron scale accuracy are more precise than what is required for these applications.

At high volumes, other microfabrication processes become more applicable for the manufacture of microfluidics. Roll-to-roll stamping, lamination, hot embossing, and injection molding of plastic components offer excellent accuracy, repeatability, and cost effectiveness once the non-recoverable engineering (NRE) costs of molds, dies, and master templates have been paid for. However, the cost of these NRE items is comparatively high, and in most circumstances, production volumes of >1 million parts are required to recover this cost.

For part volumes from 1 to 1 million, laser microfabrication offers an excellent balance between speed, cost, and accuracy for microfluidics. Laser micromachining is also unmatched in the breadth of different of materials that it can process. A single laser system can micromachine materials all the way from lightweight plastics and elastomers up through hard, durable metals and ceramics. This versatility makes laser micromaching extremely attractive for prototyping and development, as well as for small to medium run manufacturing.

The most common criticism of laser micromachining is that it is a serial, rather than batch process, and it is therefore too slow to be economical for high volume manufacturing. While certainly true in some instances, as a generalization, this is not always the case. The processing time per part is the sum of the beam exposure time plus the beam positioning time. For parts which require only minimal volumes of material to be removed, serial processes such as laser micromachining can indeed be extremely efficient and cost effective. Whereas older laser micromachining systems were often limited by clumsy beam positioning, modern systems incorporate high speed beam positioning and parts handling so that the overall processing time is limited more by the net beam exposure time, which for many applications can be fairly small. A good counter-example to the criticism of serial processing is chip resistor trimming, which is used for almost all 1% tolerance and better metal film chip resistors in the microelectronics industry today and which are produced in extremely high volumes, >10 billion/year.

Microfluidics is becoming increasingly used for miniaturized chemical analysis systems, such as the new generations of lab-on-a-chip applications which are rapidly being developed. The fundamental structure used in microfluidics is the flow channel, but integrated microfluidic systems also incorporate vias, T-junctions, sample wells, reaction chambers, mixers, and manifolds, along with some moving mechanical components such as valves, pumps, and injectors, and often some optical and electrical components for integrated control and sensing. Unlike wet and dry etching which must be carefully formulated to achieve the required material selectivity, laser micromachining can be used to process many different materials and structures at a time. For example, a laser can be used to cut a channel to one depth, cut a via to another depth, trim a metal trace, release a check valve structure, and weld two mating elements together all within the same mounting of the part. This illustrates one of the advantages that serial processing has over traditional batch processing of wafers. Another obvious advantage of serial laser processing is that no masking is required, greatly reducing the time and expense for design changes. Different parts can also be individually customized with virtually no extra tooling overhead. Microfluidics and laser micromachining are an excellent marriage of technologies which will prove essential for the rapid development of these applications.

This chapter will discuss the fundamentals of laser ablation in the microfabrication of microfluidic materials. After briefly describing the various types of lasers which are used for this purpose, the fundamental mechanisms of laser micromachining will be described, along with some data illustrating the performance of some state-of-the-art laser micromachining systems.

1.1 Lasers for micromachining

By far the most common laser used for industrial processing is the carbon dioxide (CO_2) gas laser. This popularity comes from its unique combination of high average power, high efficiency, and rugged construction. Unlike the original glass tube style gas lasers, the modern CO_2 lasers which are used for materials processing are of a hard sealed waveguide construction that use extruded aluminum RF driven electrodes to excite a $CO_2/N_2/He$ gas mixture. The lasing transitions are from asymmetric to symmetric stretch modes at 10.6 µm, or from asymmetric stretch to bending modes at 9.4 µm of the CO_2 molecule (Verdeyen, 1989). Within each of these vibrational modes there exist numerous rotational modes, and hundreds of lasing transitions can be supported by excitation into the parent asymmetrical stretch mode of the CO_2 molecules. This large number of simultaneous lasing modes along with the efficient excitation coupling through the N_2 gas is what allows CO_2 lasers to achieve power levels up to 1 kW with electrical to optical conversion efficiencies of nearly 10%. CO_2 lasers emit in the mid-infrared (MIR), most commonly at 10.6 µm, and they principally interact with their target materials via focused, radiant heating. They are used extensively for marking, engraving, drilling, cutting, welding, annealing, and heat treating an enormous variety of industrial materials (Berrie & Birkett, 1980; Crane & Brown, 1981; Crane, 1982). For micromachining applications, the long wavelength translates into a fairly large spot diameter of ~50-150 μm with a corresponding kerf width when used for through cutting.

The most common solid-state laser used in industry is the neodymium-doped yttrium-aluminum-garnet, or Nd:YAG. The YAG crystal is a host for Nd^{3+} ions, whose lasing transitions from the excited $^4F_{3/2}$ band to the energetically lower $^4I_{11/2}$ band produces emission at 1.064 μm in the near-infrared (NIR) (Koechner, 1988; Kuhn, 1998). Nearly all industrial Nd:YAG lasers are now pumped by semiconductor diode lasers, usually made of GaAlAs quantum wells and tuned to emit at ~810 nm, for optimum matching to the pertinent absorption band of Nd:YAG. Semiconductor diode pumping of Nd:YAG offers much more efficient pumping with minimal energy being lost to heat, since the diode emits only into that part of the spectrum which is needed for the pumping. However, semiconductor diode pump lasers can only be made up to ~100 W, and thus these are used only for Nd:YAG lasers of low to moderate average powers. Most industrial Nd:YAG lasers are also Q-switched, usually by means of a KD*P electrooptic intracavity modulator. When the modulator is in the non-transparent state, the pumping of the Nd:YAG rod allows the population inversion to build up to very high levels. When the modulator is rapidly switched to the transparent state, the energy stored in the inverted population is discharged at once into a single giant pulse of narrow duration and high peak power. Typical Q-switched pulse widths are in the range of ~25 ns, and with firing repetition rates of ~40 kHz, the duty cycle of a Q-switched Nd:YAG laser is ~1:1000. A ~10 W average power Nd:YAG laser can then produce pulses with peak powers of ~10 kW. This high peak power makes Q-switched Nd:YAG lasers ideally suited for nonlinear optical frequency multiplication through the use of an external cavity harmonic generating crystal such as KDP, KTP, $LiNbO_3$, or BBO. Most commonly, the 1064 nm output from the Nd:YAG is frequency doubled to produce a green output at 532 nm. The 1064 nm output can also be frequency tripled to produce 355 nm in the near ultraviolet (UVA band), or frequency quadrupled (using a sequential pair of doublers) to 266 nm in the deep ultraviolet (UVC band). All four of these commonly available Nd:YAG output wavelengths are extremely useful for micromachining purposes (Atanasov et al., 2001; Tunna et al., 2001).

Copper vapor lasers have also proven their use in high accuracy micromachining (Knowles, 2000; Lash & Gilgenbach, 1993). Similar to the Nd:YAG, they are Q-switched systems which produce high intensity pulses of typically ~25 ns at rates of 2-50 kHz and average powers of 10-100 W. Unlike the Nd:YAG, they emit directly into the green at 511 nm and 578 nm, and thus do not require a nonlinear crystal for frequency multiplication to reach these more useful wavelengths. Copper vapor lasers also have excellent beam quality and can usually produce a diffraction-limited spot on the substrate with only simple external beam steering optics. The disadvantage of copper vapor lasers is that they tend to have shorter service life and require more maintenance than Nd:YAG lasers. Frequency multiplying crystals have now become a ubiquitous feature of commercial Nd:YAG lasers, and as a result, Nd:YAGs have largely displaced the copper vapor laser for industrial micromachining applications.

Excimer lasers have also found wide use in materials processing applications. Excimer lasers operate from a molecular transition of a rare gas-halogen excited state that is usually pumped by an electric discharge. The XeCl excimer laser, which emits at 308 nm, is prototypical of these in which a pulsed electric discharge ionizes the Xe into a Xe^+ state and ionizes the Cl_2 into a Cl^- state. These two ions can then bind into a Xe^+Cl^- molecule which will loose energy through a lasing transition as it relaxes back to the XeCl state. The resulting ground state XeCl molecule readily dissociates, and these products are then recycled. Other commonly used excimer lasers are the XeF which emits at 351 nm, the KrF which emits at 249 nm, the ArF which emits at 193 nm, and the diatomic F_2 which emits at 157 nm (Kuhn, 1998). Like other laser systems which are well matched to applications in materials processing, excimer lasers produce pulses of ~50 ns with repetition rates of ~100 Hz to ~10 kHz and average powers of up to a few hundred Watts. Excimer lasers are fairly efficient in their electrical to optical conversion efficiency, but their use of highly reactive halogen gases at high pressures requires significantly more servicing and maintenance than other types. One of the most important properties of excimer lasers is their ability to create a rather large spot size which can be homogenized into a high quality flat top beam profile of up to several cm in dimension. Because of this, they have been the pre-eminent source for coherent UV radiation at moderate power levels, they can be used both as a masked or a scanned exposure source, and currently they are used extensively for UV and deep UV lithography as well as several other applications in thin film recrystallization and annealing. At higher beam intensities, they can be used for surface ablation of materials, and due to the short wavelength and short pulse width, they typically produce clean, crisp features in metals, ceramics, glasses, polymers, and composites, making them adaptable for numerous micromachining applications (Gower, 2000).

Short laser pulses, on the order of a few tens of nanoseconds, are a desirable feature for laser micromachining applications, and these can be produced with many different laser systems. As will be discussed in more detail later, the short pulse width produces nearly adiabatic heating of the substrate which allows the substrate surface temperatures to quickly reach the point of vaporization with minimal heating effects on the surrounding areas. There has been interest in laser systems which can produce even shorter pulse widths, and the foremost candidate for this has been the Ti:sapphire laser. The Ti:sapphire laser has the unique feature of being tunable over a surprisingly large fluorescence band: from ~670 nm to ~1090 nm. For efficient pumping, it needs to be optically excited in its absorption band, which is centered about 500 nm, and for which argon ion lasers and frequency doubled Nd:YAG lasers provide excellent sources (Kuhn 1998). Most Ti:sapphire lasers are configured into an optical ring resonator arrangement with a set of birefringent filters for tuning. In addition, the ring cavity usually contains a Faraday rotator and wave plates to limit the propagation to only one direction around the ring. This arrangement is well suited for wide tuning and also mode locking, through which very short pulses, on the order of a few tens of femtoseconds can be produced. Ti:sapphire lasers have thus become a key resource for spectroscopy and research on ultrafast phenomena. The Ti:sapphire laser is also capable of average powers of up to several Watts, which makes it a viable tool for micromachining. Although its operation is at longer wavelengths than those normally preferred for micromachining, its capability for tuning and producing ultrashort pulses makes it attractive for research in this area. Since it requires a pump laser of ~10 W which is already in the green, and its more complicated optical system requires more maintenance and user savvy, it is presently not a common choice for industrial micromachining applications, but this may change in the future. There are many other new laser systems under development which offer efficient generation of green light at the power levels and pulse widths required for micromachining. It is worthwhile to realize that the field of laser sources is constantly changing.

In general, the lasers best suited for micromachining are those that produce short pulses of high intensity at short wavelengths. Pulse widths of less than a microsecond are needed to allow the formed plasma to extinguish in between pulses so that subsequent pulses are not scattered and absorbed. Time for the debris plume to clear takes longer, often up to tens or hundreds of milliseconds, but its optical attenuation is usually less. Concentrating the laser radiation into short pulses of high intensity also has the benefit of more adiabatic heating of the substrate, bringing its temperature up to the vaporization point before too much of the heat can diffuse vertically and laterally away from the intended ablation zone. Shorter wavelengths generally have higher absorption coefficients in most materials, and they are thus absorbed nearer to the surface where the ablation is intended to occur. Shorter wavelengths can also be focused into a proportionally smaller diffraction-limited spot, which improves both the accuracy and precision of the ablation process. Typical working spot diameters for UV lasers in the 350 nm range are ~25 μm, although this is larger than the theoretical diffraction limit.

2. Fundamental laser micromachining processes

Laser micromachining includes a number of different processes which are differentiated by the feature geometry and the manner in which material is removed from the substrate (Ion, 2005; Schuöcker, 1999). Cutting in this context refers to using the beam to slice all of the way through a thin sheet of substrate material, leaving behind a kerf which extends completely through to the opposite side of the substrate. As is commonly the case in laser cutting of sheet metal, the material removed from the kerf is predominantly ejected out the opposite side. Ablating is usually taken to mean removal of material in a thin layer from one side only, giving only partial penetration into the thickness of the substrate, and the removed material must necessarily be ejected from the same side as which the laser is incident. In both cases, the newly removed material is ejected primarily through the kerf which has just previously been cut and which trails along behind the laser beam as it is moved along the tool path. Whereas cutting and ablating can create geometries of any shape, drilling refers to the creation of a nominally circular hole with minimal lateral translation of the beam, with either through or blind penetration. If the laser beam is held in one fixed position and pulsed, the process often termed percussion drilling, whereas if the beam is swept around in a circular pattern to first roughly remove the bulk material and then completed with a fine finishing pass to accurately define the perimeter, the process is called trepanning. Percussion drilling produces holes whose diameter is roughly the same as the diameter of the laser beam, while trepanning produces holes whose diameter is larger than the beam diameter. Because drilling does not produce a trailing kerf, all removed material must be ejected from the same side as which the laser beam was incident, and drilling is thus necessarily an ablative process, regardless of whether it creates a through or blind hole (Voisey et al., 2003).

The removal of material can involve both thermal and chemical processes, depending upon how the laser radiation interacts with the substrate. At longer wavelengths, the photon energy is insufficient to provide anything more than simple heating of the substrate. At sufficiently high intensities, however, the heating can be concentrated enough to first melt the substrate material within a localized zone, and then vaporize it in those areas where the laser intensity and subsequent heating is higher. The substrate material is thus removed via a transition to the gas phase, although the vaporized material is often subsequently ionized by the laser radiation, leading to a plasma and plume that can have the effect of occluding the incident beam. It is customary to identify three zones around the incident beam: the heat-affected zone or HAZ, the melt zone, and the vaporization zone. Some materials can pass directly from the solid phase into the vapor phase by sublimation, and for these the melt zone is absent. Both melting followed by vaporization or direct sublimation are purely thermal ablation processes.

At shorter wavelengths, the photon energy may reach the level of the chemical bond strength of the substrate. Laser radiation may then break these chemical bonds through direct photon absorption, leading to volatilization of the substrate into simpler compounds. For most organic polymers, this photolysis process produces mainly H_2O and CO_2. This occurs typically for photon energies above 3.5 eV, or for wavelengths shorter than ~350 nm, i.e. into the near UV part of the spectrum. Because the photon energy is lost to chemical bond scission, the heating effects of the beam are greatly reduced, and this regime is sometimes referred to as "cold laser machining," non-thermal ablation, or photochemical ablation. This greatly reduces the transient thermal stresses that occur as part of thermal ablation, and the result is less bowing, warping, and delamination of the substrate, as well as fewer edge melting effects which degrade feature accuracy (Yung et al., 2002). Since the peak temperature rise is greatly reduced, conductive heat flow away from the irradiation area is also reduced, and better dimensional control of the micromachined structure is obtained. There has been a general trend toward using shorter wavelength lasers for micromachining over the past two decades of development. Currently, UV lasers in the 350 to 250 nm range dominate the industrial market for the above reasons.

Thermal ablation and photochemical ablation are two ideal extremes, and laser micromachining can often involve a combination of both for any given material or composite. In addition, there are several secondary processes which can arise due to the steep temperature gradients which are produced. If the laser beam is composed of short, high-intensity pulses, as would be typical for Q-switched systems, then the adiabatic heating of the substrate can cause sufficiently high temperature gradients for which differential thermal expansion and acoustic shock can produce surface cracking or spalling of the substrate (Zhou et al., 2003). Micron-sized flakes of the substrate can be explosively ejected from this process without requiring the additional thermal energy to fully vaporize the material. This is typically more prevalent for brittle materials with low thermal conductivity, e.g. ceramics and some glasses. For materials which readily oxidize, the rapid cycle of laser heating and cooling of the melt zone can cause the formed oxide film to flake off in chips from the compressive stress that was built into the oxide during the process. This is typically more prevalent for reactive metals such as chromium, nickel, iron, and copper. Thermal spalling and oxide chipping both create debris particles which are significantly larger than the redeposition of fully vaporized substrate material. Because both thermal spalling and oxide chipping occur after the melt zone has refrozen, they leave behind a surface finish which is typically more frosted or matte in visual appearance, and microscopically cusped on a smaller scale.

Inherent to all laser micromachining is the creation of a plume of ejected material, either fully vaporized or sometimes containing micron-sized debris flakes. This plume requires time to disperse, and if the next laser pulse arrives before this takes place, the laser radiation will usually produce some degree of ionization as it is absorbed by the vapor. This ionization of the vaporized material produces a plasma which, in addition to being fairly energetic and reactive, can absorb the laser radiation further, sometimes occluding the path for the beam to reach the substrate (Eloy, 1987). This luminous plasma is what is usually responsible for the "sparkles" that mark the travel of the laser beam across the substrate. Achieving beam positioning and pulse timing to avoid the plasma and plume occlusion of the beam is a central part of tuning the recipe for any laser micromachining. This problem is generally severe in continuous wave (CW) laser micromachining, but greatly reduced for pulsed lasers which are Q-switched. While the complete plume of vaporized material usually does not have time to fully disperse in between Q-switched pulses, the more optically opaque and higher density plasma does, and laser ablation can continue onward with usually only minor attenuation. If the beam positioning is not well designed, however, the plasma and plume can become trapped into the confined spaces of the kerf, and greater time will be required for their dispersal. The most common symptom of this effect is a reduced depth of ablation for a given beam transversal rate.

2.1 Ablation process models

Laser ablation involves a complex interaction between optical, thermal, and chemical processes, but some simplifications can lead to models which can be useful for characterization, optimization, and troubleshooting of the process. Most such models start with the optics of a Gaussian beam and compute the conductive flow of heat from this source to find the temperature distribution, adding in the thermal effects which are needed to account for melting and vaporization of the substrate (Engin & Kirby, 1996; Kaplan, 1996; Olson & Swope, 1992). An idealized geometry is illustrated in Fig. 1 in which a circularly symmetric Gaussian laser beam is moved across the substrate at a constant speed v in the $+x$ direction. The beam has an average power of $P_0 = \pi r_B^2 I_0$, where I_0 is the peak intensity and r_B is the $1/e$ beam radius. The beam propagates in the $+z$ direction and meets the substrate surface in the x-y plane. The situation is more easily described by using the relative coordinate $\xi = x - vt$ which moves along with the laser beam.

The interaction of the laser beam with the substrate first involves absorption of the optical radiation and its conversion into heat for thermal (non-photo-chemical) ablation. Shorter wavelengths are absorbed more strongly at the surface with a higher absorption coefficient α, and since this is usually $\sim 10^4$ cm^{-1} or greater, the heating is effectively concentrated at the surface of the substrate. Volumetric heating effects have been considered by Zhang, et al. (2006). The surface heating density is then

$$q(\xi, y) = (1 - R)I_0 \, exp\left(-\frac{\xi^2 + y^2}{r_B^2}\right) \quad [W / m^2],$$

where R is the reflectivity loss from the surface of the substrate.

The heat transfer within the substrate is entirely by conduction, so the resulting temperature field is given by a solution to the heat conduction equation (Carslaw & Jaeger, 1959)

$$\frac{\partial T}{\partial t} - D\nabla^2 T = 0,$$

where $D = \kappa/\rho C$ is the thermal diffusivity, κ is the thermal conductivity, ρ is the mass density, and C is the specific heat capacity. The surface heating density provides a source boundary condition for the solution of the heat conduction equation. Ashby and Easterling (1984) have shown that a close analytical approximation to the solution of this problem is given by

$$T(\xi = 0, y, z, t) - T_0 = \frac{(1-R)P_0}{2\pi\kappa v \left[t(t + r_B^2 / D) \right]^{1/2}} exp\left(-\frac{(z + z_0)^2}{4Dt} - \frac{y^2}{4Dt + r_B^2} \right),$$

where T_0 is the initial temperature of the substrate, and z_0 is a parameter chosen to eliminate the surface singularity as $t \rightarrow 0$.

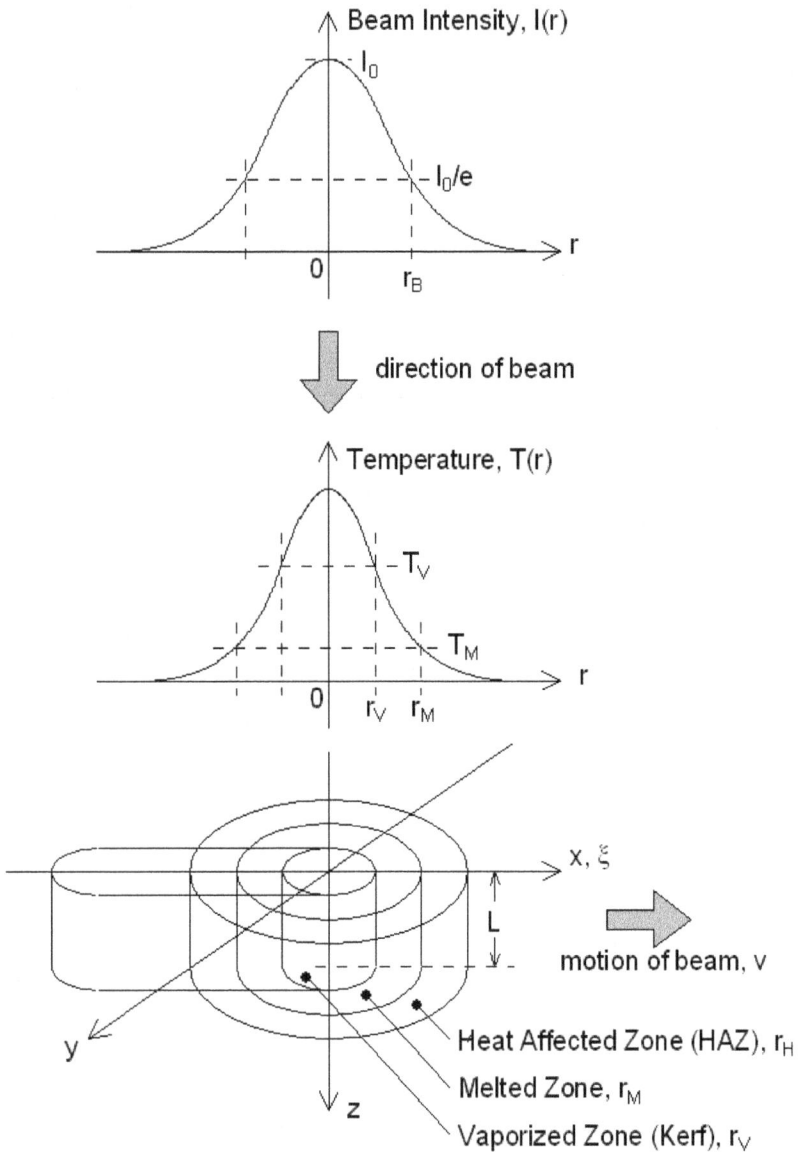

Fig. 1. Geometry and intensity and temperature profiles for laser ablation.

With a sufficiently large laser average power P_0 and a sufficiently slow beam traversal rate v, the resulting temperature field can produce first melting and then vaporization of the substrate. Three zones are commonly defined based upon the phase changes: a heat affected zone or HAZ with a radius r_H, a melted zone with a radius r_M, and a vaporized zone with a radius r_V, which forms the final kerf of width $2r_V$. For simplicity, the depth of cut is taken to be L for all three of these zones, as shown in Fig. 1. These radii are defined by the points at which the peak temperature equals the melting point T_M or the vaporization point T_V for the substrate material. It is important to recognize that these three radii are dependent upon the beam radius r_B, but are not equal to it. Similarly, the radial temperature distribution is not the same as the incident Gaussian beam shape.

In addition to simply raising the temperature of the substrate material, the incident laser power must also be used to change the phase of the material, first from solid to liquid, and then from liquid to vapor, in the case of simple thermal ablation. This energy balance is an important aspect of the ablation process model, and it can be described by the following conservation of energy relation,

$$(1 - R)P_0 = \frac{2\pi\kappa L(T_M - T_0)}{\ln(r_H / r_M)} + 2(r_M - r_V)LvE_M + \rho C(r_M - r_V)Lv(T_V - T_M) + 2r_V LvE_V,$$

where $(1 - R)P_0$ is the optical power that is absorbed by the substrate, v is the beam traversal speed, and L is the depth of the cut. The first term on the right hand side is the power required to bring the temperature of the substrate up to the melting point T_M at the inner boundary of the HAZ. The thermal conductivity is $\kappa = \rho CD$. Within the melted zone, $r_M < r < r_V$, additional power must be added for the melting phase transition, as well as to raise the temperature up to the vaporization point T_V at the inner boundary of the melted zone. The rate at which the beam sweeps out new material volume to melt is $2(r_M - r_V)Lv$ [m³/s], and the latent heat of melting is E_M [J/m³], which together give the second term. The third term on the right hand side is the power required to support the temperature difference of $T_V - T_M$ across the melted zone. Similarly, the fourth term on the right hand side is the additional power required to vaporize the material in the kerf, where E_V is the latent heat of vaporization [J/m³]. The latent heats of melting and vaporization are effectively constants which subtract from the applied optical power that falls within the melted and vaporized zones.

Most notable in the energy balance equation is the direct tradeoff that exists in the last three terms between the depth of cut L and beam traversal speed v. This makes the assumption that the incident beam does not become occluded by the features or debris that the ablation process creates. For CW laser cutting the debris plume and plasma can significantly attenuate the beam, leading to a reduction in the optical power that is available for subsequent ablation. Q-switched lasers with nanosecond pulses suffer far less from this problem, since the time between laser pulses allows the plasma time to extinguish (Chang & Warner, 1996). The above energy balance equation works fairly well for cutting depths L of up to a few beam diameters (Yuan & Das, 2007), but when the aspect ratio of the kerf becomes extreme, $L \gg 2r_V$, the sidewalls of the kerf will lead to beam reflections and scattering, and the change in depth may take the beam interaction beyond its depth of focus, both of which will have the effect of reducing the available intensity and slowing down the vertical ablation rate for deeper cuts (Bang et al., 1993). The concave bottom of laser drilled holes may also defocus the beam (Vatsya et al.,2003; Zhang ei al.,2008). This basic model has been extended to include the effects present in trepanning of holes (Zeng et al., 2005), and for trepanning with annular beam profiles (Zeng et al., 2006).

Laser ablation departs somewhat from the above model when the process involves a photochemical component. In this case, a significant fraction of the photons are absorbed directly for the process of breaking chemical bonds in the substrate, and these photons do not produce direct substrate heating, as would be the case for pure thermal ablation. Process models for this situation must break the photon flux into a thermally absorbed portion and a photochemical portion. The thermal portion behaves as per the above model description, while the photochemical portion creates volatilized products in proportion to the energy density of the specific bonds which are broken. Simple energy balance arguments are useful for predicting the photochemical ablation rate, under the assumption that any left-over energy that does not directly produce photochemical ablation is directed toward substrate heating of the same region. The relative split of the incident photon flux between thermal and photochemical ablation is usually taken to be proportional to the relative absorption coefficients of the two processes. However, it must be cautioned that the appropriate absorption coefficients are themselves temperature dependent and proper modeling of the optical absorption becomes a central problem in any multi-physics simulation of ablation.

2.2 Optical absorption

The most important principle of laser micromachining is that the laser output wavelength must be one which is strongly absorbed by the material to be processed. If the material is highly transparent at the wavelength of the laser, then no optical absorption and energy transfer will take place. For semiconductors and other crystalline materials, this normally means that the photon energy must be greater than the energy bandgap. For polymers and other amorphous materials, the photon energy must be greater than the energy difference between the lowest unoccupied molecular orbital (LUMO) and the highest occupied molecular orbital (HOMO). For both of these cases, that usually entails a laser emitting in the visible or UV with photon energies of ~1 eV or greater.

2.2.1 Absorption of laser radiation

The primary interaction between laser radiation and a solid is photochemical excitation of electrons from their equilibrium states to some excited states by the absorption of photons. Some of these transitions are schematically shown in Fig. 2. Interband transitions take place when photon energy is larger than bandgap of the material. In this process, electron-hole pairs are generated. The free electrons may jump back from conduction band to valence band through thermal (dashed lines) or photochemical processes. If the photon energy is less than bandgap of the material, the energy can be absorbed by defect levels in the banbgap or produce Intraband transitions. Both transitions will induce thermal processes as electrons jump back to valence band. With higher laser light intensities, multi-photon absorption is favored, because the probability of non-linear absorption increases strongly with laser intensity. The coherent multi-photon transitions would generate electron-hole pairs similar to interband transitions. (Linde et al., 1997)

Thus, the initial electronic excitation is followed by complex secondary processes, which can be classified into thermal and photochemical processes. The type of interaction between laser radiation and the material depends on laser parameters (wavelength, pulse duration, and fluence) and on the properties of the materials (Baeuerle, 2000; Mai & Nguyen, 2002).

Laser ablation (material removal) can be analyzed on the basis of photothermal (purely pyrolytic), photochemical (purely photolytic), and photophysical processes, in which both thermal and non-thermal mechanisms contribute to the overall ablation rate.

2.3 Thermal process

The thermal transition of the electrons can be described by the relaxation time τ_T as shown in Fig. 2. When τ_T is smaller than the time required for desorption of species from the surface, τ_R, a photothermal process occurs. Thus, the photothermal ablation is based on the excitation energy being instantaneously transformed into heat. Due to the rapid dissipation of the excitation and ionization energy from the electrons to the lattice, the material surface is heated rapidly and vaporized explosively with or without surface melting. This regime applies to pulsed laser ablation by infrared- (IR-) and visible- (VIS-) laser radiation, and to most cases of ultraviolet- (UV-) laser radiation with nanosecond and longer pulses. These result in relatively high ablation rates and a rough surface finish (Baeuerle, 2000; Ehlich & Tsao, 1989; Luft et al., 1996; Schubart & Otto, 1997).

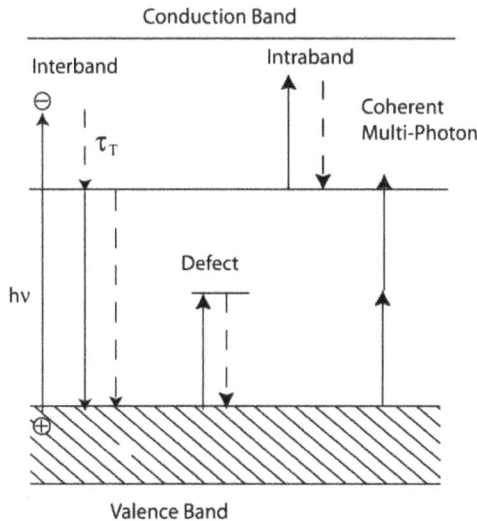

Fig. 2. Schematic of different types of electronic excitation in a solid.

With moderate-to-high laser fluences, and pulse lengths of nanoseconds, screening of the incident radiation by the vapor and plasma plume becomes important. Screening of the incident laser light by absorption and scattering within the vapor plume diminishes the intensity that reaches the substrate. The ablation rate depends on photon energy, laser fluence, spot size and material properties and is in a range between 0.1 and several µm/pulse to be considered as a useful machining method. Additionally, with shorter wavelengths of laser radiation, the laser-plasma interaction becomes less pronounced (Baeuerle, 2000).

2.4 Photochemical process

If $\tau_T \gg \tau_R$, the laser excitation can result in direct bond scission, the electrons freed from the broken bonds will be desorbed from the surface, and the process will be photochemical in nature. With purely photochemical (non-thermal) processes, the temperature of the system remains essentially unchanged under laser irradiation. The ablation rate is relatively slow ($\leq$ 1μm/pulse), but high surface quality can be achieved because of the absence of surface melting and explosive evaporation of the material (Baeuerle, 2000; Mai & Nguyen, 2002).

3. Nd: YAG 266 nm and 355 nm laser micromachining

3.1 Laser ablation settings

Chen and Darling (2005, 2008) have reported sysmatic studies of laser micromachining using Nd:YAG 266 nm and 355 nm lasers recently. An Electro-Scientific Industries (ESI) model 4440 laser micromachining system with a Light Wave Enterprises 210 diode-pumped frequency-tripled (355 nm) and a Photonics Industries diode-pumped frequency-quadrupled (266 nm) Nd:YAG laser were used to micromachine the samples. Output powers of the 355 nm laser were 4.8 W at repetition rate of 10 kHz and 3.0 W at repetition rate of 20 kHz, and that of the 266 nm laser was 0.5 W at repetition rate of 5 kHz. The stage was moved up and down to adjust the z-axis to focus and de-focus the laser beam on the samples. The x-y stage allowed scan speeds from 0 to 250 mm/s. The laser scan speed and repetition rate were adjusted to control the total energy of micromachining, and the focus/defocus was adjusted by moving z-axis stage up to control the laser fluences of the laser spots as shown in Table 1.

The microfluidic materials, such as sapphire, silicon and Pyrex, were micromachined by both 266 nm and 355 nm Nd:YAG lasers. A series of 1 mm × 1 mm square cavities were created by laser micromachining with various laser machining conditions. Fig. 3 and Fig. 4 depict the typical laser micromachining cavities of sapphire and silicon. The ablation square cavities were inspected by an SEM and were measured for depth using an optical microscope and a scanning profilometer. Fig. 5 shows typical measurement data using a Tencor/KLA P-15 profilometer. A certain amount of solidified molten silicon remains in the ablation area after laser machining (Dauer et al., 1999). Thus, silicon wafers were cleaned and etched using a 22 wt% KOH solution at 75°C for 4 minutes to clear the ablation debris.

Z-position[a] (μm)	Fluence[b]		
Repetition Rate (Hz)	10k (355 nm)	20k (355 nm)	5k (266 nm)
0	866	271	50.93
300	96.24	30.01	24.02
600	34.65	10.83	13.93
900	17.68	5.52	9.08
1200	10.69	3.34	6.39
1500	7.16	2.24	4.73
1800	5.13	1.60	3.65
2100	3.85	1.20	2.90

[a]Laser focus at 0 z position, the z position shows the distance of the stage moving up;
[b]Laser fluence was calculated by energy per pulse/ spot size area; fluence unit = J/cm^2

Table 1. The fluences of the laser versus z-stage positions

Fig. 3. The SEM image of Nd:YAG laser micromachining on sapphire

Fig. 4. SEM image of Nd:YAG laser micromachining on silicon.

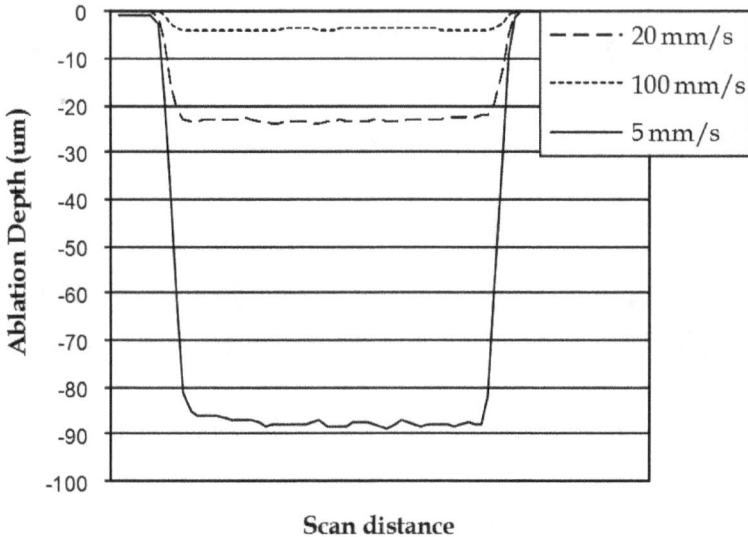

Fig. 5. Depth profiles for 355 nm Nd:YAG laser micromachining of sapphire with fluence of 9.27 J/cm² and virious scan speeds.

3.2 Laser micromachining ablation rate

The ablation rates of the laser micromachining were calculated as:

$$\frac{\text{Total removed volume of the material}}{\text{The number of total pulses} \times \text{spot size area}}$$

Figures 6 - 8 show the plots of ablation rates as a function of laser fluences with various laser scan speeds for sapphire, silicon and Pyrex using both Nd:YAG 266 nm and 355 nm lasers. It is observed that in the cases of both sapphire and Pyrex, the 266 nm laser provides higher ablation rates than the 355 nm laser under the same micromachining conditions. On the other hand, Fig. 7 (silicon) shows the varied ablation rates of Nd:YAG 355 nm laser micromachining using 20 mm/s and 50 mm/s scan speeds. The varied result is caused by the plume screening effect on the slower scan speed laser micromachining condition . In this case, the dwell time of laser light on the surface of the silicon is longer than the time for vapor/plasma formation, which attenuates the intensity of incident laser radiation.

The threshold fluences of laser micromachining of sapphire and silicon were calculated as shown in Table 2. All samples either did not exhibit fixed laser ablation threshold values or showed surface melting phenomena. Those results indicate that a thermal process was engaged in the laser micromachining of all the materials micromachined by both 266 nm and 355 nm Nd:YAG lasers. In general, the ablation rates using Nd:YAG 266 nm laser are higher than using 355 nm laser. This is due to the 266 nm laser producing a greater photochemical component.

		Sapphire		Silicon	
	Scan speeds	5 mm/s	10 mm/s	20 mm/s	50 mm/s
Threshold Fluence	266nm Nd:YAG	0.4	0.4	0.96	0.95
J/cm²	355nm Nd:YAG	1.19	1.10	1.31	1.29

Table 2. The threshold fluences of laser micromachining of sapphire and silicon

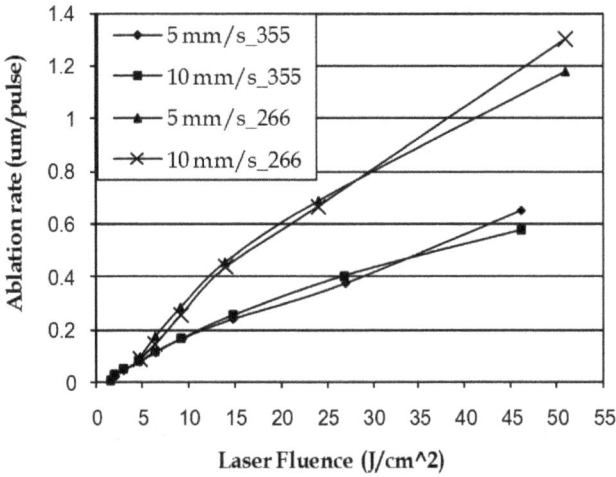

Fig. 6. The ablation rates for laser micromachining versus laser fluence for sapphire with different cutting speeds using 266 nm and 355 nm Nd:YAG lasers.

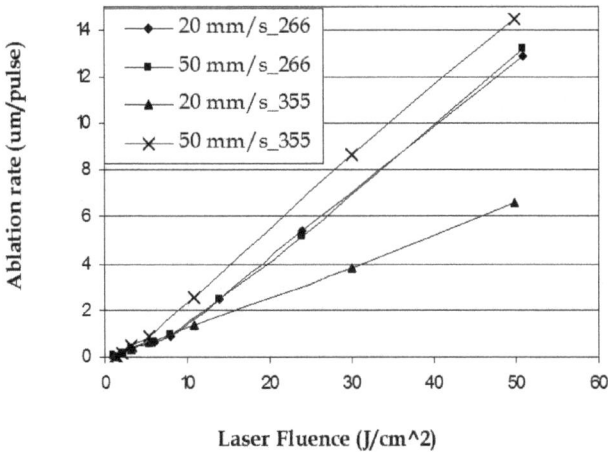

Fig. 7. The ablation rates of laser micromachining versus laser fluences for silicon with different cutting speeds using 266 nm and 355 nm Nd:YAG lasers.

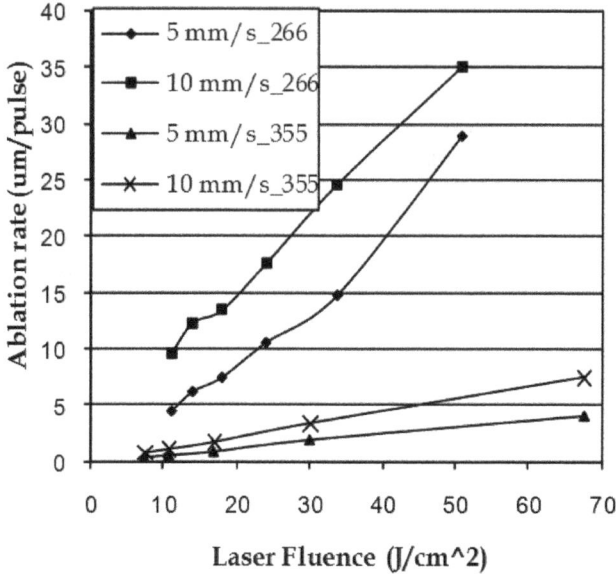

Fig. 8. The ablation rates for laser micromachining versus laser fluences for Pyrex with different cutting speeds using 266 nm and 355 nm Nd:YAG lasers.

3.3 The ablation efficiency

The ablation efficiency was calculated by dividing the ablation rate by the energy per pulse to normalize the ablation rate performed by the 355 nm and 266 nm Nd:YAG lasers. Figures 9 - 11 show the plots of ablation efficiency as a function of laser fluence with various scan speeds using both lasers. The results indicate that at high laser fluences, the ablation efficiencies of the 266 nm laser are better than that of the 355 nm laser for all three materials.

Figure 10 (silicon) shows that the ablation rate of 266 nm Nd:YAG laser micromachining is slower than 355 nm laser micromachining under 50 mm/s scan speed after normalizing the ablation rate by energy per pulse. The result points out that at the laser fluences higher than 10 J/cm², the ablation efficieny of the 266 nm laser is 1.5 times faster than that of the 355 nm laser at the scan speed of 50 mm/s, and 3.2 times faster in the case of 20 mm/s as shown in Table 3.

	Sapphire	Silicon		Pyrex
Ablation Efficiency		50 mm/s	20 mm/s	
266 nm/355 nm	9	1.5	3.2	13

Table 3. The comparison of Nd:YAG 266 nm and 355 nm laser ablation efficiencies to sapphire, silicon and Pyrex with laser fluence larger than 10 J/cm².

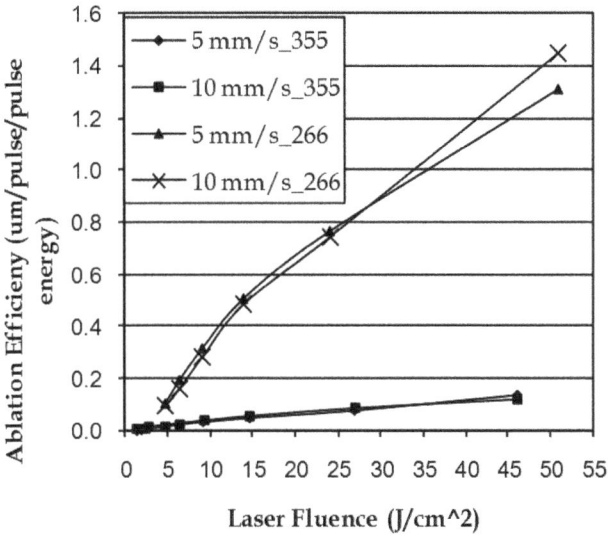

Fig. 9. Laser ablation efficiency versus laser fluences for sapphire under different scan speeds using the 266 nm and 355 nm Nd:YAG lasers.

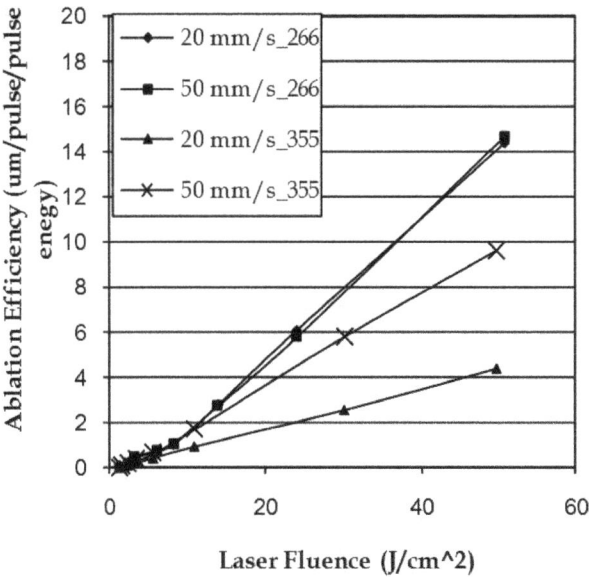

Fig. 10. Laser ablation efficiency versus laser fluence for silicon under different scan speeds using the 266 nm and 355 nm Nd:YAG lasers.

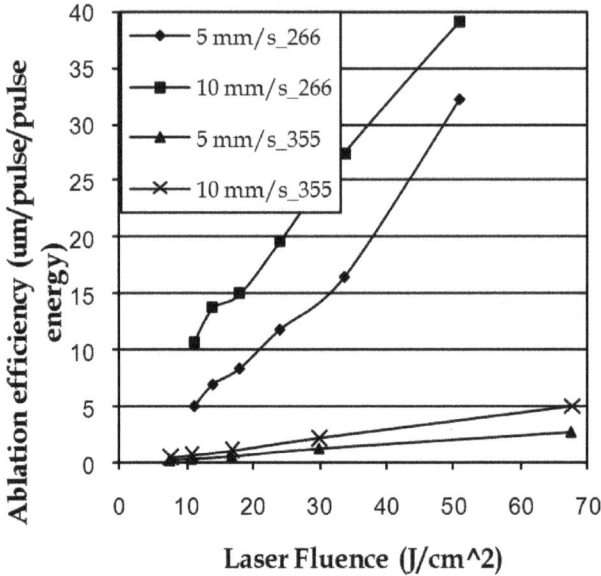

Fig. 11. Laser ablation efficiency versus laser fluence for Pyrex under different scan speeds using the 266 nm and 355 nm Nd:YAG lasers.

3.4 The ablation precision of laser micromachining

By computing the average ablation depths and standard deviation, the depth of laser micromachining can be characterized as:

$$\text{Average depth (mean)} \pm \text{standard error } (=2.58 \times \text{standard deviation/} \sqrt{\text{sample size}}));$$

which give 99% of the cutting depths falling into this range (Lindgren et al., 1978), and the laser machining precision is defined as,

$$\text{Precision} = 2 \times \text{standard error} / \text{average depth}$$

Figure 12 shows the plot of laser machining precision as a function of laser fluence using Nd:YAG 266 nm and 355 nm lasers with different scan speeds. The results portray the Nd:YAG 266 nm laser providing better precision than the 355 nm laser, and Nd:YAG laser micromachining more generally providing better precision in the order of sapphire, silicon and then Pyrex.

4. CO_2 laser cutting of microfluidic plastic laminates

CO_2 lasers have become the most used laser system for industrial fabrication and materials processing. This is due to a combination of their relatively low cost, high optical power and efficiency, and robust operation over a long service life. They are routinely applied to an extremely wide range of material processing, including scribing, marking, drilling, cutting, and heat treating of metals, ceramics, and polymers. CO_2 laser processing has also been

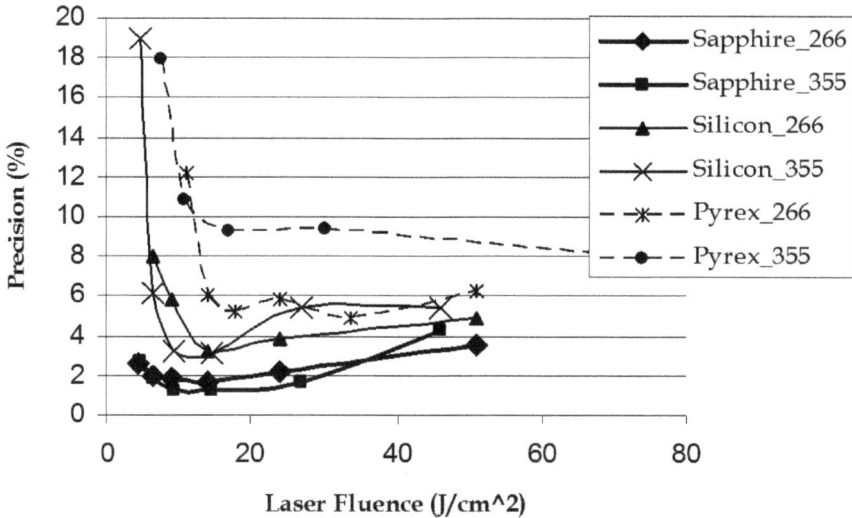

Fig. 12. Laser micromachining precision versus laser fluences for sapphire, silicon and Pyrex using the 266 nm and 355 nm Nd:YAG lasers.

extensively applied to the field of microfluidics, principally in the form of through-cutting of plastic laminates. A great many applications for microfluidics demand disposable cartridges for the liquid contacting elements of the system. Disposable cartridges, in turn, demand extremely low cost materials and fabrication methods, often in the range of pennies per part, to be competitive in the marketplace. One approach, which has gained great popularity over the past decade, is the construction of microfluidic cartridges from a series of laser-cut plastic laminates which are aligned and bonded together. This method of fabrication offers enormous flexibility in both the design of the microfluidic plumbing as well as the materials which are used to create it.

One example of a fairly advanced microfluidic cartridge created as a bonded stack of laser-cut plastic laminates is shown in Fig. 13. (Lafleur, 2010). As illustrated, this type of microfluidic cartridge can utilize both thick, rigid layers as well as thinner, flexible layers in its construction, allowing channel thicknesses from a few mils up to several mm to be created. The layers can be aligned and bonded together using a variety of techniques, including heat fusing, heat staking, solvent welding, or through the use of adhesives which are either applied directly, or which can be a pressure-sensitive adhesive which comes on one or both sides of a given layer. The cartridge shown in Fig. 13 only uses 6 layers, but cartridges employing over 20 layers are becoming more routine (Lafleur, 2010). Common structural materials for plastic laminate microfluidics include polymethyl methacrylate (PMMA), polyethylene (PE), polycarbonate (PC), and acetate. In addition, semi-permeable membranes such as Nafion and nitrocellulose are frequently employed. As is true for other types of microfluidic systems, the control of surface hydrophobicity / hydrophilicity is of paramount concern, and plays a predominant role in the materials selection.

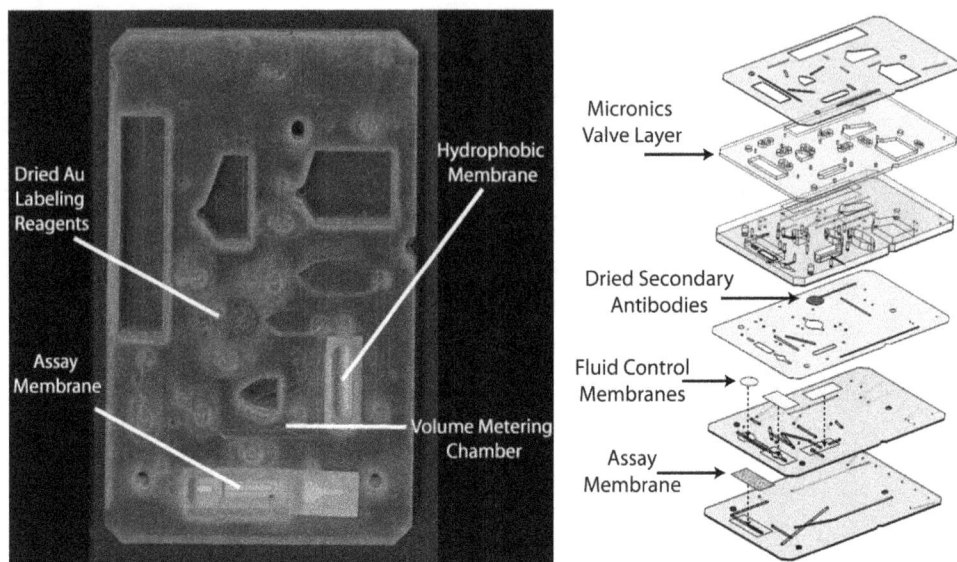

Fig. 13. A laser-cut plastic laminate microfluidic cartridge for carrying out an immunoassay. From Lafleur (2010).

5. Discussion

The laser ablation processes, thermal and photochemical, are determined by the materials properties. Figure 14 depicts the absorption coefficients of transparent materials, sapphire and Pyrex, and Table 4 shows some physical properties of those three materials.

	Eg (eV)	Melting temp. (C)	Bond strength (kJ/mol)	Absorption Coefficient@ 266nm(cm⁻¹)	Absorption Coefficient@ 355nm(cm⁻¹)	Evaporation Temp.* (C)
Sapphire	7.8	2054	511 ± 3	5.19	4.74	1800
Silicon	1.12	1414	326.8 ± 10	2.0E6	1.07E6	1350
Pyrex	7.8	821	799.6 ± 11.3	14.7	1.93	--

* Rough estimates of source evaporation temperatures are commonly based on the assumption that vapor pressures of 10^{-2} Torr must be established to produce efficient source removal rates (Maissel & Glang, 1970).

Table 4. Some physical properties of sapphire, silicon, and Pyrex (Chen & Darling, 2005, 2008)

In general, the laser ablation rates of sapphire, silicon, and Pyrex micromachined by near UV (355 nm) and mid-UV (266 nm) nanosecond pulsed Nd:YAG lasers, are higher using the 266 nm laser than the 355 nm laser in the absence of plume screening effects. Under those high laser fluency micromachining conditions, non-linear optical phenomena such as multi-photon process become important, and the 266 nm laser (with photon energy = 4.66 eV) has a higher probability to induce photochemical process than the 355 nm laser (with photon energy = 3.50 eV). Therefore, the ablation rates increase more in the cases of wide bandgap materials, such as sapphire and Pyrex, than the increase in the case of narrow bandgap material, like silicon as laser fluence increasing.

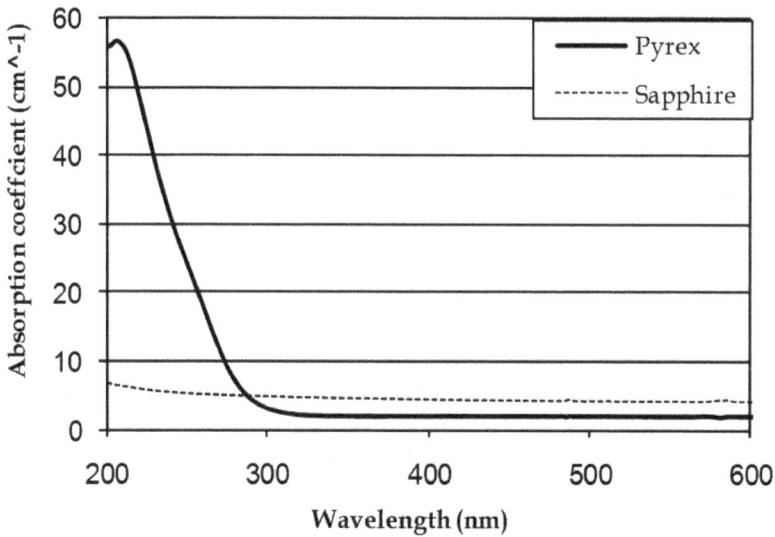

Fig. 14. The absorption coefficients versus wavelength for the transparent materials tested.

Sapphire has relatively the same level of absorption at 266 nm and 355 nm, however, the 266 nm laser provides a higher ablation efficiency at a given laser fluence than the 355 nm laser caused by higher photochemical process contributing to the overall ablation. Therefore, 266 nm laser micromachining on sapphire would provide not only slighly better absorption but also higher probability of photochemical process than 355 nm laser. In the case of silicon with its narrow band gap and high absorption at both wavelengths, the ablation efficiencies are not much different between the 266 nm and 355 nm lasers.

Pyrex has a low melting temperature, a high bond strength, a low absorption coefficient, and a wide energy band gap, as shown in Table 4. This implies that a predominantly thermal process was engaged in the laser micromachining of Pyrex by the 266 nm and 355 nm lasers. However, Pyrex shows better ablation efficiency using 266 nm laser due to more photochemical process at the higher absorption coefficient and higher energy (Mai & Nguyen, 2002; Baeuerle, 2000; Lim & Mai, 2002; Craciun & Craciun, 1999; Craciun et al., 2002; Hermanns, 2000).

Laser micromachining of plastic laminates for microfluidics nearly always involves through-cutting of each layer. CO_2 laser systems do not offer sufficient beam control to allow accurate machining to a prescribed depth, nor would the inhomogeneity of the plastic films support this type of machining. During the laser micromachining, plastic laminates are most often supported on mesh or grille working platens to allow the beam and the ablation debris to completely pass through to the other side without obstruction. Very thin, fragile or flexible materials, such as nitrocellulose membranes, are usually supported by a sacrificial backing piece, and for this situation, the laser micromachining reverts back to pure surface ablation with the debris exiting from the same side as which the laser was incident. The greatest issue with CO_2 laser through-cutting of plastics is the degree of edge melting that occurs along the kerf. While the vaporization temperatures for most plastics are comparatively low, so are the melting temperatures, and the CO_2 laser beam is both broad in diameter and deeply penetrating, all of which can combine to easily cause run-away heating of the areas surrounding the desired kerf. This is particularly a problem in CW CO_2 systems. The most common approach to combating this problem is to tune the beam traversal speed to a fairly high value which produces a shallow depth of cut, and then to scan back and forth repeatedly until the full depth of cut is achieved. The time between successive passes is chosen to be greater than the time required for the substrate to cool back down to a stable working point. Through cutting of laminates does offer the advantage that larger cavities and channels can be created by simply tracing the beam around their edges and dropping out the waste as one single piece, as opposed to scanning back and forth to ablate away the entire volume. This conserves laser beam time, minimizes heating, and creates finished parts faster, with the only negative feature being the need to reliably capture the waste pieces so that they do not get caught in the remainder of the manufacturing process.

Nearly all of the materials used for plastic laminate microfluidics can also be readily photochemical ablated by UV lasers, usually producing harmless H_2O and CO_2 gas as by products. UV laser cutting of plastics is a premier method that gives the best geometrical accuracy due to the smaller beam spot and the photochemical ablation process which produces significantly less edge melting along the kerf. However, CO_2 lasers still dominate the market for this type of machining as a result of their much lower cost and ease of use as compared to UV laser systems.

6. Conclusion

This chapter discusses the fundamentals of laser ablation in the microfabrication of microfluidic materials. The removal of material involves both thermal and chemical processes, depending upon how the laser radiation interacts with the substrate. At longer wavelengths and low laser fluencies, the thermal process dominates. While the photon energy of the laser radiation is sufficiently high, the laser radiation can provide heating, with or without melting the substrate material, and then vaporize it. At shorter wavelengths, the ablation process shifts to photochemical. The photon energy of laser radiation reaches the level of the chemical bond strength of the substrate, and then breaks these chemical bonds through direct photon absorption, leading to volatilization of the substrate into simpler compounds.

In the cases of the ablation rates of sapphire, silicon, and Pyrex, micromachined by near UV and mid-UV nanosecond pulsed Nd:YAG lasers. All three materials have higher ablation efficiencies using the 266 nm laser than the 355 nm laser due to better absorption and higher probability of photochemical process using 266 nm laser. The ablation efficiencies are increased more for the case of high melting temperature or/and finite absorption materials such as sapphire and Pyrex. The increase is less for narrow band gap or/and high absorption materials such as silicon.

Laser systems can micromachine materials all the way from lightweight plastics and elastomers up through hard, durable metals and ceramics by carefully selecting laser wavelengths, pulse duration, and fluencies. This versatility makes laser micromaching extremely attractive for prototyping and development, as well as for small to medium run manufacturing.

7. References

Ashby, M. F. & Easterling, K. E. (1984). The transformation hardening of steel surfaces by laser beams – I. Hypo-eutectoid steels. *Acta Metal.*, Vol. 32, No. 11. pp. 1935-1948, ISSN: 0001-6160

Atanasov, P. A., Eugenieva, E. D., & Nedialkov, N. N. (2001). Laser drilling of silicon nitride and alumina ceramics: A numerical and experimental study. *J. Appl. Phys.*, Vol. 89, No. 4. pp. 2013-2016, ISSN: 0021-8979

Baeuerle, D. (2000). *Laser Processing and Chemistry* (3rd Ed.), Springer, ISBN 3-540-66891-8, New York, United States of America

Bang, S. Y., Roy, S. & Modest, M. F. (1993). CW laser machining of hard ceramics–II. Effects of multiple reflections. *Int. J. Heat Mass Transfer*, Vol. 36, No. 14, pp. 3529-3540, ISSN: 0017-9310

Berrie, P. G. & Birkett, F.N. (1980). The drilling and cutting of polymethyl methacrylate (Perspex) by CO_2 laser. *Opt. Lasers Eng.*, Vol. 1, No. 2, pp. 107-129, ISSN: 0143-8166

Carslaw, H. S. & Jaeger, J. C. (1959). *Conduction of Heat in Solids* (2nd Ed.), Oxford University Press, ISBN 0-19-853303-9, Oxford, UK

Chang, J. J. & Warner, B. E. (1996). Laser-plasma interaction during visible-laser ablation of methods. *Appl. Phys. Lett.*, Vol. 69, No. 4, pp. 473-475, ISSN: 0003-6951

Chen, T.-C. & Darling, R.B. (2005). Parametric studies on pulsed near ultraviolet frequency tripled Nd:YAG laser micromachining of sapphire and silicon. *J. Mater. Sci. Technol.*, Vol. 169, No. 2, pp. 214–218. ISSN: 1005-0302

Chen, T.-C. & Darling, R.B. (2008). Laser micromachining of the materials using in microfluidics by high precision pulsed near and mid-ultraviolet Nd:YAG lasers. *J. Mat. Proc. Tech.*, Vol. 198, No. 1-3, pp. 248-253, ISSN: 0924-0136

Craciun, V. & Craciun, D. (1999). Evidence for volume boiling during laser ablation of single crystalline targets. *Appl. Surf. Sci.*, Vol. 138–139, pp. 218–223, ISSN: 0169-4332

Craciun, V., Bassim, N., Singh, R.K., Craciun, D., Hermann, J., & Boulmer-Leborgne, C. (2002). Laser-induced explosive boiling during nanosecond laser ablation of silicon. *Appl. Surf. Sci.*, Vol. 186, No. 1-4, pp. 288–292, ISSN 0169-4332

Crane, K. C. A. & Brown, J. R. (1981). Laser-induced ablation of fibre/epoxy composites. *J. Phys. D: Appl. Phys.*, Vol. 14, No. 12, pp. 2341-2349, ISSN: 0022-3727

Crane, K. C. A. (1982). Steady-state ablation of aluminium alloys by a CO_2 laser. *J. Phys. D: Appl. Phys.*, Vol. 15, No. 10, pp. 2093-2098, ISSN: 0022-3727

Dauer, S., Ehlert, A., Buttgenbach, S. (1999). Rapid prototyping of micromechanical devices using a Q-switched Nd:YAG laser with optional frequency doubling. *Sens. Actuators A*, Vol. A76, No. 1-3, pp. 381–385, ISSN: 0924-4247

Ehrlich, D.J. & Tsao, J.Y. (Eds.). (1989). *Laser Microfabrication: Thin Film Processes and lithography*, Academic Press, ISBN-10: 0122334302 Salt Lake City, USA

Eloy, J.-F. (1987). *Power Lasers*, Halsted Press / John Wiley & Sons, ISBN 0-470-20851, New York, New York

Engin, D. & Kirby, K. W. (1996). Development of an analytical model for the laser machining of ceramic and glass-ceramic materials. *J. Appl. Phys.*, Vol. 80, No. 2. pp. 681-690, ISSN: 0021-8979

Gower, M. C. (2000). Industrial applications of laser micromachining. *Optics Express*, Vol. 7, No. 2, pp. 56-67, ISSN: 1094-4087

Hermanns, C. (2000). Laser cutting of glass, *Proc. SPIE 4102 international symposium on Inorganic Optical Materials II*, pp. 219-226, ISBN: 9780819437471, San Diego, [California], USA, August 2000

Ion, J. C., (2005). *Laser Processing of Engineering Materials*, Elsevier Butterworth-Heinemann, ISBN 0-7506-6079-1, Oxford, UK / Burlington, Massachusetts

Kaplan, A. F. H. (1996). An analytical model of metal cutting with a laser beam. *J. Appl. Phys.*, Vol. 79, No. 5, pp. 2198-2208, ISSN: 0021-8979

Knowles, M. R. H. (2000). Micro-ablation with high power pulsed copper vapor lasers. *Optics Express*, Vol. 7, No. 2, pp. 50-55, ISSN: 1094-4087

Koechner, W. (1988). *Solid-State Laser Engineering* (2nd Ed.), Springer-Verlag, ISBN 0-387-18747-2, New York, New York

Kuhn, K. J. (1998). *Laser Engineering*, Prentice Hall, ISBN 0-02-366921-7, Upper Saddle River, New Jersey

Lafleur, L. K. (2010). *Design and Testing of Pneumatically Actuated Disposable Microfluidic Devices for the DxBox: A Point-of-Care System for Multiplexed Immunoassay Detection in the Developing World*, Ph.D. Dissertation, The University of Washington, Seattle, Washington

Lash J. S. & Gilgenbach, R. M. (1993). Copper vapor laser drilling of copper, iron, and titanium foils in atmospheric pressure air and argon. *Rev. Sci. Inst.*, Vol. 64, No. 11, pp. 3308-3313, ISSN: 0034-6748

Lim, G. C., & Mai, T.-A. (2002). Laser micro-fabrications: present to future applications, *Proc. SPIE 4426 Second International Symposium on Laser Precision Microfabrication*, pp. 170-176, ISBN: 9780819441379, Singapore, May 2001

Lindgren, B.W., McElrath, G.W., Berry, D.A. (1978). *Introduction to Probability and Statistics*. (4th Ed.) Macmillan Publishing Co., ISBN: 0023709006, Basingstoke UK

Luft, A., Franz, U., Emsermann, A., Kaspar, J. (1996). A study of thermal and mechanical effects on materials induced by pulsed laser drilling. *Appl. Phys. A*, Vol. 63, No. 2, pp. 93–101. ISSN: 0947-8396

Maissel, L.I., & Glang, R. (1970). *Handbook of Thin Film Technology*, McGraw-Hill, ISBN-10: 0070397422, New York

Mai, T.-A., & Nguyen, N.-T. (2002). Fabrication of micropumps with Q-switched Nd: YAG, *Proc. SPIE 4426 Second International Symposium on Laser Precision Microfabrication*, pp. 195-202, ISBN: 9780819441379, Singapore, May 2001

Olson, R. W. & Swope, W. C. (1992). Laser drilling with focused Gaussian beams. *J. Appl. Phys.*, Vol. 72, No. 8, pp. 3686-3696, ISSN: 0021-8979

Schuöcker, D. (1999). *High Power Lasers in Production Engineering*, Imperial College Press / World Scientific, ISBN 981-02-3039-7, London / Singapore

Tunna, L., Kearns, A., O'Neill, W., & Sutcliffe, C. J. (2001). Micromachining of copper using Nd:YAG laser radiation at 1064, 532, and 355 nm wavelengths. *Optics & Laser Tech.*, Vol. 33, No. 3, pp. 135-143, ISSN: 0030-3992

Vatsya, S. R., Bordatchev, E. V. & Nikumb, S. K. (2003). Geometrical modeling of surface profile formation during laser ablation of materials. *J. Appl. Phys.*, Vol. 93, No. 12, pp. 9753-9759, ISSN: 0021-8979

Verdeyen, J. T. (1989). *Laser Electronics* (2nd Ed.), Prentice Hall, ISBN 0-13-523630-4, Englewood Cliffs, New Jersey

Voisey, K. T., Kudasia, S. S., Rodden, W. S. O., Hand, D. P., Jones, J. D. C., & Clyne, T. W. (2003). Melt ejection during laser drilling of metals. *Mat. Sci. Eng.*, Vol. A356, No. 1-2, pp. 414-424, ISSN: 0921-5093

von der Linde, D., Sokolowski-Tinten, K., & Bialkowski, J. (1997). Laser-solid interaction in the femtosecond time regime. *Appl. Surf. Sci.*, Vol. 109/110, pp. 1–10, ISSN:0169-4332.

Yuan, D. & Das, S. (2007). Experimental and theoretical analysis of direct-write laser micromachining of polymethyl methacrylate by CO_2 laser ablation. *J. Appl. Phys.*, Vol. 101, No. 2, pp. 024901-1-6, ISSN: 0021-8979

Yung, K. C., Mei, S. M., & Yue, T. M., (2002). A study of the heat-affected zone in the UV YAG laser drilling of GFRP materials. *J. Mat. Proc. Tech.*, vol. 122, pp. 278-285, ISSN: 0924-0136

Zeng, D., Latham, W. P., & Kar, A. (2005). Two-dimensional model for melting and vaporization during optical trepanning. *J. Appl. Phys.*, Vol. 97, No. 10, pp. 104912-1-7, ISSN: 0021-8979.

Zeng, D., Latham, W. P., & Kar, A. (2006). "Shaping of annular laser intensity profiles and their thermal effects for optical trepanning. *Opt. Eng.*, Vol. 45, No. 1, pp. 014301-1-9, ISSN: 0091-3286

Zhang, C., Salama, I. A., Quick, N. R., & Kar, A. (2006). One-dimensional transient analysis of volumetric heating for laser drilling. *J. Appl. Phys.*, vol. 99, No. 11, pp. 113530-1-10, ISSN: 0021-8979

Zhang, C., Quick, N. R. & Kar, A. (2008). A model for self-defocusing in laser drilling of polymeric materials. *J. Appl. Phys.*, Vol. 103, No. 1, pp. 014909-1-8, ISSN: 0021-8979

Zhou, M., Zeng, D. Y., Kan, J. P., Zhang, Y. K., Cai, L., Shen, Z. H., Zhang, X. R., & Zhang, S. Y., (2003). Finite element simulation of the film spallation process induced by the pulsed laser peening. *J. Appl. Phys.*, vol. 94, no. 5, pp. 2968-2975, ISSN: 0021-8979

3

Miniature Engineered Tapered Fiber Tip Devices by Focused Ion Beam Micromachining

Fei Xu, Jun-long Kou, Yan-qing Lu and Wei Hu
College of Engineering and Applied Sciences and
National Laboratory of Solid State Mi-crostructures,
Nanjing University, Nanjing,
P. R. China

1. Introduction

Optical fibers have been the basis of the modern information technology since Kao and Hockham proposed glass waveguides as a practical medium for communication in 1965. A lot of different optical fiber active/passive devices including couplers, interferometers, gratings, resonators and amplifiers have been widely employed for applications on telecommunications and sensing networks (Agrawal, 2002). For a number of applications, it is important to reduce the device's size. Small size is often attractive for particular sensing applications because of some benefits such as fast response to detecting small objection with little perturbation on the object being measured. There are two steps to obtain fiber devices as small as possible. First, it is to taper or etch the fiber and reduce its diameter. A subwavelength-scale microfiber is the basic element of miniature fiber devices and sub-systems (Tong et al., 2003; Brambilla et al., 2004, 2005, 2010). The second is to engineer the microfiber to realize miniature version of conventional fiber devices. There are various fabrication methods to engineer the microfiber, such as CO_2 laser, femtosecond (fs) laser, HF acid etching, arc splicing and focused ion beam (FIB). Most of these techniques have the difficulties in carving the microfiber freely because of the resolution. The latest progress in FIB technique has opened a new widow for ultra-small size fiber devices. So far, FIB is the most flexible and powerful tool for patterning, cross-sectioning or functionalizing a subwavelength circular microfiber due to its small and controllable spot size and high beam current density.

FIB systems have been produced commercially for approximately thirty years, primarily for large semiconductor manufacturers. FIB systems operate in a similar fashion to a scanning electron microscope (SEM) except, rather than a beam of electrons and as the name implies, FIB systems use a finely focused beam of ions that can be operated at low beam currents for imaging or high beam currents for site specific sputtering or milling (http://en.wikipedia.org/wiki/Focused_ion_beam). The fine and controllable ion spot size and high beam current density are perfect for micro- and nano-fabrications with high spatial resolution (~ 10 nm). As a result, FIB has recently become a popular candidate for fabricating high-quality micro-devices or high-precision microstructures. Originally, FIB processing was used for mask repair (Liang et al., 2000), integrated circuit chip repair/modification (Liu et al., 2006), cross-sectional imaging of critical parts of semiconductor devices and sample preparation for transmission electron microscopy (Daniel et al., 1998; Hopman et al., 2008; Jeon et al., 2010). Besides these applications, FIB milling can also be used to assist carbon nanotube growth and manipulation (Hofmann et al., 2005; Deng et al., 2006), pattern magnetic data storage media (Terris et al., 2007) and structure hard-to-etch materials like SiC or LiNbO$_3$. In the field of optoelectronics, there have been extensively studies toward utilizing the FIB as a machining tool to fabricate planar micro-optical components with low surface roughness for integrated optical circuits, for example, the end facet mirrors, ring resonators, gratings and photonic crystals (Hopman et al., 2008). Obviously, FIB processing can and in fact has been widely applied to fabricate microfiber based devices to reduce the size of fiber devices as much as possible.

In this chapter, we will review several kinds of ultra-small engineered tapered fiber tip (TFT) devices including interferometers and gratings by FIB micromachining and their characteristics and sensing applications.

2. Fabrication and measurement

Standard optical TFT is an optical microfiber with only one output or input end and a taper transition. The taper transition is connected to untapered fiber at the extremities which can easily be connected to other fiber optic components. The taper is etched or pulled from a standard single mode fiber when heated by a CO$_2$ laser, electrical microheater or a small flame. Since the TFT is for analyte detecting rather than launching the light, it should be short enough in order to be rigid. However, too short and sharp shape results in high losses due to the poor 'adiabaticity' of the taper profile which couples light to lossy unbound modes (Love et al., 1991). During the last decade, much work has been carried out to study and optimize TFT profiles for telecom devices. Technology development allows manufacturing tapers with diameters well below 100 nm and it is possible to tailor the taper shape to an ideal profile (Brambilla & Xu, 2007). The quickest and simplest way to manufacture short TFT relies on using a commercially available pipette puller. This method is often used to manufacture fibre tips for optical tweezers and scanning near-field optical microscopy (SNOM) tips. In this chapter, we make TFTs using a commercial pipette puller (model P-2000, Sutter Instrument). The P-2000 is a microprocessor-controlled CO$_2$ laser-based micropipette puller. The bare fibre is held on two puller stages. The P-2000 can also be used to pull tubes and optical fibres to extremely small diameters. The pipette puller has five parameters which can be adjusted to achieve the wanted profile. The fabrication process is simple, convenient and extremely fast, which usually takes less than 0.3 second. The obtained TFT is then checked under a high-magnification optical microscope. Figure 1 shows a microscope image of a typical TFT with a sharp profile.

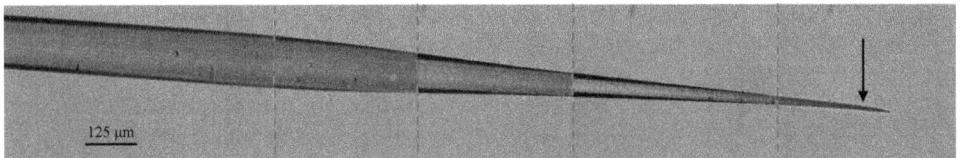

125 μm

Fig. 1. Microscope image of a typical TFT, five photographs separated by four dashed vertical lines are used to show the whole profile of the TFT. The black arrow indicates the milling location (Kou et al., 2010b). Reprinted with permission. Copyright 2010 Optical Society of America

The TFT is then coated with a thin metal layer such as aluminium (Al) or gold (Au, for exciting surface waves discussed in Section 4.2). The coating thickness is around 30 ~ 150 nm and the metal is deposited on only one side of the taper. The metal Al is used as a conductive layer to prevent gallium ion accumulation in the FIB micromachining process. Then, the Al-coated TFT is placed stably in the FIB machining chamber (Strata FIB 201, FEI Company, Ga ions) using conductive copper tape. We generally use a 30.0 kV gallium ion beam with current 60 ~ 300 pA. This enables us to make structures with high accuracy and sharp end-faces. We mill the structures from the taper end with small diameter to that with bigger diameter, because the milled part becomes non-conductive when the metal is removed by the beam. The total micro-machining process takes about 15 ~ 30 minutes depending on the size of the machined structures. Finally, the TFT is immersed in hydrochloric acid for about 15 ~ 30 minutes to totally remove the Al layer before cleaned with deionized water. In our experiment, the cavity or grating is made from a two-step process. Because there are some remains adhering onto the surfaces of the cavity after the first milling step, a second step under the same or smaller beam current is used to improve the surface smoothness.

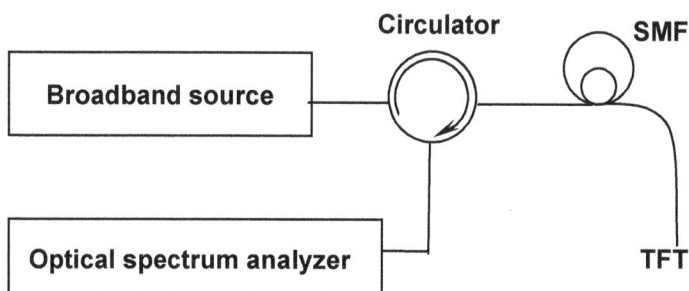

Fig. 2. Experimental setup of an FPMI.

In this chapter, we mainly consider the reflected signals. The reflective spectral response of these TFT based devices are measured with a broadband source (1525 ~ 1610 nm) and an Ando AQ6317B optical spectrum analyzer (OSA) through a circulator, as shown in Fig. 2. The TFTs before milling display an ignorable reflection of less than - 100 dB over the whole broadband spectrum.

3. FIB machined micro-cavity TFT interferometers

Optical fiber interferometers have been extensively used in various sensing applications due to its advantages of versatility, linear response and relatively simple structure. In the past decades, a lot of efforts have been made to develop intrinsic and extrinsic interferometers, especially the micro-cavity Fabry-Perot interferometers (MCFPIs). MCFPIs with tens-of-micrometer-length cavity are attractive because of the small size, large free spectrum range (FSR) and high sensitivity. The cavity can be assembled by splicing two single mode fibers (SMFs) to a hollow-core fiber (Sirkis et al., 1993), inserting a silica SMF and a multi-mode fiber into a glass capillary (Bhatia et al., 1996), or splicing a SMF and an index-guiding photonic crystal fiber together (Villatoro et al., 2009). Although many progresses have been made, people are still pursuing new micro cavity fabrication techniques to improve the cavity length precision, structure accuracy and the process repeatability. Femtosecond laser technology thus was proposed recently showing great success in micromachining fiber devices. MCFPIs can be quickly fabricated by milling a small-open hole in a SMF for liquid and gas sensing (Rao et al., 2007). However, even the fs-laser machined MCFPIs still show low fringe visibility of several dBs in liquids due to the rugged surfaces inside the cavity; what's more, it is difficult to focus the laser spot to a sub-wavelength scale due to the diffraction limit. Thus the micromachining accuracy is limited and the size of the micro-cavity is large (tens of micrometers). The latest progress in FIB technique opens a new widow of opportunity for ultra-small size cavity (Kou et al., 2010a, 2010b). Microcavities with nanometer-scale accuracy in a subwavelength microfiber could be fabricated by FIB, which is relatively difficult for fs laser approach. There are several typical geometries which can be realized by FIB machined-TFT as shown in Fig. 3. Among them, an open-notch in one side is the most preferred and easiest to be fabricated.

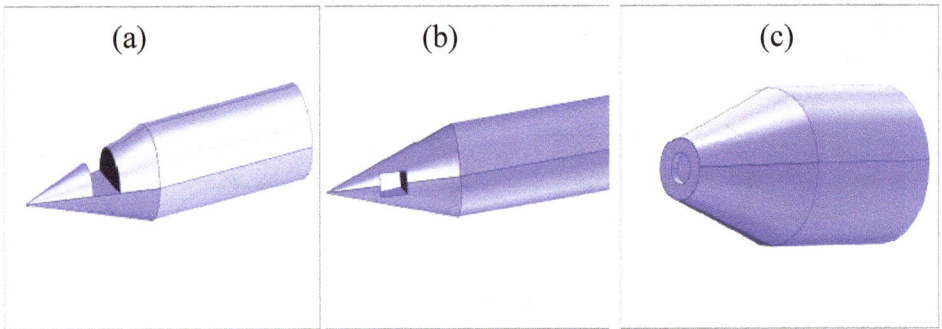

Fig. 3. Illustration of several typical geometries which can be realized by FIB machining, (a) a side open notch, (b) a hole in the middle and (c) a hole in the tip end and parallel to the fiber axis.

For geometry (a), due to the low reflectivity of the air-glass interfaces, multiple reflections have negligible contributions to the optical interference. However, a TFT consists of a SMF and a MMF in nature, without splicing. It may hold both the original single core mode and the multi-modes in the cladding at different positions. As shown in Fig. 4, we only consider two reflections I_1 and I_2 at the two end-faces, respectively. The fundamental LP_{01} mode can be coupled to high-order LP_{0m} mode in the taper transition or be excited to high-order LP_{0m} mode at the end-faces. I_1 or I_2 possibly includes LP_{01} or LP_{0m} mode (Kou et al., 2010b). We also break the cavity and measure the reflection I_1 at end-face 1; flat reflective spectrum without obvious interference fringes is observed. Accordingly, a reasonable assumption is to consider only one dominated mode in I_1 (LP_{01} mode) and I_2 (LP_{01} or LP_{0m} mode excited when inputting I_2 into end-face 1) (Kou et al., 2010b). We call this kind of device as a hybrid FP modal interferometer (FPMI). The interference spectrum can be modelled using the following two-beam optical interference equation (Kou et al., 2010b):

$$I = I_1 + I_2 + 2\sqrt{I_1 I_2}\, \cos(\delta + \varphi_0) \tag{1}$$

The phase difference between two modes in I_1 and I_2 is (Kou et al., 2010b)

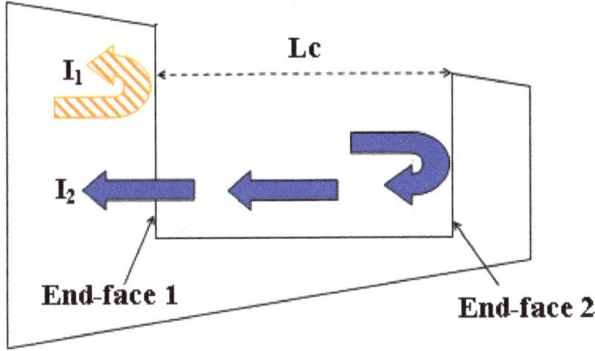

Fig. 4. Illustration of the FPMI. I_1 and I_2 are the reflections at end-face 1 and end-face 2 respectively; L_c is the length of the cavity. When I_2 enters end-face 1, the fundamental mode is possible to be excited to a higher-order mode (Kou et al., 2010b). Reprinted with permission. Copyright 2010 Optical Society of America

$$\begin{cases} \delta = \delta_1 + q\delta_2 = (2\pi / \lambda)(\Delta_1 + q\Delta_2) \\ \Delta_1 = 2n_c L_c \\ \Delta_2 = \int (n_2(r) - n_1(r))dz(r) \end{cases} \tag{2}$$

and FSR is (Kou et al., 2010b)

$$\text{FSR} = 2\pi\lambda/\delta \tag{3}$$

where q = 0 (for LP_{01} in I_2) and 1 for (LP_{0m} in I_2); Δ_1 (δ_1) and Δ_2 (δ_2) are the optical path length difference (the phase difference) owing to the micro-cavity and the modal difference in the taper transition, respectively; $n_1(r)$ and $n_2(r)$ are the effective index of LP_{01} and LP_{0m} modes, respectively, functions of local radius r(z) of the TFT at position z, which can be calculated by three-layer model of finite cladding step-profile fiber with the TFT profile r(z) which can be obtained from the microscope figure of the TFT (Kou et al., 2010b).

3.1 FIB machined FPMI for temperature sensing
The FPMI can be applied as a high-temperature sensor. Its extremely small size and especially unique structure offer great potentials for fast-response high temperature sensing particularly in small and harsh area with high temperature gradient, such as micro-flame and high temperature gas-phase/liquid-phase flow in microfluidics channel. Figure 5 shows an SEM picture of a FPMI with a micro-notch cavity from the side view and cross section after cleaving the TFT at the cavity. The end-face is very sharp and smooth. The cavity is 4.4 µm long and 5 µm high, located at the position with the local radius r = 4.6 µm.

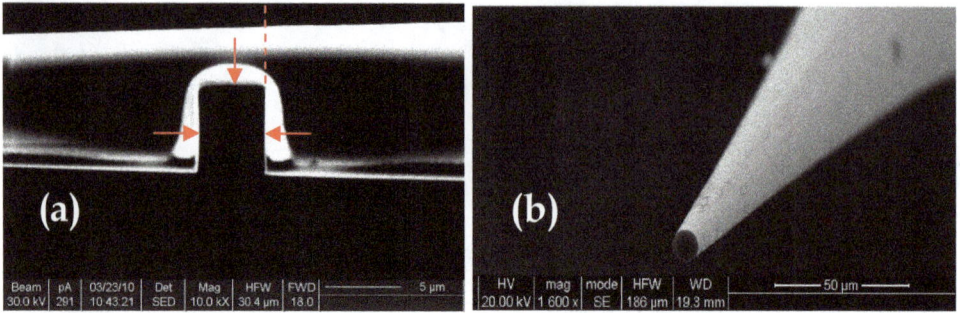

Fig. 5. SEM image (a) of the micro-notch cavity from the side view: three arrows show the edges of the cavity at the fiber tip, (b) of the cross section with the fiber tip cleaved at the position indicated in (a) by a dash line (Kou et al., 2010b). Reprinted with permission. Copyright 2010 Optical Society of America

The reflective spectral response of this FPMI device is measured with the setup as shown in Fig. 2. The TFT without a cavity displays an ignorable reflection of less than - 100 dB over the whole broadband spectrum. Hence, the detected signal is the light reflected only at the two end-faces of the micro-cavity, and the reflection at the tip end is negligible. The interference spectra of the FPMI device at room temperatures (19 °C) are shown in Fig. 6. The spectra indicates a free spectral range (FSR) of ~ 11 nm and a fringe visibility of ~ 11 dB around 1550 nm, which is larger than some other MCFPI sensors (Choi et al., 2008), and enough for sensing application. δ_1 is ~ 12π and δ_2 ~ 295π for LP$_{03}$ mode, and FSR ~ 10 nm, in good agreement with what we obtain in the experiment. In our calculation, λ = 1530 nm, L_c = 4.4 µm and n_c = 1 (Kou et al., 2010b).

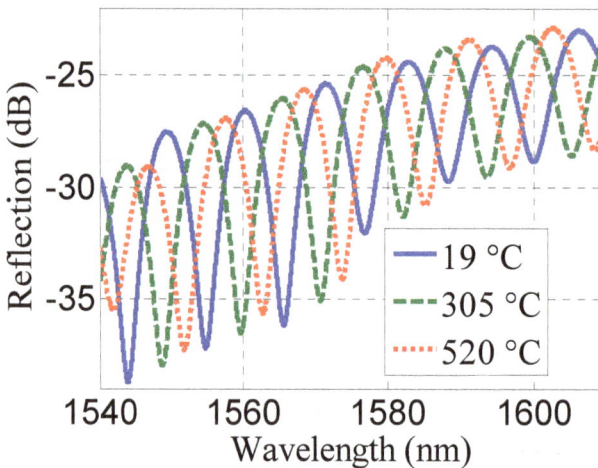

Fig. 6. Interference spectra of the FPMI device in air at different temperatures (Kou et al., 2010b) Reprinted with permission. Copyright 2010 Optical Society of America

We characterize the thermal response of the FTMI device by heating it up in a micro-furnace (FIBHEAT200, Micropyretics Heaters International Inc.) and temperature ranging from room temperature (19 °C) to 520 °C is measured by a thermocouple (TES-1310, Type K, TES Electrical Electronic Corp.). The spectrum and temperature were recorded when both of them are stable for several minutes (Kou et al., 2010b).

The temperature sensitivity S_T is defined as the interference wavelength shift divided by the corresponding temperature change. S_T depends on temperature through the thermal expansion and/or thermo-optics effect (Choi et al., 2008; Kou et al., 2010b):

$$
\begin{cases}
S_T = \dfrac{d\lambda}{dT} = \dfrac{2\pi}{\delta}\left(\dfrac{d\Delta_1}{dT} + \dfrac{d\Delta_2}{dT}\right) = \dfrac{2\pi}{\delta}\left(2\alpha_T L_c + \dfrac{d\Delta_2}{dT}\right) \\
\dfrac{d\Delta_2}{dT} \approx \int \left[\dfrac{\partial(n_1 - n_2)}{\partial n}\sigma_T + \dfrac{\partial(n_1 - n_2)}{\partial r}\alpha_T\right]dz
\end{cases}
\tag{4}
$$

where σ_T (1.1×10^{-5} /°C) is the thermo-optics coefficient and α_T (5.5×10^{-7} /°C) is the thermal expansion coefficient. There are two contributions from temperature change: the temperature-induced length variation in the cavity, and the temperature-induced index variation and taper volume variation in taper transition. The first one is less than 1 pm/°C and ignorable, it agrees with the fact that those previous micro-cavity FP interferences in SMF by femtosecond laser machining are temperature-insensitive; the second one is about 10 ~ 20 pm/°C and dominates in temperature sensing (Kou et al., 2010b).

Figure 7 displays the measured interferometer wavelength shifts ($\Delta\lambda$) and error on temperature (T). As the temperature increases, the interferometer wavelength shifts to longer wavelength. A third-order polynomial was used to fit the wavelength shifts across the entire calibration range. The average sensitivity of the device is ~ 17 pm/°C, which is very close to the theoretical result. Higher sensitivity can be obtained by optimizing the profile of the SMF-TT or using special fiber taper with higher thermo-optics coefficient (Kou et al., 2010b).

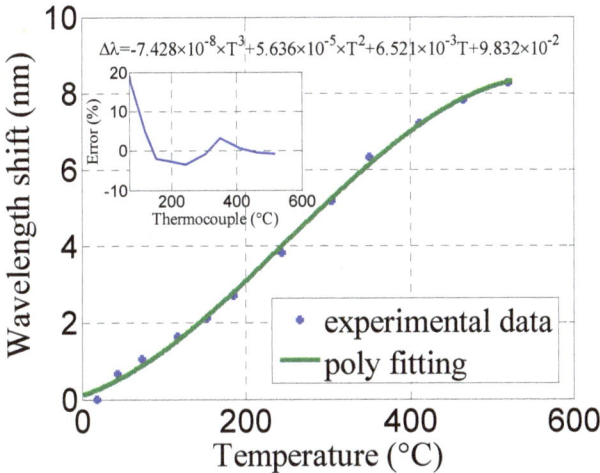

Fig. 7. Dependence of the measured wavelength shift on temperature. The asterisk represents the measured results while the solid line is the fitting result. The inset shows the dependence of error on temperature (Kou et al., 2010b). Reprinted with permission. Copyright 2010 Optical Society of America

3.2 FIB machined FPMI for refractive index sensing

The FPMI also can be employed as a refractive index sensor. Figure 8 shows the SEM picture of another FPMI with a micro-notch cavity from the side view. The cavity is 3.50 µm long and 2.94 µm high, located at the position with the local radius r = 2.4 µm.

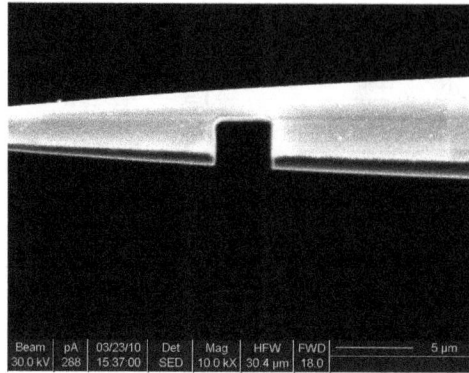

Fig. 8. SEM image of the micro-notch cavity from the side view (Kou et al., 2010a). Reprinted with permission. Copyright 2010 Optical Society of America

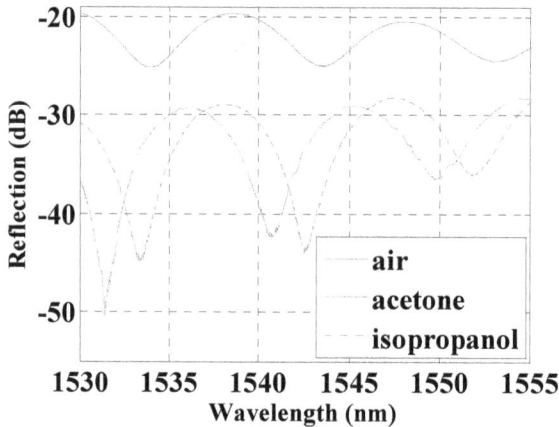

Fig. 9. Interference spectra of the MPRI device in air (solid line), acetone (dashed line) and isopropanol (dotted line), at room temperature (25 °C) (Kou et al., 2010a). Reprinted with permission. Copyright 2010 Optical Society of America

The reflective Interference spectrum is measured with the same setup as shown in Fig. 2. Figure 9 shows the interference spectra of the MPRI in air, acetone and isopropanol at room temperature (25 °C). The interference spectrum indicates a fringe visibility maximum of ~ 20 dB, which is much higher than those of typical MCFPIs in liquids.

The performance of resonant or interferometer refractive index sensors can be evaluated by using the sensitivity S_R, which is defined as the magnitude in shift of the resonant wavelength divided by the change in refractive index of the analyte. The sensitivity was measured by inserting the sensor in mixtures of isopropanol and acetone. These solutions were chosen with the objective of simulating aqueous solutions, having a refractive index in the region around 1.33 at a wavelength of 1550 nm. The ratio was increased by adding small calibrated quantities of isopropyl to the solution at a position far from the sensor. The refractive indexes of pure isopropyl and acetone at 1550 nm are 1.3739 and 1.3577 respectively (Wei et al., 2008).

Figure 10 displays the shifted spectral wavelength as a function of the liquid mixture refractive index. The asterisks represent the measurement results while the solid line is the best-fitting. As the refractive index increases, the spectrum shows a red-shift. The sensitivity of the device is 110 nm/RIU (refractive index unit) according to Fig. 11. Higher sensitivity can be obtained by optimizing the profile of the microfiber taper probe. Due to its small size, fiber-probe structure, all fiber connection, linear response, low-cost, easy fabrication and high sensitivity, MPRI devices are promising in various chemical and biological applications. It even may offer fantastic potentials sensing inside sub-wavelength liquid droplets, bubbles or biocells because of its unique probe structure and possible smaller size (Kou et al., 2010a).

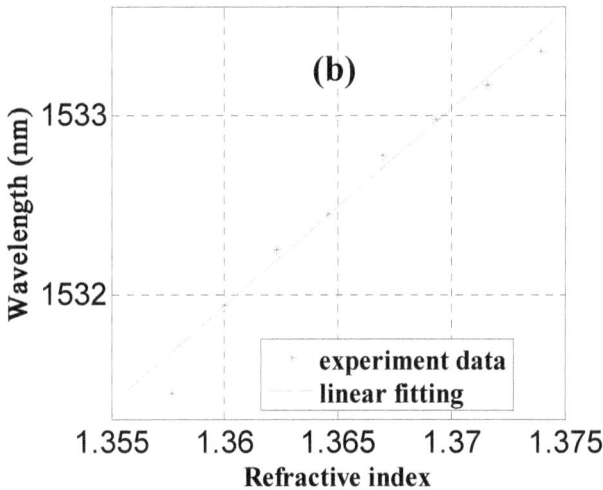

Fig. 10. The shifted spectral wavelength as a function of the liquid mixture refractive index. The asterisks represent the measured results while the solid line is the fitting results (Kou et al., 2010a). Reprinted with permission. Copyright 2010 Optical Society of America

4. FIB machined TFT micro-grating

Since their discovery in 1978 (Hill et al., 1978), optical fiber gratings have found a variety of applications in telecom and sensing because of their relatively low cost, inherent self-referencing and multiplexing/demultiplexing capabilities. Over the last two decades, fiber gratings including fiber Bragg gratings (FBGs) and long-period gratings (LPGs) have been manufactured mainly by modifying the core refractive index using interferometric or point-by-point techniques; most of interferometric techniques use a phase mask and an ultraviolet (UV) laser (Hill et al., 1993) (typically excimer or frequency doubled Ar+ ion) or femtosecond lasers (near IR or UV). Gratings based on surface etched corrugations have also been demonstrated in etched fibers using photolithographic techniques (Lin & Wang, 2001). However all these gratings fabricated in thick fibers have weak refractive index modulations ($\Delta n_{mod} \sim 10^{-4} - 10^{-3}$) and the related grating lengths are of the order of several millimeters. To reduce the grating length, strong refractive index modulations ($\Delta n_{mod} > 10^{-2}$) are necessary. Strong Δn_{mod} can be obtained by alternating layers of different materials, one of which can be air. Although this process in normal optical fibers imposes the removal of large amounts of material (the propagating mode is confined at a depth > 50 µm from the fiber surface), in fiber tapers and tips it requires the removal of small amounts of matter because the propagating mode is confined to the silica/air interface. A few techniques have been proposed for the fabrication of gratings in microfibers, including photorefractive inscription using CO_2 lasers (Xuan et al., 2009), femtosecond lasers (Martinez et al., 2005; Xuan et al., 2010) and wrapping a microfiber on a microstructured rod (Xu et al., 2009, 2010). None of them produced strong and short Bragg gratings. In some cases extra polymer coatings are needed, while in others the use of CO_2 lasers implies that the grating length is still long (it only can be used to write long period gratings (LPG) or high-order FBG). As a consequence, devices based on gratings tend to have a sizeable length: typically FBGs have lengths in the order of few millimetres. FIB technique provides a powerful way to mill the microfiber with directly and flexibly without a mask and realize compact micro-gratings (tens of micrometers) with colourful structure in the surface. Such small size and unique structure grating offer great potentials for a lot of sensing applications such as high temperature and refractive index sensing with the advantages of fast-response, ability to work in harsh environments and occupying little space.

4.1 FIB machined micro-grating for temperature sensing

Figure 11 shows an SEM micrograph of an ultra-short second-order TFT micro-grating (TFT-MG). The grating has 11 shallow corrugations with period $\Lambda = 1.1$ µm, providing a total length of ~ 12 µm, two orders of magnitude shorter than FBGs fabricated in conventional optical fiber. Each notch is ~ 1.6 µm deep and ~ 0.6 µm long. The average radius at the position where the notches are located is ~ 2.7 µm. The Bragg wavelength of the grating can be calculated from $\lambda_g = 2n_{eff}\Lambda/m$, where n_{eff} is the mode effective refractive index in the equivalent unperturbed geometry, Λ is the period and m is the Bragg order. Unlike conventional circularly-symmetric FBGs, this nanostructured TFT-MG has asymmetric periodic corrugations. The modal field and n_{eff} in the nano-structured TFT can be derived analytically from the bare TFT using the method developed by W. Streifer, which considers an equivalent structure, obtained by shifting the boundary between air and silica to compensate for the different geometry. Figure 12 shows the cross-sections of an un-etched fiber, an etched fiber and the equivalent unperturbed geometry with this method, which shifts the boundary between air and silica to compensate for the different geometry (Streifer et al., 1975, 1978), respectively. The effective groove height h_{eff} of the equivalent unperturbed geometry satisfies (Kou et al., 2011a):

$$\begin{cases} (1-\tau)(\theta_g - \sin\theta_g \cos\theta_g) = \theta_{eff} - \sin\theta_{eff} \cos\theta_{eff} \\ \theta_g = \arccos\{(r - h_g)/r\} \\ \theta_{eff} = \arccos\{(r - h_{eff})/r\} \end{cases} \qquad (5)$$

where τ is the grating duty cycle, h_g is the groove height and r is the fiber radius, respectively. In our device, r = 2.7 μm, τ = 0.33 and h_g = 1.6 μm, we find h_{eff} = 1.2 μm by solving Eq. 5 and n_{eff} = 1.428 by utilizing a finite element method. Thus, the Bragg wavelength is 1571 nm. It agrees well with the following experimental results.

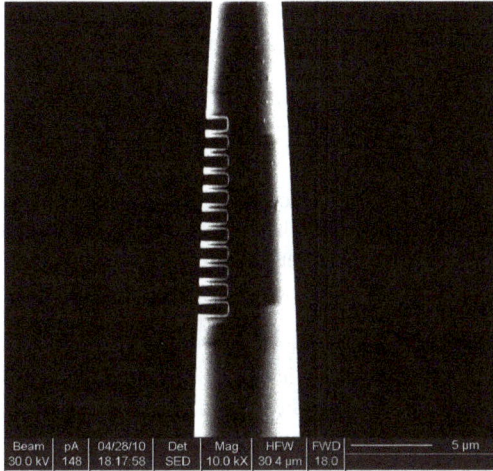

Fig. 11. SEM picture of the nanostructured TFT-MG. The grating has 11 notches and a total length of ~ 12 μm. The notch length and depth are ~ 0.6 μm and ~ 1.6 μm, respectively. The grating period is Λ ~ 1.1 μm (Feng et al., 2011).

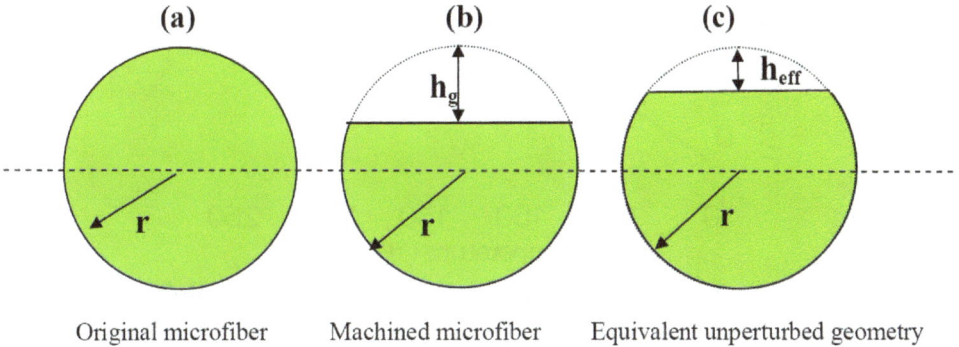

Fig. 12. The cross-sections of un-etched fiber (a), etched fiber (b) and equivalent unperturbed geometry (c), respectively. h_g is the groove height and h_{eff} is effective height (Kou et al., 2011a).

The reflective spectral response of the TFT-MG in Fig. 10 is measured with the setup shown in Fig. 2. We characterize the thermal response of the TFT-MG by heating it up in a micro-furnace from room temperature (20 °C) to 230 °C. The spectrum and temperature are recorded when both of them are stable for several minutes.

The interference spectra of the TFT-MG at different temperatures (23 °C, 47°C, 104°C, 153°C and 228 °C) are shown in Fig. 13. The Bragg wavelength is ~ 1570 nm, in agreement with our theoretical calculation. The spectra indicate a reflection peak-to-trough ratio around 1570 nm of ~ 10 dB at the Bragg wavelength which is achieved with as few as 11 periods and is similar with or even better than some other long length fiber gratings, enough for sensing applications.

The temperature sensitivity S_T is defined as (Kou et al., 2011a):

$$S_T = \frac{d\lambda_g}{dT} = \frac{2}{m}\left(\sigma_T\Lambda\frac{\partial n_{eff}}{\partial n_{silica}} + \Lambda\alpha_T n_{eff} + r\Lambda\alpha_T\frac{\partial n_{eff}}{\partial r}\right) \tag{6}$$

where σ_T (1.4×10-5 /°C) is the thermo-optic coefficient and α_T (5.5×10-7 /°C) is the thermal expansion coefficient. According to our calculations, the first part is about 15 ~ 20 pm/°C and dominates in temperature sensing. Thermal expansion effect (the second and third parts of Eq. 6) contributes little to the total sensitivity (< 6%), mainly due to the low thermal expansion coefficient of silica. Moreover, in the first part of Eq. 2, $\partial n_{eff}/\partial n_{silica}$ is nearly 1 and does not change much with the microfiber diameter, which means that the most efficient method to increase thermal sensitivity is to use fiber with higher thermo-optics coefficient.

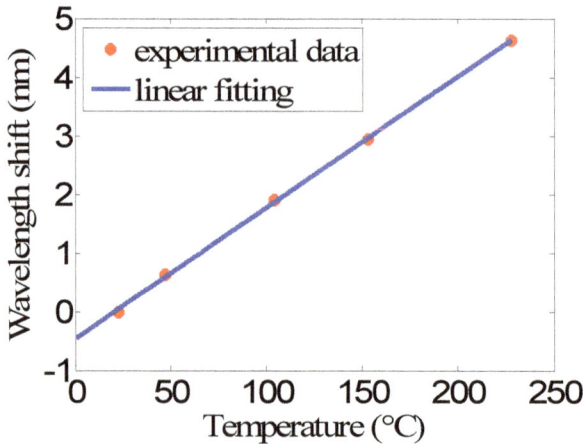

Fig. 13. Dependence of the measured wavelength shift on temperature. The asterisk represents the measured results while the solid line is the linear fitting result (Feng et al., 2011).

Figure 13 displays the measured resonant wavelength shifts ($\Delta\lambda$) on temperature (T). As the temperature increases, the interference wavelength shifts to longer wavelength. The average sensitivity of the device is ~ 22 pm/°C, which is very close to the theoretical result, higher than or similar with previous fiber grating sensors. Higher sensitivity can be obtained by use special fiber taper with higher thermo-optics coefficient.

First-order micro-grating (m = 1) with smaller period also can be fabricated in TFT by FIB. Figure 14 shows the SEM photography of a first-order TFT-MG from the side view. The grating has shallow corrugations of period Λ = 600 nm with 61 periods. The total length is about 36.6 μm, which is extremely short. Every groove is 200 nm in depth, located at the position with the local radius around r = 3.25 μm. The resonant spectra of the TFT-MG at different temperatures are shown in Fig. 15. The Bragg wavelength is ~ 1550 nm, with excited higher order mode as deduced from our theoretical calculation. The spectra indicate an extinction ratio of ~ 11 dB at the Bragg wavelength which is achieved with a 36.6 um long Bragg grating. The average temperature sensitivity of the device from room temperature to around 500 °C is ~ 20 pm/°C as shown in Fig. 15 (b), which is similar with the second-order TFT-MG. It is reasonable because the main thermal contribution is from the thermo-optic effect (Kou et al., 2011a).

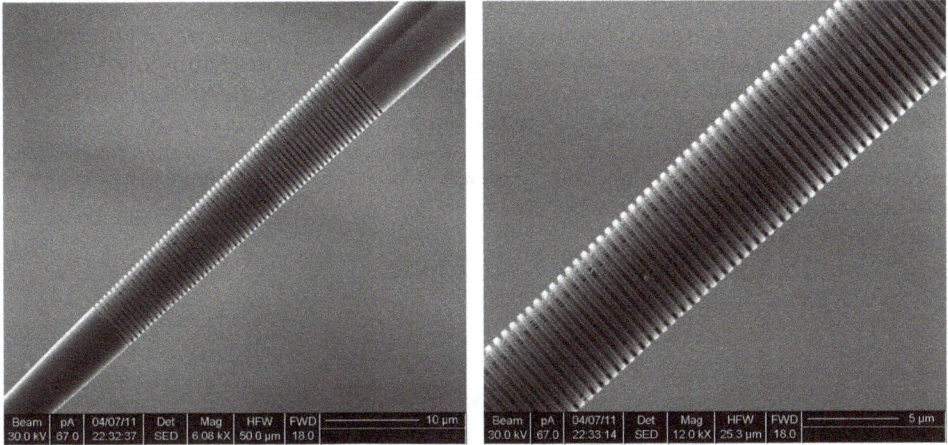

Fig. 14. Left: FIB picture of the TFPG with 61 periods (~ 36.6 μm in length and Λ = 600 nm). Right: magnified picture of the grating (Kou et al., 2011a). Reprinted with permission. Copyright 2010 Optical Society of America

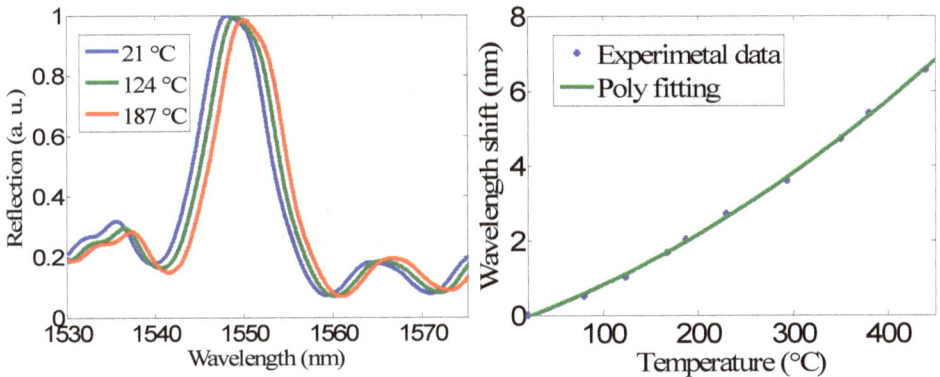

Fig. 15. (a) Reflection spectra of the first-order TFT-MG in air at different temperatures. (b) Dependence of the measured wavelength shift on temperature. The asterisk represents the measured results while the solid line is the linear fitting result (Kou et al., 2011a). Reprinted with permission. Copyright 2010 Optical Society of America

4.2 FIB machined metal-dielectric-hybrid micro-grating for refractive index sensing

Conventional FBGs have been extensively developed to measure the temperature, pressure or stress. But it is scarcely used to measure the environmental refractive index variation because there is almost no evanescent field penetrating outside of a standard 125 μm diameter FBG. TFT-MG may overcome the drawback with the available evanescence field interacting with the outer environments. The sensitivity of a pure-silica TFT-MG with the diameter of several micrometers is about tens of nm/RIU. By inducing metal-cladding, more cladding modes are possible to be excited and higher sensitivity can be obtained, which is so called grating-assisted surface plasmon-polariton (SPP)-like grating sensor (Nemova & Kashyap, 2006).

Figure 16 shows the SEM picture of a metal-dielectric-hybrid TFT-MG (MD-TFT-MG) by FIB milling. The fabrication process is similar with those mentioned ones above. But the fiber tip is coated with a gold layer with thickness of 30 nm on one side by magnetron sputtering and it is kept all the way throughout the experiment. We choose gold due to its relatively low absorption in the infrared and inertness to oxidation when exposed in air. Then a grating is fabricated by FIB milling at the fiber tip with local radius of ~ 3 μm. The grating has shallow corrugations of period Λ = 578 nm with 17 periods. The total length is about 10 μm, which is extremely short with local radius of ~ 3 μm.

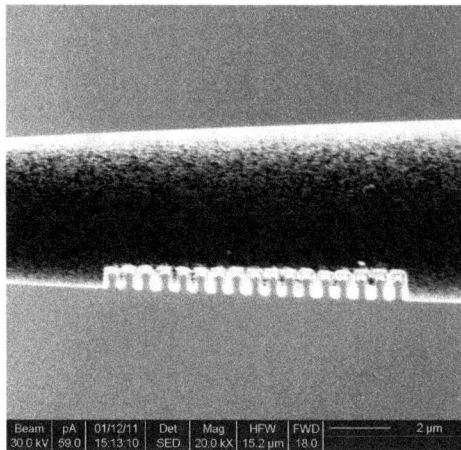

Fig. 16. SEM picture of the metal-dielectric-hybrid fiber tip grating (~ 10 μm in length and Λ = 578 nm). Right: magnified picture of the grating (Kou et al., 2011b).

Optical characterization of the MD-TFT-MG in Fig. 16 is performed using the same setup as shown in Fig. 2. Figure 17 shows the reflection spectra of the MD-TFT-MG in air, acetone, and isopropanol, respectively. The extinction ratio is about ~ 10 dB. There are several valleys and peaks with different characteristics in the spectral range of ~ 100 nm. They shift when the outer environment changes from acetone to isopropanol. However, these valleys and peaks show larger shifts at longer wavelengths, while those at shorter wavelength region shift much less and almost stop at specific wavelengths. This unique response to outer liquid refractive index comes from the fact that the reflected light can be coupled to different modes. In the micrometer-diameter metal-dielectric-hybrid TFT, several modes are probably excited with similar propagation constant because of the metal cladding. Some modes are well confined in the tip and have negligible field overlap with the liquid while some modes are not. The different valleys and peaks correspond to the coupling between these different forward and backward propagating modes, with different response properties for the outer environment.

The reflection resonant condition for the grating is:

$$\frac{2\pi}{\lambda_g}[n_f + n_b] = \frac{2\pi}{\Lambda} \tag{7}$$

where n_f and n_b are the effective indices of the forward and backward modes, respectively. For simplicity, we assume a theoretical model to explain our experimental results which is simple and not perfectly matched with the experiment but can give the fundamental mechanism of the device. Within the model, the microfiber is 6 μm in diameter with uniform metal cladding (20 nm in thickness). However, the real device is much more complicated, with nonuniform metal cladding and diameter. And if an asymmetrical mode field lies mainly near the grating, leading to a larger modal overlap with the grating, it may result in a higher sensitivity. Figure 3 shows the calculation on the effective index of one cladding mode and one core mode as a function of outer liquid refractive index n_l. Due to the existence of the metal layer, the cladding mode has a larger effective index (corresponding to long resonant wavelength) than that of the core mode (corresponding to short resonant wavelength) and has a larger overlap with the taper surface and the outside environment, leading to a higher sensitivity to the surrounding medium which is in coincidence with the spectra of Fig. 2.

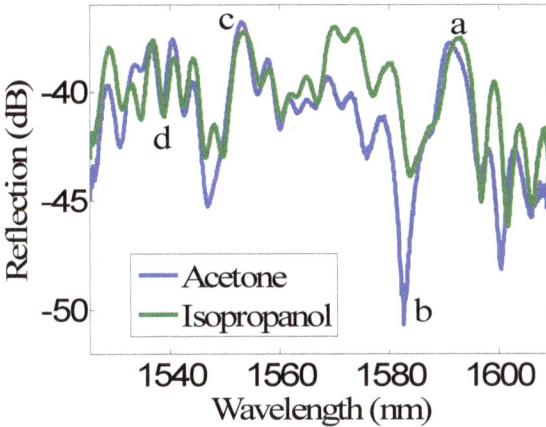

Fig. 17. Measured reflection spectra of the FTG when immersed in acetone and isopropanol (Kou et al., 2011b).

The performance of resonant refractive index sensors can be evaluated by using sensitivity S, which is defined as the magnitude in shift of the resonant wavelength divided by the change in refractive index of the analyte. In our experiment, the sensitivity is measured by inserting the sensor in a beaker containing mixtures of isopropanol and acetone, where the isopropanol component has the following ratios: 0, 1/7, 2/7, 3/7, 4/7 5/7, 6/7, and 1 (Kou et al., 2011b).

Figure 18 displays measured resonant wavelength shifts of several peaks and valleys and fitting of this FTG on the liquid refractive index (a, b, c, d as marked in Fig. 2, a and c are peaks; b and d are valleys). As the refractive index increases, the resonant wavelength shifts to longer wavelength. The sensitivities of different modes change severely. It can be as high as 125 nm/RIU (peak a) or as low as 7 nm/RIU (valley d). For peak a (or valley b), both the resonant wavelength and sensitivity are larger than those of peak c (or valley d). According to our theoretical calculation, we believe peak a (or valley b) corresponds to cladding mode while peak c (or valley d) is core mode. The smallest sensitivity can be further decreased to nearly zero by optimizing the tip grating profile and metal coating. Because of many different properties on the outer liquid refractive index, the metal-dielectric-hybrid FTG can be applied as a multi-parameter sensor and the index-insensitive channel can be used to simultaneously measure temperature, pressure, and so on (Kou et al., 2011b).

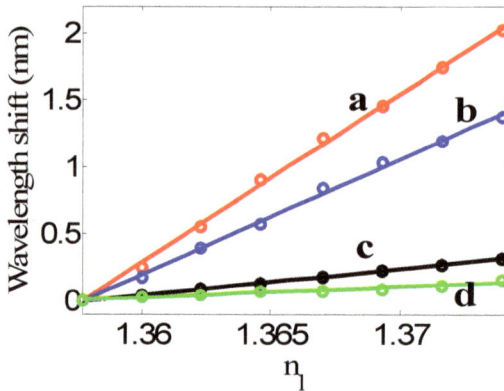

Fig. 18. Dependence of wavelength shift on outer liquid refractive index n_1. The asterisks represent the experimental results with the solid line of linear fitting (Kou et al., 2010b).

5. Conclusion

In this chapter, FIB machined TFT based micro-devices including interferometers and gratings are demonstrated. Being a very flexible, mask-less, direct write process, FIB milling is perfect for carving nanoscale geometries precisely in microfibers. Various miniature fiber devices can be realized and they show great potential in sensing with the unique geometry and size. The sensitivity such as of temperature or refractive index can't increase too much because it mainly depends on the fiber materials and size. But the ultra-small size is attractive for some special application, in particular for detecting small-size objects. Some novel geometry is possible to be realized in microfiber such as an inline-microring, a slot-microfiber etc.

6. Acknowledgment

This work is supported by National 973 program under contract No. 2010CB327803, 2012CB921803 and 2011CBA00200, NSFC program No. 11074117 and 60977039. The authors also acknowledge the support from the Priority Academic Program Development of Jiangsu (PAPD), and the Fundamental Research Funds for the Central Universities.

7. References

Agrawal, G. P. (2002). Fiber-optic communication systems. New York, Wiley-Interscience.

Bhatia, V., Murphy, K. A., Claus, R. O., Jones, M. E., Grace, J. L., Tran, T. A. & Greene, J. A. (1996). Optical fibre based absolute extrinsic Fabry-Perot interferometric sensing system. *Measurement Science & Technology*, Vol. 7, No. 1, (1996), pp. 58-61

Brambilla, G. (2010). Optical fibre nanowires and microwires: a review. *Journal of Optics*, Vol. 12, No. 4, (2010), pp. 043001

Brambilla, G., Finazzi, V. & Richardson, D. (2004). Ultra-low-loss optical fiber nanotapers. *Optics Express*, Vol. 12, No. 10, (2004), pp. 2258-2263

Brambilla, G., Koizumi, E., Feng, X. & Richardson, D. J. (2005). Compound-glass optical nanowires. *Electronics Letters*, Vol. 41, No. 7, (2005), pp. 400-402

Brambilla, G. & Xu, F. (2007). Adiabatic submicrometric tapers for optical tweezers. *Electronics Letters*, Vol. 43, No. 4, (2007), pp. 204-206

Choi, H. Y., Park, K. S., Park, S. J., Paek, U.-C., Lee, B. H. & Choi, E. S. (2008). Miniature fiber-optic high temperature sensor based on a hybrid structured Fabry-Perot interferometer. *Optics Letters*, Vol. 33, No. 21, (2008), pp. 2455-2457

Daniel, J. H., Moore, D. F. & Walker, J. F. (1998). Focused ion beams for microfabrication. *Engineering Science and Education Journal*, Vol. 7, No. 2, (1998), pp. 53-56

Deng, Z. F., Yenilmez, E., Reilein, A., Leu, J., Dai, H. J. & Moler, K. A. (2006). Nanotube manipulation with focused ion beam. *Applied Physics Letters*, Vol. 88, No. 2, (2006),

Feng, J., Ding, M., Kou, J.-l., Xu, F. & Lu, Y.-q. (2011). An optical fiber tip micro-grating thermometer. *IEEE Photonics Journal*, Vol. 3, No. 5, (2011), pp. 810-814,

Hill, K. O., Fujii, Y., Johnson, D. C. & Kawasaki, B. S. (1978). Photosensitivity in Optical Fiber Waveguides - Application to Reflection Filter Fabrication. *Applied Physics Letters*, Vol. 32, No. 10, (1978), pp. 647-649

Hill, K. O., Malo, B., Bilodeau, F., Johnson, D. C. & Albert, J. (1993). Bragg Gratings Fabricated in Monomode Photosensitive Optical-Fiber by UV Exposure through a Phase Mask. *Applied Physics Letters*, Vol. 63, No. 3, (1993), pp. 424-424

Hofmann, S., Cantoro, M., Kaempgen, M., Kang, D. J., Golovko, V. B., Li, H. W., Yang, Z., Geng, J., Huck, W. T. S., Johnson, B. F. G., Roth, S. & Robertson, J. (2005). Catalyst patterning methods for surface-bound chemical vapor deposition of carbon nanotubes. *Applied Physics A: Materials Science & Processing*, Vol. 81, No. 8, (2005), pp. 1559-1567

Hopman, W. C. L., Ay, F. & Ridder, R. M. d. (2008). Focused ion beam milling strategy for sub-micrometer holes in silicon. Workshop FIB for Photonics, Eindhoven,the Netherlands. http://en.wikipedia.org/wiki/Focused_ion_beam

Jeon, J., Floresca, H. C. & Kim, M. J. (2010). Fabrication of complex three-dimensional nanostructures using focused ion beam and nanomanipulation *Journal of Vacuum Science & Technology B*, Vol. 28, No. 3, (2010), pp. 549-553

Kou, J.-l., Feng, J., Wang, Q.-j., Xu, F. & Lu, Y.-q. (2010a). Microfiber-probe-based ultrasmall interferometric sensor. *Optics Letters*, Vol. 35, No. 13, (2010a), pp. 2308-2310

Kou, J.-l., Feng, J., Ye, L., Xu, F. & Lu, Y.-q. (2010b). Miniaturized fiber taper reflective interferometer for high temperature measurement. *Optics Express*, Vol. 18, No. 13, (2010b), pp. 14245-14250

Kou, J.-l., Qiu, S.-j., Xu, F. & Lu, Y.-q. (2011a). Demonstration of a compact temperature sensor based on first-order Bragg grating in a tapered fiber probe. *Optics Express*, Vol. 19, No. 19, (2011a), pp. 18452-18457

Kou, J.-l., Qiu, S.-j., Yuan, Y., Zhao, G., Xu, F. & Lu, Y.-q. (2011b). Miniaturized Metal-dielectric-hybrid Fiber Tip Grating for Refractive Index Sensing. *IEEE Photonics Technology Letters*, Vol. 23, No. 22, (2011), pp. 1712-1714 ,(2011b),

Liang, T., Stivers, A., Livengood, R., Yan, P. Y., Zhang, G. J. & Lo, F. C. (2000). Progress in extreme ultraviolet mask repair using a focused ion beam. *Journal of Vacuum Science & Technology B*, Vol. 18, No. 6, (2000), pp. 3216-3220

Lin, C. Y. & Wang, L. A. (2001). A wavelength- and loss-tunable band-rejection filter based on corrugated long-period fiber grating. *IEEE Photonics Technology Letters*, Vol. 13, No. 4, (2001), pp. 332-334

Liu, K., Soskov, A., Scipioni, L., Bassom, N., Sijbrandij, S. & Smith, G. (2006). Electrical breakthrough effect for end pointing in 90 and 45 nm node circuit edit. *Applied Physics Letters*, Vol. 88, No. 12, (2006), pp. 124104

Love, J. D., Henry, W. M., Stewart, W. J., Black, R. J., Lacroix, S. & Gonthier, F. (1991). Tapered single-mode fibres and devices. I. Adiabaticity criteria. *Optoelectronics, IEE Proceedings Journal*, Vol. 138, No. 5, (1991), pp. 343-354

Martinez, A., Khrushchev, I. Y. & Bennion, I. (2005). Thermal properties of fibre Bragg gratings inscribed point-by-point by infrared femtosecond laser. *Electronics Letters*, Vol. 41, No. 4, (2005), pp. 176-178

Nemova, G. & Kashyap, R. (2006). Fiber-Bragg-grating-assisted surface plasmon-polariton sensor. *Optics Leters*, Vol. 31, No. 14, (2006), pp. 2118-2120

Rao, Y.-J., Deng, M., Duan, D.-W., Yang, X.-C., Zhu, T. & Cheng, G.-H. (2007). Micro Fabry-Perot interferometers in silica fibers machined by femtosecond laser. *Optics Express*, Vol. 15, No. 21, (2007), pp. 14123-14128

Sirkis, J. S., Brennan, D. D., Putman, M. A., Berkoff, T. A., Kersey, A. D. & Friebele, E. J. (1993). In-line fiber ealon for strain measurement. *Optics Letters*, Vol. 18, No. 22, (1993), pp. 1973-1975

Streifer, W. & Hardy, A. (1978). Analysis of two-dimensional waveguides with misaligned or curved gratings. *IEEE Journal of Quantum Electronics*, Vol. 14, No. 12, (1978), pp. 935-943

Streifer, W., Scifres, D. & Burnham, R. (1975). Coupling coefficients for distributed feedback single- and double-heterostructure diode lasers. *IEEE Journal of Quantum Electronics*, Vol. 11, No. 11, (1975), pp. 867-873

Terris, B., Thomson, T. & Hu, G. (2007). Patterned media for future magnetic data storage. *Microsystem Technologies*, Vol. 13, No. 2, (2007), pp. 189-196

Tong, L. M., Gattass, R. R., Ashcom, J. B., He, S. L., Lou, J. Y., Shen, M. Y., Maxwell, I. & Mazur, E. (2003). Subwavelength-diameter silica wires for low-loss optical wave guiding. *Nature*, Vol. 426, No. 6968, (2003), pp. 816-819

Villatoro, J., Finazzi, V., Coviello, G. & Pruneri, V. (2009). Photonic-crystal-fiber-enabled micro-Fabry?Perot interferometer. *Optics Letters*, Vol. 34, No. 16, (2009), pp. 2441-2443

Wei, T., Han, Y., Li, Y., Tsai, H.-L. & Xiao, H. (2008). Temperature-insensitive miniaturized fiber inline Fabry-Perot interferometer for highly sensitive refractive index measurement. *Optics Express*, Vol. 16, No. 8, (2008), pp. 5764-5769

Xu, F., Brambilla, G., Feng, J. & Lu, Y.-Q. (2010). A Microfiber Bragg Grating Based on a Microstructured Rod: A Proposal. *IEEE Photonics Technology Letters*, Vol. 22, No. 4, (2010), pp. 218-220

Xu, F., Brambilla, G. & Lu, Y. (2009). A microfluidic refractometric sensor based on gratings in optical fibre microwires. *Optics Express*, Vol. 17, No. 23, (2009), pp. 20866-20871

Xuan, H., Jin, W. & Liu, S. (2010). Long-period gratings in wavelength-scale microfibers. *Optics Letters*, Vol. 35, No. 1, (2010), pp. 85-87

Xuan, H., Jin, W. & Zhang, M. (2009). CO_2 laser induced long period gratings in optical microfibers. *Optics Express*, Vol. 17, No. 24, (2009), pp. 21882-21890

4

Microwave Meta-Material Absorbers Utilizing Laser Micro-Machining Technology

Hongmin Lee
Kyonggi University,
Korea

1. Introduction

Recently, artificially structured electromagnetic (EM) materials have become an extremely active research area because of the possibility of creating materials which exhibit novel EM responses not available in nature. This includes negative refractive index (NRI), super-lens, cloaking, and more generally, coordinating transformation materials. For the most part, these composites, often called meta-materials (MTMs). The double negative (DNG) MTM structure was realized in 2000 by appropriately depositing SRRs and thin-wires on dielectric substrate. Since then, most of reported designs have a 1D or 2D geometry that responds only to one (two) electrical and magnetic components of the electromagnetic fields. Much of the work in MTM has been focused on the real parts of permittivity and permeability to enable the creation of a negative refractive index material. However, they can be manipulated to create a high performance absorber. According to effective medium theory, MTMs can be represented by the complex electric permittivity ε_{eff} (= ε' + $j\varepsilon''$) and magnetic permeability μ_{eff} (= μ' + $j\mu''$). By varying the dimensions of electric and magnetic components, it is possible to adjust permittivity and permittivity independently. Additionally, by tuning the electric and magnetic resonances a MTM can be impedance matched to free space, resulting reflectivity R = 0. The additional multiple layers or metallic back-plate will also ensure transmission T = 0. As a result, 100 % absorbance A (= 1– R – T) is theoretically possible.

The microwave absorbers are used in military application to reduce the radar cross-section (RCS) of a conducting object and electromagnetic (EM) interference among microwave components. One of the earliest approaches for the design of EM absorber structure is based on the use of Salisbury screen. This type of absorber needs the resistive sheet and a metallic ground plane. The metallic backing plays two main roles; 1) it is used to avoid power transmission on the other side of the absorber, 2) it cancels out a reflected component that combined with the impinging wave. Recently, advancement in absorber technology has been obtained by using artificially MTMs to create a high-performance absorber for the microwave and terahertz frequency regime. In practice, it is difficult to make the absorber's electrical size small enough at low frequencies. For the design of compact microwave absorbers made by MTM complimentary pairs, we need to choose proper unit cell structures which are characterized by oppositely signed values of real parts of permittivity and permittivity. However, the absorbers are usually made with metallic backing plates in order to avoid power transmission on the absorbers' other side, which may represent many problems for stealth applications. In order to design a metallic backplane-less absorber with double- negative MTM unit cell structure, we may refer to the sketch shown in Fig. 1. The two metallic pattern layers separated by dielectric spacer can be placed either orthogonal to EM wave propagation direction or parallel. If the two metallic pattern layers are placed orthogonal to EM wave propagation direction, as shown in Fig. 1(a), the radar cross section (RCS) of the object may increase at frequencies other than aimed design frequency bands.

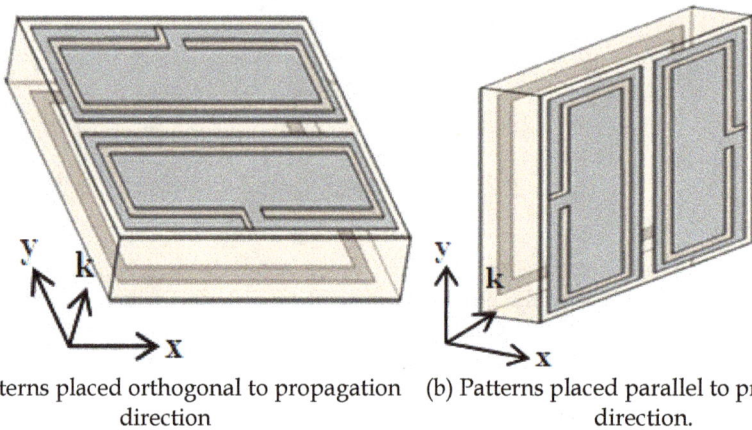

(a) Patterns placed orthogonal to propagation direction   (b) Patterns placed parallel to propagation direction.

Fig. 1. Sketch describing the geometry of MTM absorber unit cell (where, k is the wave propagation direction).

In order to avoid this problem, the two metallic pattern layers would need to be paced parallel to EM wave propagation direction, as shown in Fig. 1(b). In this study, the prototype absorber resonator structures were fabricated using both of a surface micromachining process technique and a standard photolithography technique.

2. Design of a miniaturized meta-material microwave absorber

2.1 Double negative unit cell design

The practical implementation of a double negative MTM unit cell involves the proper choice of both the structures with the negative real part of the permittivity and the negative real part of the permeability. A single unit cell of the proposed absorber consisted of distinct metallic elements, as shown in Fig. 2(a) and 2(b). The electric responses were provided by two symmetrically placed open complimentary split ring resonators (OCSRRs), as shown in Fig 2(a). We have constructed NRI MTM unit cell using open complimentary split ring resonator (OCSRR) and split ring resonator (SRR) arrangement. The OCSRR has been derived from two former planar resonant structures: the open split ring resonator (OSRR) and the complimentary split ring resonator (CSRR). As compared to SRR and CSRR, the electrical size of OCSRR is smaller and it can be modeled as an open parallel resonant circuit. The OCSRR is modified CSRR structure exhibiting negative permittivity and the SRR structure exhibits negative permeability. Each unit cell is printed on the two side of a FR-4 substrate. We use a double-layered structure with a SRR and two OCSRRs which are put on top of each other to make a miniaturized MTM absorber unit cell for 2 GHz frequency band.

The magnetic responses were provided by two spirals, as shown in Fig. 2(b). We created electromagnetic responses by the OCSRRs with two spirals in a parallel plane separated by a lossy dielectric substrate, as shown in Fig. 2(c). The absorber unit cell is made of a FR-4 substrate whose relative dielectric constant is ε_r = 4.4, and loss angle tangent tan δ = 0.025, and thickness t = 0.8 mm. The metal for metallic patterns is a copper whose conductivity is σ = 5.8 $\times$ 10^{7}S/m. By changing the geometry and the separation between the OCSRRs and the spirals the electromagnetic responses are tuned to match the impedance to free space and minimize the transmission at the aimed design frequency. Computer simulations for one unit cell are carried out using the commercial finite-difference time domain solver Microwave Studio by CST. The program simulated a single unit cell with appropriate boundary conditions, as shown in Fig. 2(c). The perfect electric conductor (PEC) boundary conditions are applied to the top and bottom walls of the waveguide, where as perfect magnetic conductor (PMC) boundary conditions are applied to the side walls of the waveguide. The other two opposite sides of the waveguide is assigned as waveguide ports. The total dimension of a cell is 7.3 mm $\times$ 7 mm $\times$ 0.8 mm. A single unit cell is placed inside a waveguide, and a vertically polarized transverse electromagnetic (TEM) wave is incident normally on the front side of port 1, as shown in Fig. 2(c). The scattering parameters of this MTM unit cell were then simulated, and the absorbance was calculated using the equation A = 1- $|S_{11}|^2$- $|S_{21}|^2$. The simulated magnitudes of S_{11} and S_{21} parameters are plotted in Fig. 3(a). We observe that both the reflection and transmission are very low at the resonance frequency of 2.43 GHz, which indicates a strong absorption of the EM wave energy.

In order to express the effective permittivity and permeability of artificial material in terms of the scattering parameters, they are conventionally retrieved from scattering parameters of a unit cell under plane wave excitation [11]. The impedance parameters and ABCD parameters for two- port network can be calculated from scattering parameters using simple transformation. Then the Bloch-Floquet theorem was used to calculate the Bloch impedance Z_B, and complex propagation constant γ.

$$\gamma = \cos^{-1}((Z_{11}+Z_{22})/2Z_{21})/p. \tag{1}$$

Where p is the size of the MTM unit cell, and the Bloch impedance Z_B can be expressed as

$$Z_B = B/(e^{j\gamma p}-A). \tag{2}$$

Where, the parameter A is the voltage ratio between two ports with open-circuit at port 2 and the parameter B is the trans-admittance with short-circuit at port 2 can be expressed using impedance parameters as

$$A = Z_{11}/Z_{22}, B = (Z_{11}Z_{22} - Z_{21}^2)/Z_{21}. \tag{3}$$

The effective permittivity ε_{eff} and permeability μ_{eff} can then easily calculated from Bloch impedance and propagation constant with the free space wave number k_0, and wave impedance Z_0 of the empty waveguide, respectively.

$$\mu_{eff} = (\gamma Z_B)/(k_0 Z_0). \tag{4}$$

$$\varepsilon_{eff} = (\gamma Z_0)/(k_0 Z_B). \tag{5}$$

The extracted frequency dependence of the effective parameter results are plotted in Fig. 3. The real and imaginary components ε_{eff} (= ε' - jε'') and μ_{eff} (= μ' - jμ'') are plotted in Fig. 3(b) and (c), respectively. There is a frequency interval, in which one effective parameter is negative ε' for OCSRRs, μ' for sipral). Note that both the real components of the effective permittivity and permeability (ε' and μ') are negative, and the imaginary components (ε'' and μ'') are positive at the aimed design frequency of 2.43 GHz. This meets the general condition for the power flow and the phase velocity to be oppositely directed to the power flow, which is written as;

$$\varepsilon'\mu'' + \mu'\varepsilon'' > 0. \tag{6}$$

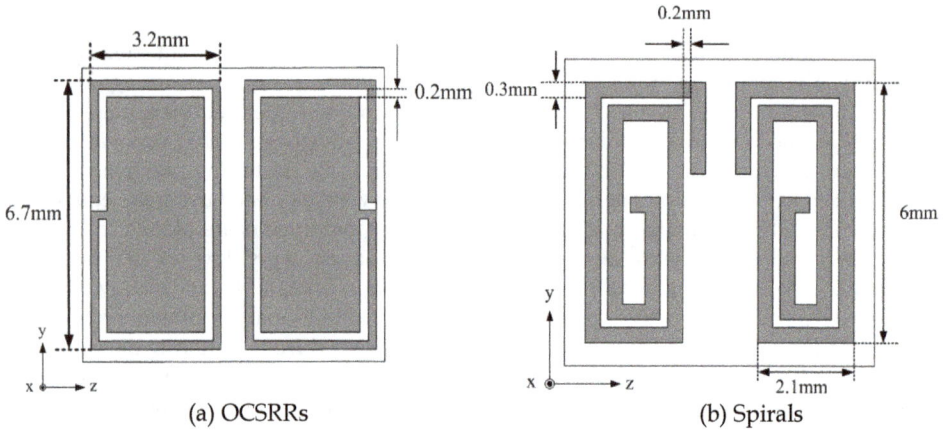

(a) OCSRRs   (b) Spirals

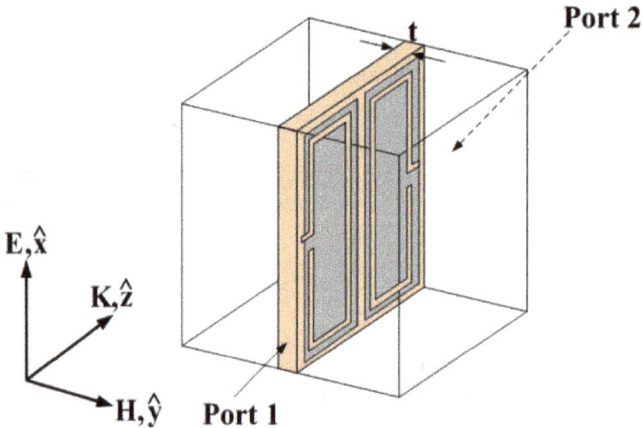

(c) Single unit cell showing the direction of propagation of incident electromagnetic wave (substrate thickness t = 0.8 mm).

Fig. 2. Schematic of optimum absorber unit cell and simulation setup.

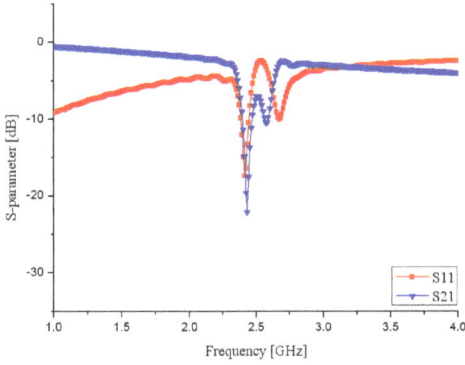

(a) Simulated S-parameters of one unit cell

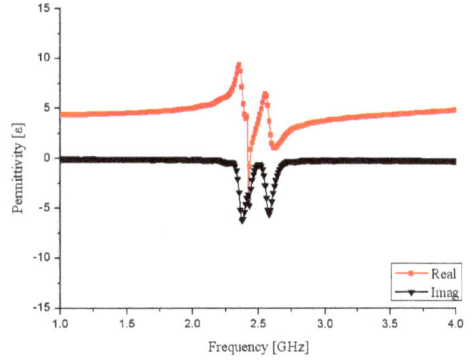

(b) Effective permittivity

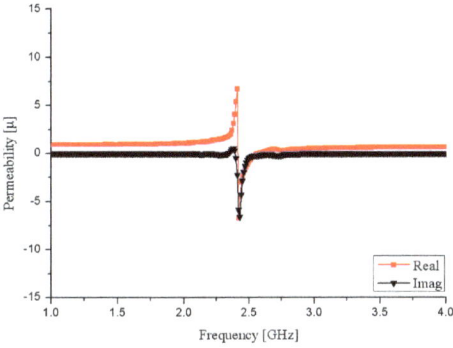

(c) Effective permeability

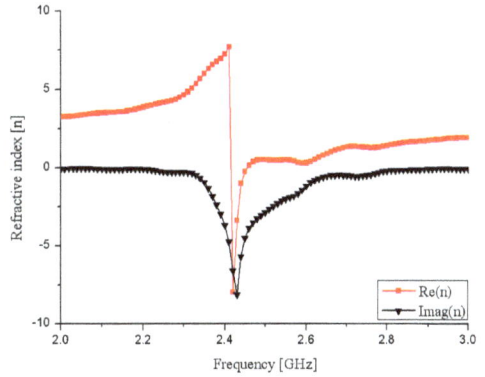

(d) Refractive index

Fig. 3. Simulated results for the single absorber unit cell.

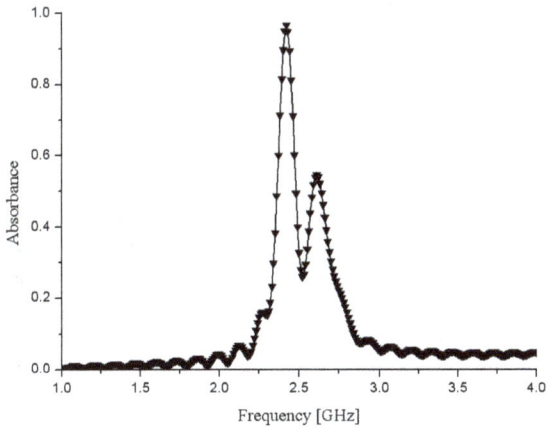

Fig. 4. Simulated absorbance of the metamaterial absorber cell.

(a) f = 2.43 GHz

(b) f = 2.54 GHz

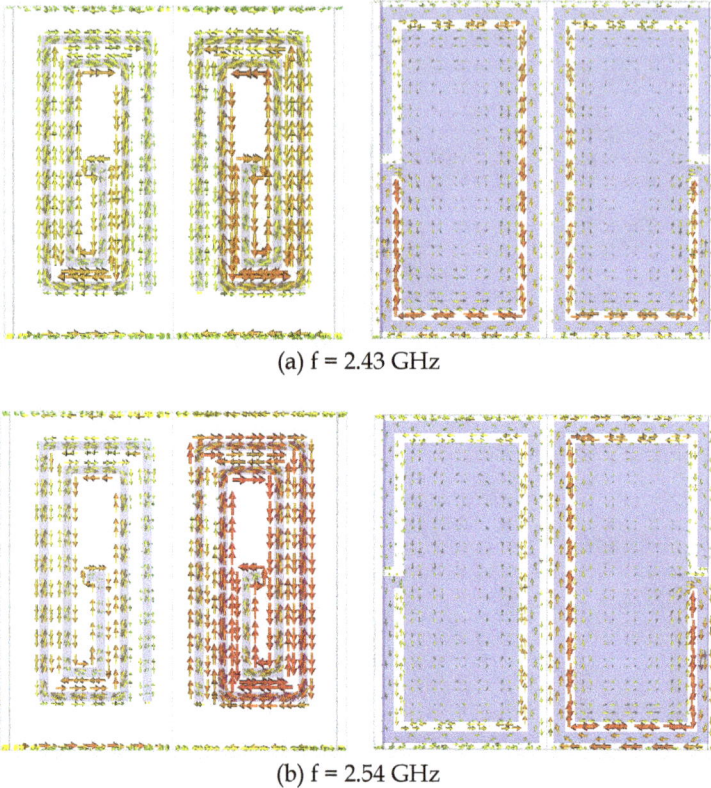

Fig. 5. Simulated surface current densities in the spirals and OCSRRs.

As a result, the unit cell can be regarded as a double negative metamaterial unit cell over the frequency range 2.43-2.45 GHz. As shown in Fig. 3(c), the imaginary part of the refractive index is large ($n'' \approx 8$) in the left-handed frequency region which means strong absorption of the EM wave energy. The simulated absorbance curve over a broader frequency range is plotted in Fig. 4. The maximum absorbance peak is 96% at 2.43 GHz, there is a secondary absorbance peak at approximately at 2.54 Hz. In order to understand the nature of this absorbance, the simulated surface current densities in the top resonator structure of spiral and the lower resonator structure of OCSRRs for 2.43 and 2.54 GHz resonances are shown in Fig. 5, respectively. For the 2.43 GHz resonance, we observe that the counter-circulating currents flow on both the spirals provide magnetic resonance, and the stronger current density takes place in both the shorted-end of the left-side OCSRRs, which provide electric resonance. In contrast, the 2.54 GHz resonance is determined by the magnetic response associated with a circulating current flowing on the right-side spiral and the electric resonance associated with the shorted-end of the right-side OCSRR. Fig. 6 shows the simulated S-parameters for the different horizontal spacing lengths g between the absorber cells and the simulated results are list in the Table 1. When the spacing between two cells is 6 mm, the arrayed cell shows good impedance matched to free space impedance and maximum absorbance.

(a) Unit cells array

(b) $|S_{11}|$

(c) $|S_{21}|$

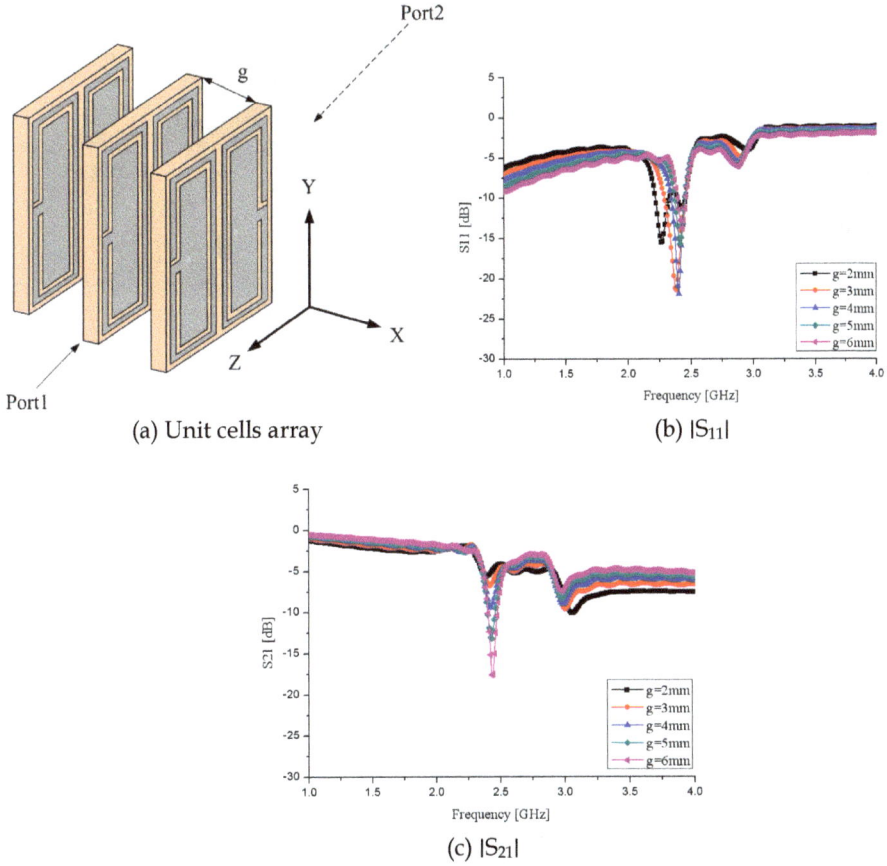

Fig. 6. Simulated results for the different horizontal spacing lengths g between the absorber cells.

g [mm]	S11 [dB]	S21 [dB]
2	-11.4	-5.0
3	-12.5	-6.6
4	-13.3	-9.4
5	-13.2	-13.2
6	-13.8	-17.6

Table 1. The summary of the simulated results for the unit cells array.

2.2 Experimental results

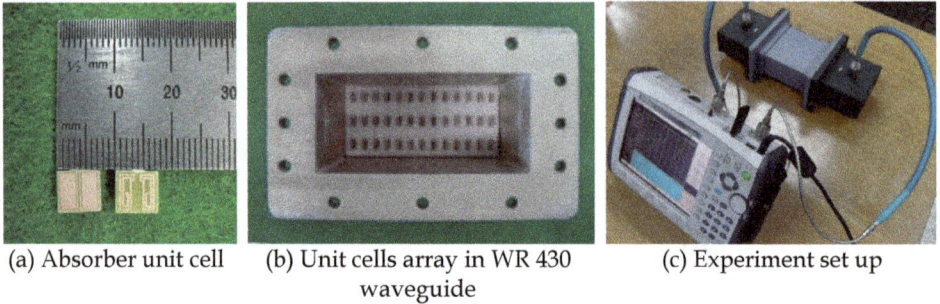

| (a) Absorber unit cell | (b) Unit cells array in WR 430 waveguide | (c) Experiment set up |

Fig. 7. Photographs of the fabricated prototype absorber unit cell and unit cells array.

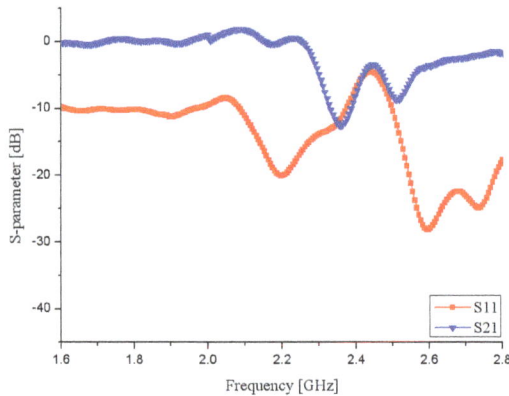

Fig. 8. Measured S-parameters of the planar arrayed (13 × 3) unit cells.

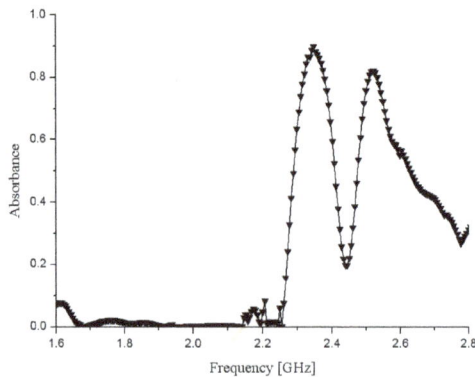

Fig. 9. Measured absorbance curve.

In order to verify a new type of the backplane-less absorber was designed without the resistive sheet. We fabricate a proto-type unit cell on a FR-4 substrate (ε_r = 4.4, tan δ = 0.025, and thickness t = 0.8 mm) using standard photolithography technique, and an experiment was carried out. The photographs of the fabricated two-layer metallization MTM absorber sample are shown in Fig. 7 (a). In order to verify the effectiveness of the double-negative MTM absorber cells without a metallic backing plate, a planar array of absorber unit cells (13 × 3) was mounted on a polystyrene foam substrate with a relative permittivity of 1.02. The inter-element spacing between two unit cells was set to 6 mm. As shown in Fig. 7 (b), this planar absorber structure was placed inside a rectangular waveguide (WR430), and measurements were taken with the waveguide over the frequency range from 1.6 GHz to 2.8 GHz. At these frequencies, the EM wave propagation was confined to the TE_{10} mode. The measured magnitudes of S_{11} and S_{21} parameters for the planar arrayed (13 × 3) unit cells are plotted in Fig 8. This results show that the backplane-less planar absorber structure is able to achieve the well matched broadband absorber. However, the experimental transmission minimum frequency of 2.43 GHz is shifted approximately 90 MHz lower compared to the simulated transmission minimum frequency. By using the measured S_{11} and S_{22} parameters, the calculated corresponding absorbance A (= 1- $|S_{11}|^2$- $|S_{21}|^2$) was plotted in Fig. 9. It exhibits a maximum absorbance of 90% at 2.34 GHz, with a full width at half maximum (FWHM) of 120 MHz over the frequency range from 2.29 to 2.41 GHz. This broader bandwidth performance is likely due to a mutual coupling effect between the cells. Although, the field distribution of the TE_{10} mode wave is different compared to the plane wave, the proposed planar absorber structure shows the possibility of the planar-type absorber without metallic backing plates or a resistive sheet.

A new idea for a backplane-less planar-type microwave absorber based on the arrayed double-negative MTM unit cells has been presented. By properly arrayed a single unit cell with a SRR and two OCSRRs configuration, has been shown to effectively absorb most of the impinging power. The two main advantages of the proposed absorber are: 1) the absence of a metallic back plate or a resistive sheet and 2) the reduced thickness (close to $0.06\lambda_0$) in the EM wave propagation direction. The total miniaturized MTM absorber unit cell size was 7.4 mm × 7 mm × 0.8 mm. The proposed layout can be easily extended to work for more compact and thin backplane-less planar absorber designs.

3. Design of a metamaterial absorber with an enhanced operating band-width

3.1 Choice of the resonant inclusion on the edge-coupled SRRs

We present the employment of the modified microwave meta-material absorber structure in order to achieve a further enhancement of the operating bandwidth. The SRRs can be used as electrically resonant particles which exhibit a strong resonant permittivity, and can also be used as the magnetic resonant particles which exhibit a strong resonant permeability. Fig. 10 shows the various types of the edge-coupled SRRs with different electromagnetic orientations. The electric field polarization is kept along the x-axis, and magnetic field polarization is kept along the y-axis. In principle, when the SRRs excited purely by magnetic field normal to the SRRs plane with electric field perpendicular to the gap (Fig. 10(b)), it exhibits a magnetic response only. SRRs excited by electric field parallel to the gap and the magnetic field lying completely in the SRRs plane (Fig. 10(d)), it exhibits an electric response only. Furthermore, the SRRs can exhibit both a purely magnetic and electric response to different electromagnetic orientation, as shown in Fig. 10(a). This results in a complicated bi-anisotropic EM behaviour. When the SRRs excited by in-plane magnetic field with electric field perpendicular to the gap (Fig. 10(c)), there is no magnetic or electric response.

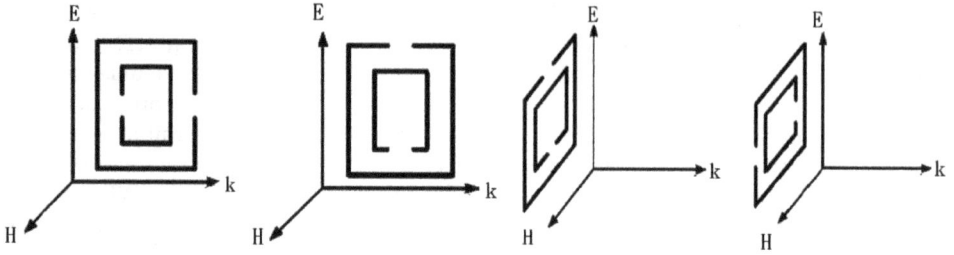

Fig. 10. The various types of the edge-coupled SRRs with different electromagnetic orientations.

3.2 Double negative unit cell design

The two type of the proposed absorber unit cells consisted of distinct metallic elements are shown in Fig. 11 (type 1) and Fig. 12 (type 2). In the absorber unit cell type 1, the purely electric and magnetic responses were provided by the two symmetrically placed OCSRRs and an SRR, respectively, as shown in Fig 11. In the absorber unit cell type 2, the purely electric responses were provided by the two OCSRRs and the electromagnetic responses were provided by the 90⁰ rotated SRR, as shown in Fig 12.

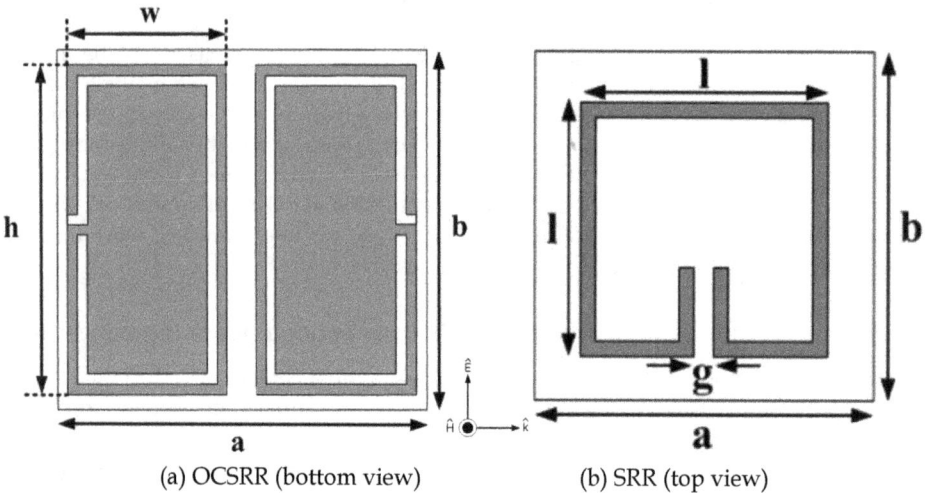

(a) OCSRR (bottom view)   (b) SRR (top view)

Fig. 11. Geometry of the absorber unit cell (type 1).

We created electromagnetic responses by combining the OCSRRs with an SRR in a parallel plane separated by a lossy dielectric substrate. The absorber unit cells are made of a FR-4 substrate whose relative dielectric constant is $\varepsilon_r = 4.4$, and loss angle tangent tan δ = 0.025, and thickness t = 1.8 mm. The metal for metallic patterns is a copper whose conductivity is σ = 5.8 × 10⁷S/m. By changing the geometry and the separation between the OCSRRs and SRR, the electromagnetic responses are tuned to match the impedance to free space and minimize the transmission at the aimed design frequency. The geometrical dimensions are: a = 7.4 mm, b = 7 mm, w = 3.2 mm, h = 6.4 mm l = 4.6 mm, and g = 0.4 mm. Computer simulations for one unit cell are carried out using the commercial finite-difference time

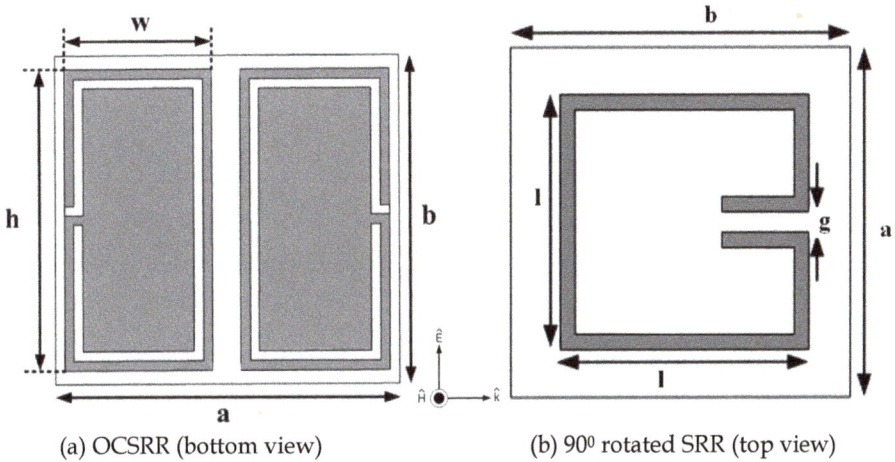

(a) OCSRR (bottom view)   (b) 90⁰ rotated SRR (top view)

Fig. 12. Geometry of the absorber unit cell (type 2).

domain solver Microwave Studio by CST. The scattering parameters of theses MTM unit cells were then simulated, and the absorbance was calculated using the equation $A = 1 - |S_{11}|^2 - |S_{21}|^2$. The calculated absorbance is plotted in Fig. 13. In case of type 1, the maximum absorbance peak is 93% at 2.47 GHz, there is a weak secondary absorbance peak 40% at 2.98 Hz. In case of type 2, the maximum absorbance peak is 84% at 2.6 GHz, there is a strong secondary absorbance peak 86% at 2.97 Hz. When three cells are arrayed in vertical direction with a configuration of type 2, it exhibits near same absorbance characteristics. In order to understand the nature of this absorbance, the simulated surface current densities in the top resonator structure of a SRR and the lower resonator structure of OCSRRs are shown in Fig. 13 and Fig. 14, respectively.

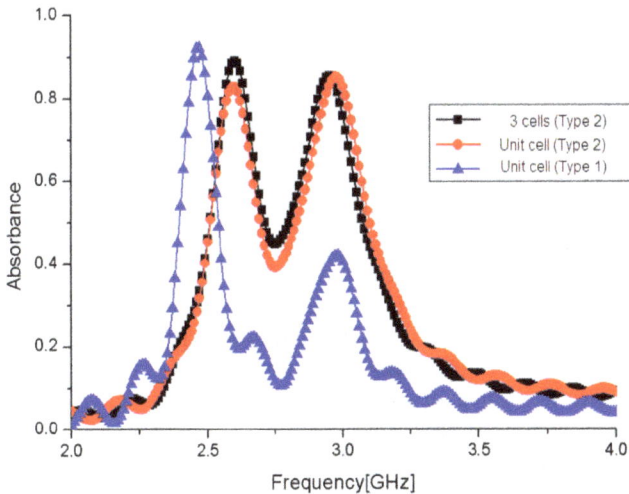

Fig. 13. Calculated absorbance of the absorber unit cells with a different configuration.

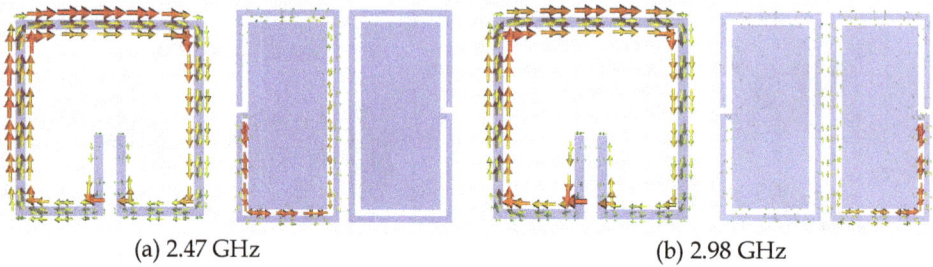

(a) 2.47 GHz   (b) 2.98 GHz

Fig. 14. Simulated surface current densities (type 1).

(a) 2.6 GHz   (b) 2.97 GHz

Fig. 15. Simulated surface current densities (type 2).

For the 2.47 GHz resonance in type 1, we observe that the counter-circulating currents flow on the SRR provide magnetic resonance, and the stronger current density takes place in the shorted-end of the left-side OCSRR, which provide electric resonance. In contrast, the 2.98 GHz resonance is determined by the magnetic response associated with a circulating current flowing on the SRR and electric resonance associated with the shorted-end of the right-side OCSRR. For the 2.6 GHz resonance in type 2, we observe that a dipole resonance currents and weak circulating flow on the SRR, and the stronger current density takes place in the shorted-end of the right-side OCSRR. The 2.97 GHz resonance is determined by the magnetic response associated with a circulating current flowing on the SRR and electric resonance associated with the shorted-end of the left -side OCSRR.

Fig. 16 shows the sketch of the planar arrayed (13 × 4) structures of MTM absorber unit cells of type 2. The unit cells are vertically arrayed without a gap length and the horizontally arrayed with a spacing lengths g = 6 mm. A vertically polarized transverse electromagnetic (TEM) wave impinges upon this structure with the incident angle from 0^0 to 45^0 and the scattering parameters were then simulated. The simulated S-parameters are plotted in Fig. 17. We observe that the planar arrayed unit cell shows the first and secondary absorbance peak at 2.93, and 3.27 GHz, respectively. Due to a mutual coupling effect between the cells, the two maximum absorbance frequencies are shifted to a higher frequency band.

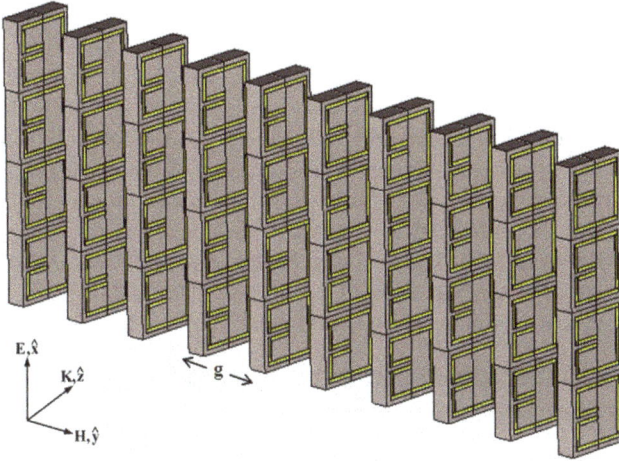

Fig. 16. The planar arrayed (39 × 39) unit cells of type 2.

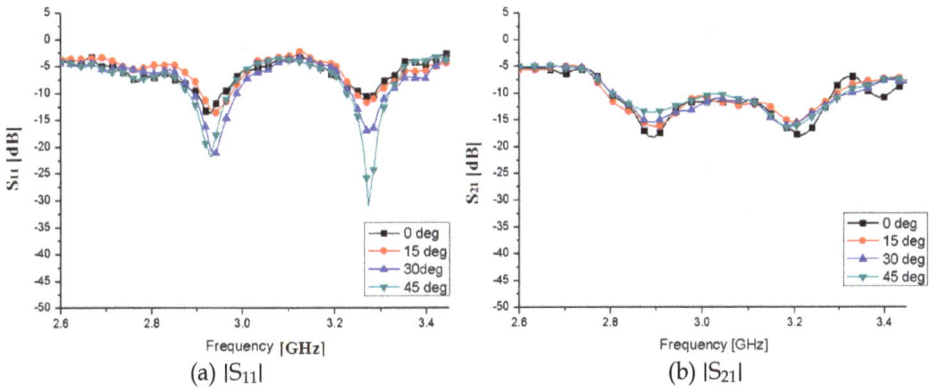

(a) |S_{11}|

(b) |S_{21}|

Fig. 17. The simulated S-parameters.

3.3 Experimental results

In order to verify a new type of the backplane-less absorber was designed without the resistive sheet. We fabricate a proto-type unit cell on a FR-4 substrate (ε_r = 4.4, tan δ = 0.025, and thickness t = 1.8 mm) using standard photolithography technique, and an experiment was carried out. The photographs of the fabricated two-layer metallization MTM absorber sample are shown in Fig. 18 (a). In order to verify the effectiveness of the double-negative MTM absorber cells without a metallic backing plate, a planar array of absorber unit cells (39 × 39) was mounted on acryl test-zig, as shown in Fig. 18(b). The inter-element spacing between two unit cells was set to 7 mm and the total size of the planar absorber was 300 m × 300 mm, and measurements were taken over the frequency range from 2.6 to 3.4 GHz. The calculating absorbance by using the measured magnitudes of S_{11} and S_{21} parameters for the planar arrayed absorber cells are plotted in Fig. 18 (c). With the increasing incidence angle from 0^0 to 45^0 the absorbance curves have little change and the two absorbance peaks remain more than 90% for different incident angles. For normal incident waves, the experimental data exhibits a maximum absorbance of 93% at 2.93 GHz, with a full width at half maximum (FWHM) of 250 MHz, and a secondary peak absorbance 87% at 3.26 GHz with a FWHM of 260 MHz. By properly arrayed a single unit cell with a 90^0 rotated SRR and two OCSRRs configuration, has been shown to effectively absorb most of the impinging power. The two main advantages of the proposed absorber are: 1) the absence of a metallic back plate or a resistive sheet and 2) the enhanced absorbance operating band by using electromagnetic responses of the 90^0 rotated SRR. The thickness of the planar MTM absorber structure in the EM wave propagation direction was shown 7.4 mm ($\approx \lambda_0$) at 2.6 GHz. The proposed layout can be easily extended to work for more compact and thin backplane-less planar absorber designs.

(a) One period of the arrayed unit cells.

(b) A planar arrayed MTM absorber

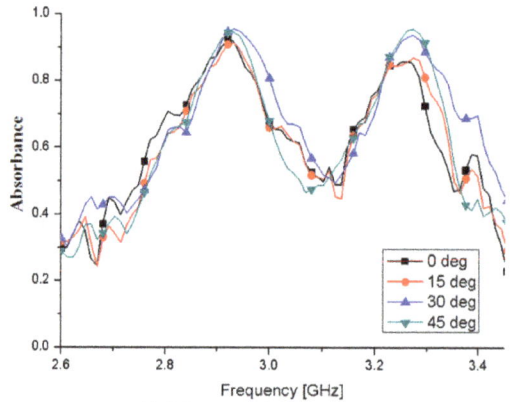

(c) Measured absorbance

Fig. 18. Photographs of the fabricated absorber and the calculated results of the absorbance.

4. Design of a meta-material absorber applications

4.1 Three-dimensional resonator for EM waves absorption

In recent years, as the use of mobile phones has become widespread, the EM waves shielding technology is a hot issue for the intelligent building construction. In this work, a three-dimensional resonators made of cubic arrangements of the OCSRRs and a metal strips is proposed. By installing these spatial band-stop cube resonators inside the concrete walls, the incident electromagnetic waves into the concrete walls can be blocked effectively over the specific frequency band. For the practical applications it is desirable to have a less sensitive to the incident angle and polarization of the impinging waves. However, it is very difficult to implement in practices due to its size and surrounding environments of the concrete. In order to solve this problem, two OCSRRs were used as basic elements and a metal strip conductor was connected between these OCSSRs. As a result, a three-dimensional resonators (7 mm $\times$ 7 mm $\times$ 7 mm) which acts individually interacting with

(a) Geometry of one unit cell

(b) Electric field distribution

(c) Surface current distribution

(d) Simulated S-parameters of one unit cell

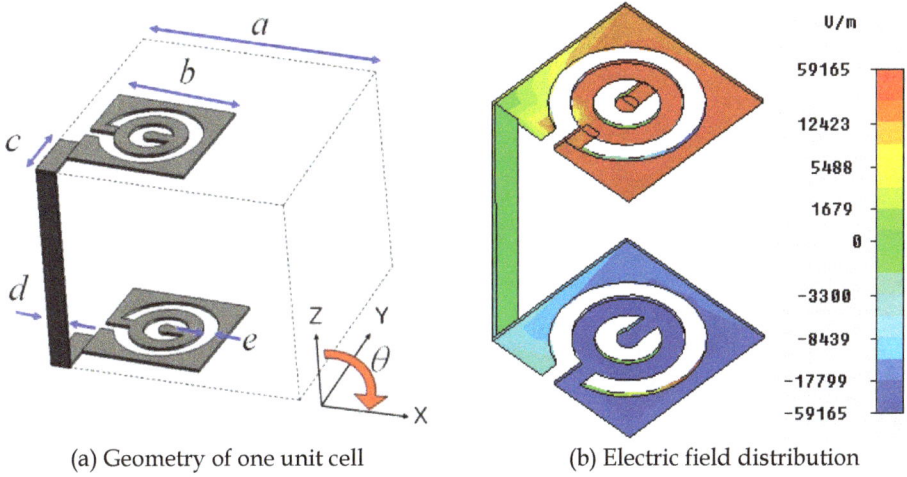

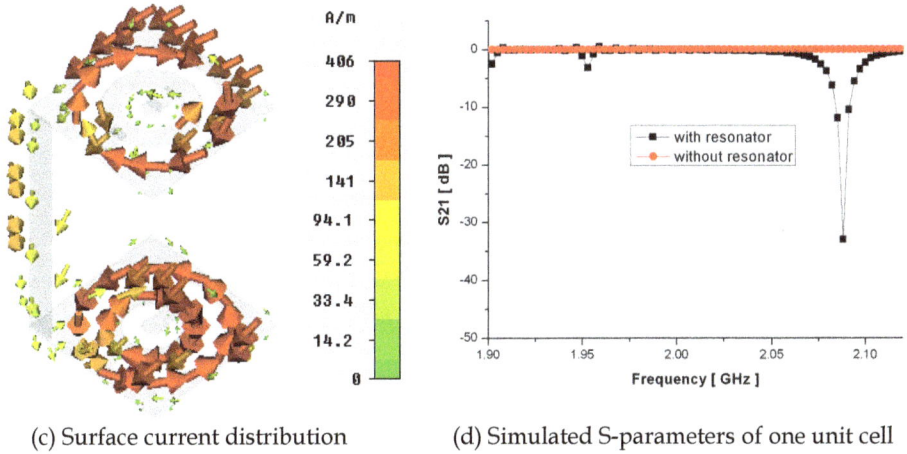

Fig. 19. Geometry of one unit cell and the simulated results.

the incoming electromagnetic waves at 2 GHz frequency band. The geometry of one unit element and the simulated results are shown in Fig. 19. The geometrical dimensions are: a = 7 mm, b = 3.8 mm, c = 2 mm, d = 0.85 mm, and e = 0.45 mm. When the unit element is placed within a rectangular waveguide (109 mm × 54.5 mm × 150 mm), the simulated transmission coefficient through with/without a resonator is shown in Fig. 19 (b). The resonance frequency of the proposed resonator shows 2.08 GHz and the stop-band width (≤ -10dB) is 19 MHz. The resonance frequency can be adjusted by the variation of the capacitance and inductance, which depend on the physical length of the slot width, ring spacing, ring radius, and the height of metal strip. The unit cell of the three-dimensional resonator (7 mm × 7 mm × 7 mm) is shown in Fig. 20. On each side of the six sides of the cube structures with a metallic patterns of a resonator and a lossy dielectric material (ε_r = 6.0) is filled inside of the cube.

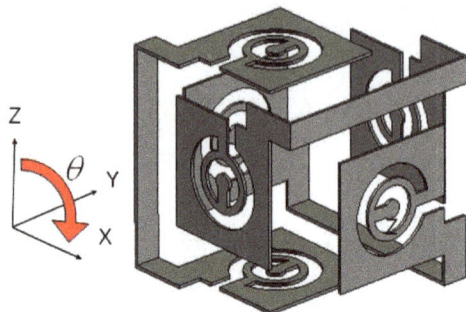

Fig. 20. Geometry of tree-dimensional resonator.

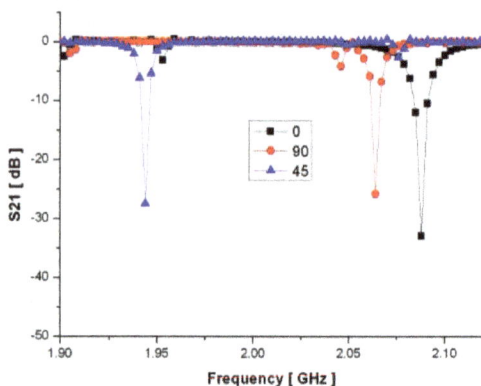

Fig. 21. Transmission coefficient variation for various rotated angle.

The simulated transmission coefficient variation of a resonator by the various rotated angles (θ) of a cube is shown in Fig. 21. The transmission coefficient is degenerated by the cross polarized effect and the performance of a resonator is determined by the direction of incident waves. The cubic resonator shows stop-band characteristics for a selected set of rotated angles of 0º, 45º and 90º, but its resonance frequency is changed due to anisotropic nature of the structures. The photographs of a fabricated three-dimensional resonator cubic resonator and a concrete block sample are shown in Fig. 22. The resonator structures were fabricated with a surface micromachining process on flexible polyimide substrate. A 200 μm-thick copper was film was e-beam evaporated on the polyimide substrate to form the ground plane. For the metallic patters, direct laser micro-machining technology was used, and another 10 μm-thick polyimide was coated on the top of the metallic patterns. A 'T'-shaped metallic patterns for covering the surfaces of a cubic resonator and a photograph of the metallic-zig structure for the concrete blocks fabrication are shown in Fig. 23. In order to check the stop-band characteristics of the proposed resonator, three concrete block samples with a different cubic resonator arrangement were loaded into the jig and tested.

In a measurement system, a Wiltron 360B vector network analyzer was connected to the measurement jig through an R-band waveguide to coaxial connector. A WR-430 standard waveguide, ranging from about 1.7 GHz to about 2.6 GHz, was used as a measurement jig. The WR-430 standard waveguide had a size of 10.922 cm × 5.461cm. Fig. 24 shows the measured transmission coefficient through the waveguide loaded with the cubic resonators array structures. We observed that the stop band-width becomes wider in comparison to the response of one cubic resonator. When nine cubic resonators are embedded into a concrete block (109 mm × 54.5 mm × 150 mm), the resonance frequency is about 2.05GHz, which is lower than simulated frequency and the stop-band width ($\leq -10dB$) is 150 MHz. The proposed concrete block structure can be applied to the intelligent building construction for EM waves shielding.

Fig. 22. Photographs of the fabricated three-dimensional resonator and a concrete block sample.

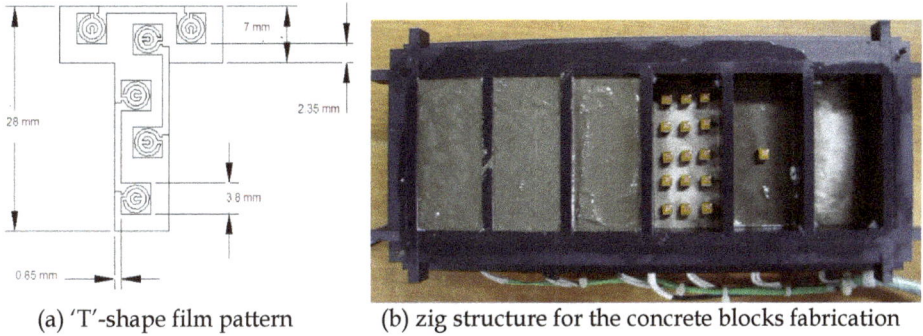

(a) 'T'-shape film pattern   (b) zig structure for the concrete blocks fabrication

Fig. 23. Geometry of 'T'-shape film pattern and zig structure for fabricating the concrete blocks with cubic resonators.

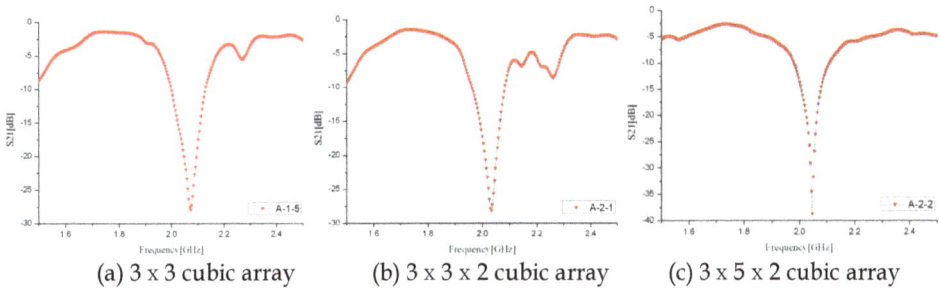

(a) 3 x 3 cubic array   (b) 3 x 3 x 2 cubic array   (c) 3 x 5 x 2 cubic array

Fig. 24. Measured transmission coefficient for the different cubic resonators array.

4.2 Isolation improvement of two closely spaced loop antennas using double negative MTM absorber unit cells

A standard method in improving isolation is to limit the radiation in the propagating direction and to increase the distance between the antennas. However, this method is not highly efficient because the area allocated to the antennas in mobile device is very small. In order to improve the isolation performance several studies have been done on spatial diversity and MIMO systems. A multiple slits etched on a single ground plane were used to isolate the feeding port of two closely-packed antennas. Another technology is to link the two antennas with a so-called neutralization line in order to increase the port-to-port isolation. However, the aforementioned techniques are only suitable for mobile diversity and MIMO antennas with a common ground plane. This study presents a new method to improve the isolation between two antennas without a shared ground plane. In order to increase the isolation between the two loop antennas, double negative MTM absorber unit cells were used. The MTMs are artificial composite structures that exhibit a homogeneous effective permittivity ε and permeability μ, which become negative over an operating frequency range. Due to the unique characteristics of the MTMs and their integration, a number of the potential MTM applications have been researched in many small antenna applications. A MTM absorber consists of a structure with negative permittivity and another structure with negative permeability, resulting in a negative refractive index. We use a double-layered structure with a SRR and two OCSRRs which are put on top of each other in order to make a miniaturized thin MTM absorber unit cell for the 2 GHz frequency band.

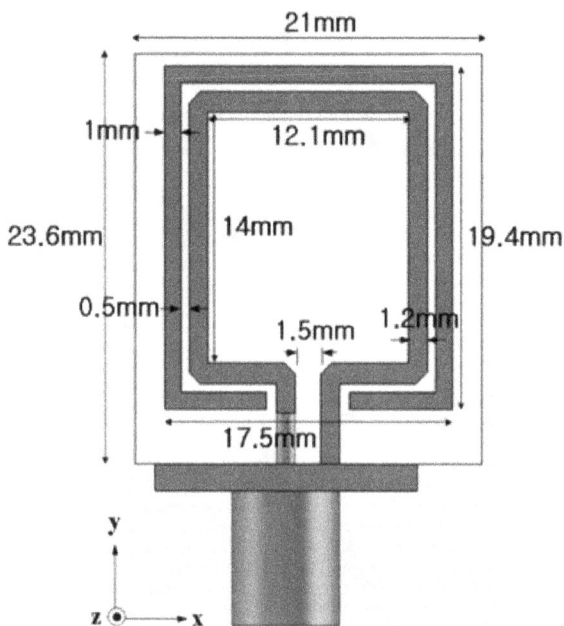

Fig. 25. Geometry of the proposed loop antenna.

The geometry of the proposed loop antenna is shown in Fig. 25. It was fed by a 50Ω coaxial connector and designed on a FR-4 dielectric substrate (relative permittivity = 4.6, thickness = 2 mm). The proposed antenna consists of a small square loop and a SRR shaped loop. The single loop antenna occupied 21 mm × 23.6 mm × 2 mm; the total length of the inner loop was 54 mm long ($\approx$ 0.5 λ_g at 2.5 GHz). The input impedance of an electrically small loop antenna has a large inductive component. In order to match the input impedance, it needs a capacitive element. A small capacitive split ring resonator (SRR) structure was used as a matching element in this design. The two vertical loop arms have the same currents with a null in middle of the top loop section. The currents are seen to be equivalent to that of a pair of parallel monopole antennas driven in-phase with a spacing of approximately 1/8 λ_g. The geometry of the proposed MTM absorber unit cell is shown in Fig. 27. Two OCSRRs and a SRR structure were placed on both sides of the FR-4 dielectric substrate (relative permittivity = 4.6, thickness = 2 mm). Two OCSRRs are located with even symmetry on one side of the FR-4 substrate. We constructed an MTM unit cell using an OCSRR and SRR arrangement. The proposed absorber unit cell based on the double negative MTM structure was shown to reach a thickness on the order of 0.059 λ_0 at the frequency of 2.54 GHz. The frequency characteristics of the unit cell structure were simulated by using a periodic boundary condition (PBC) method; the dimensions of the unit cell were 7.4 mm × 7 mm × 2 mm. The simulated scattering parameters for the proposed unit cell are plotted in Fig. 28. The performance of the proposed absorber shows a maximum absorbance of 93% at 2.54 GHz, and the full width at half maximum absorbance ($\geq$ 50%) was at 200 MHz.

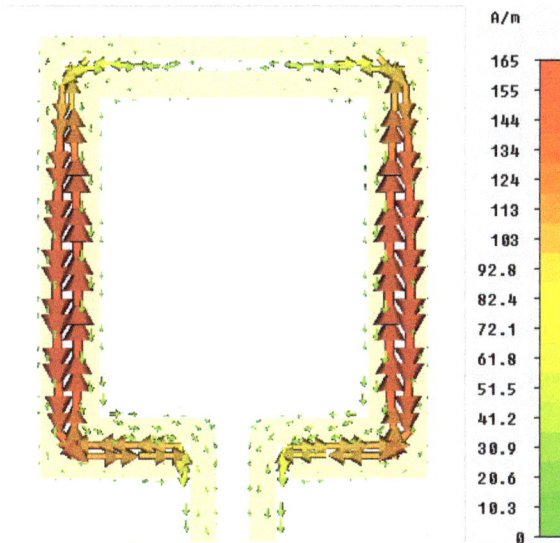

Fig. 26. The simulated surface current distributions.

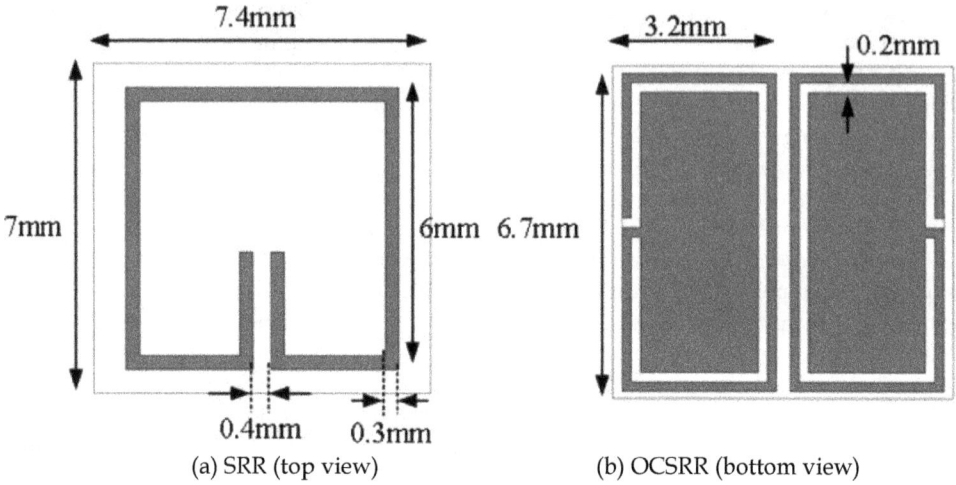

(a) SRR (top view)   (b) OCSRR (bottom view)

Fig. 27. Geometry of the proposed MTM absorber unit cell.

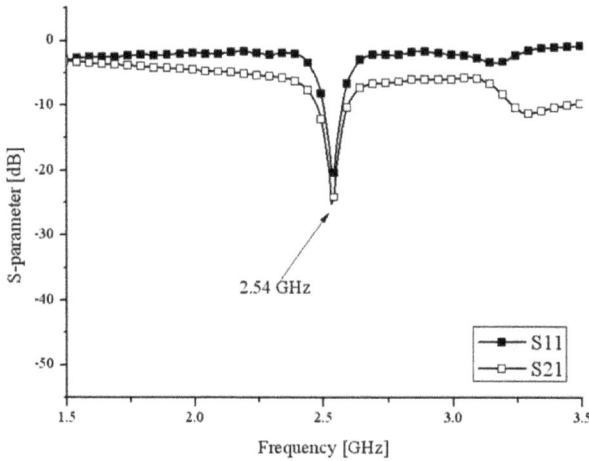

Fig. 28. Simulated S-parameters for the proposed MTM absorber cell.

The geometry of the two proposed loop antennas with the three MTM absorber cells is shown in Fig. 29. It consists of two loop antennas and three MTM absorber cells; the total antenna size was 51 mm × 23.6 mm × 2 mm. When the inter-element spacing between two loop antennas is set to 13.5 mm ($\approx \lambda_g/4$), the maximum isolation of -23 dB between the two antennas was obtained at 2.54 GHz. When three cells were placed between the antennas, the resonant frequency changed slightly, but the port-to-port isolation $|S_{21}|$ is efficiently improved by about 10 dB within the operating bandwidth. The simulated magnetic field distribution for the antenna with the MTM absorber cells is shown in Fig. 30. When a signal with a frequency of 2.54 GHz was applied to port 1, most of the magnetic fields are induced onto the two vertical loop's arms near port 1. When the MTM absorber cells were placed in the orthogonal direction against the magnetic fields, most of the magnetic fields were not coupled to the loop antenna at port 2. As a result, the port-to-port isolation between the two loop antennas was improved by using the MTM absorber cells between the two antennas. In order to verify the simulations, two loop antennas with three double negative MTM absorber cells were fabricated on a FR-4 substrate with the relative dielectric constant 4.6 and a thickness of 2 mm; two 50Ω SMA connecters were installed.

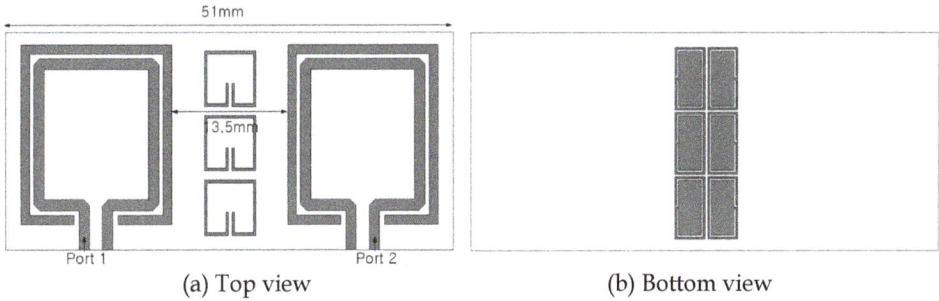

(a) Top view   (b) Bottom view

Fig. 29. Geometry of the proposed antenna with three absorber cells.

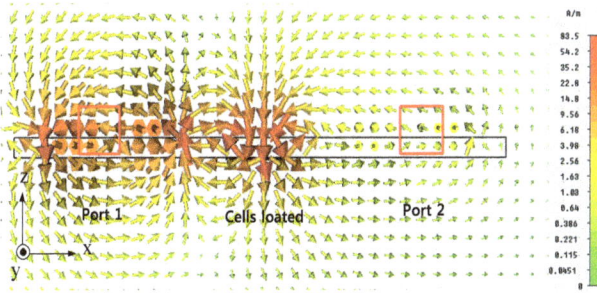

Fig. 30. The simulated magnetic field distributions of the antenna when a signal was fed to port 1(f = 2.54GHz).

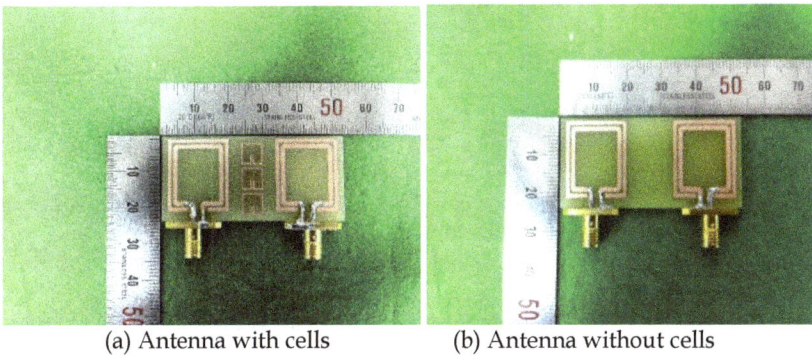

(a) Antenna with cells   (b) Antenna without cells

Fig. 31. Photographs of the fabricated antenna

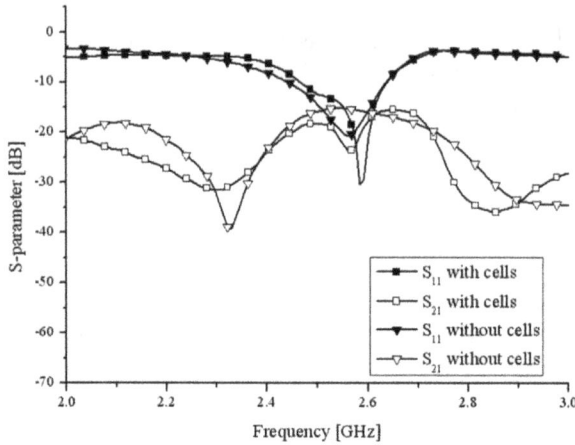

Fig. 32. Comparison of the measured return loss and isolation for the proposed antenna with/without cells.

The photographs of the prototype fabricated antennas are shown in Fig. 31. Comparisons of the measured return loss and isolation results for the proposed antennas are shown in Fig. 32. The measured S-parameters were similar to the simulated results. When three absorber cells were placed between the antennas, the isolation $|S_{21}|$ shows a 10 dB improvement when compared to the isolation of the antennas without the absorber cells. A maximum isolation of -24 dB was achieved at 2.54 GHz. The operating bandwidth of the proposed antenna was slightly lower due to the absorber cells between the two loop antennas.

A new method to improve isolation between two closely spaced small antennas with ground plane is presented. The proposed antenna structures are composed of two small loop antennas and three double negative MTM absorber unit cells exhibiting negative permittivity and permeability at the aimed frequency of 2.54 GHz. The proposed absorber cell consists of a double-layered structure with a SRR and two OCSRRs. They are put on top of each other has shown to effectively absorb most of the impinging power. Both of them have same resonance frequency of 2.54 GHz. In order to obtain high isolation between the two closely spaced antennas, the proposed absorber cells can be effectively used. This technique can be applied for isolation improvement in multi antenna systems.

5. Conclusions

Much of the work in MTM has been focused on the real parts of the permittivity and permeability to enable the creation of a NRI material for various applications including electromagnetic wave cloaking, perfect lenses, and microwave components. By varying the dimensions of the electric and magnetic components, it is possible to adjust the permittivity and permeability independently. Additionally, by tuning the electric and magnetic resonances, an MTM can be impedance-matched to free space. As a result, 100 % absorbance is theoretically possible. By properly arrayed a single unit cell, it is possible to make the absorber's electrical size small enough. Since from the practical point of view, it would be rather difficult to employ a photo-etching technology for the fabrication of the absorber cells, operated at the higher microwave frequency and terahertz frequency regime.

In this work, the authors present the design and fabrication of the two-dimensional and three-dimensional microwave meta-material absorber unit cells by using a standard photolithography technique and a micro-machining technology, respectively. Moreover, a new type MTM absorber structure was proposed. By properly arrayed a single unit cell with a 90^0 rotated SRR and two OCSRRs configuration, has been shown to effectively absorb most of the impinging power. The two main advantages of the proposed absorber are: 1) the absence of a metallic back plate or a resistive sheet and 2) the enhanced absorbance operating band by using electromagnetic responses of the 90^0 rotated SRR. The thickness of the planar MTM absorber structure in the EM wave propagation direction was shown 7.4 mm ($\approx \lambda_0$) at 2.6 GHz. The proposed layout can be easily extended to work for more compact and thin backplane-less planar absorber designs.

6. Acknowledgment

This research was supported Basic Science Research Program through National Research Foundation of Korea (NRF) funded by the Ministry of Education, Science and Technology (No. 2010-0011646)."

7. References

R. L. Fnate and M. T. McCormack (1968). Reflection properties of the Salisbury screen, *IEEE Trans. Antennas Propagation*, vol. AP-30, no. 10, pp. 1443-1454.

N. I. Landy, S. Sajuyigbe, J. J. Mock, D. R. Smith, and W. J. Padilla (2008). Perfect metamaterial absorber, *Physics Review. Lett.*, vol. 100, pp. 274021-4.

H. Tao, N. I. Landy, C. M. Bingham, X. Zang, R. D. Averitt, and W. J. Padilla (2008). A metamaterial absorber for the terahertz regime: design, fabrication and characterization, *Optical Express*, vol. 16, pp. 7181-7188.

Y. Cheng and H. Yang (2010). Design, simulation, and measurement of metamaterial absorber, *Microwave Optical Technology Lett.*,vol. 52, pp. 877-880.

H. Tao, C. M. Bingham, D. Pilon, K. Fan, A. C. Strkwerda, D. Shrekenhammer, W. J. Padilla, X. Zhang, and R. D. Averitt (2010). A dual band terahertz metamaterial absorber, *Journal of Applied Physics D*,vol. 43, pp. 225102-225106.

M. H. Li, H. L. Yang, and X. W. Hou (2010). Perfect metamaterial absorber with dual bands, *Progress in Electromagnetic Research*,vol. 108, pp. 37-49.

Y. Cheng, H. Yang, Z. Cheng, and N. Wu (2011). Perfect metamaterial absorber based on a split-ring-cross resonator, *Journal of Applied Physics A*, vol. 102, pp. 99-103.

K. B. Alici, F. Bilotti, L. Vegni, and E. Ozbay (2010). Experimental verification of metamaterial based subwavelength microwave absorbers, *Applied Physics*, vol.108, pp. 0831131-0831136.

A. Velez, F. Aznar, J. Bonache, J. M. Velazquez-Ahumada, and F. Martin (2009). Open complimentary split ring resonators (OCSRRs) and their application to wideband CPW band pass filters, *IEEE Microwave & Wireless Component Letter*, vol.19, pp. 197-199.

D. R. Smith, D. C. Vier, T. Koschny, and C. M. Soukoulis (2005). Electromgnetic parameter retrieval from inhomogeneous metamaterials, *Physics Review*, vol. E71, pp. 0316171-10.

R. A. Depine and A. Lakhtakia (2004). A new condition to identify isotropic dielectric-magnetic materials displaying negative phase velocity, *Microwave Optical Technology Lett.,*vol. 41, pp. 315-316.

J. Lee and S. Lim (2011). Bandwidth enhanced and polarization-insensitive metamaterial absorber using double resonance, *Electron Lett.,* vol. **47**, pp. 8-9.

M. A. Jensen J. W. Wallace (2004). A review of antennas and propagation for MIMO wireless communications, *IEEE Trans. Antennas Propagation,* vol. 52, pp. 2810-2824.

C. Chiu, C. Chen, D. Murch, and R. Rowell (2007). Reduction of mutual coupling between closely-packed antennas elements, *IEEE Trans. Antennas Propagation,* vol. 55, pp. 1732-1738.

Y. Shin and S. Park (2007). Spatial diversity antenna for WLAN application, *Microwave Optical Technology Lett.,* vol. 49, pp. 1290-1294.

A. Diallo, C. Luxey, P. Le Thuc, and G. Kossiavas (2008). Enhanced two-antenna structures for universal mobile telecommunication system diversity terminals, *Microwaves, Antennas and Propagation, IET,*vol. 2, pp. 93-101.

A. Chebihi, C. Luxey, A. Diallo, P. Le Thuc, and R. Staraj (2008). A noval isolation technique for closely spaced PIFAs UMTS mobile phones, *IEEE Trans. Antennas Propagation,* vol. 7, pp. 665-668.

N. Engheta and R. W. Ziolkowski (2006). Metamaterials: physics and engineering explorations, Wiley-Inter-sciencce, NJ.

F. Falcone, T. Lopetegi, J. D. Baena, R. Marques, and M. Sorolla (2004). Effective negtive-ε stop-band microstrip lines based on complementary split ring resonators, *IEEE Microwave & Wireless Component Lett.,* vol. 14, pp. 280-282.

D. R. Smith, D. C. Vier, T. Koschny, and C. M. Soukoulis (2005). Electromgnetic parameter retrieval from inhomogeneous metamaterials, *Physics Review,* vol. E71, pp. 0316171-10.

5

Laser Ablation for Polymer Waveguide Fabrication

Shefiu S. Zakariyah

Advanced Technovation Ltd,
Loughborough Innovation Centre, Loughborough,
UK

1. Introduction

An increase in interconnection density, a reduction in packaging sizes and the quest for low-cost product development strategy are some of the key challenges facing micro-opto-electronics design and manufacture. The influence of high-density, small-sized products has placed significant constraints on conventional electrical connections prompting various fabrication methods, e.g. photolithography, being introduced to meet these challenges and ameliorate the rapidly changing demand from consumers. While high-power solid state lasers are fundamental to large scale industrial production, excimer laser on the other hand has revolutionised the manufacturing industry with high precision, easy 3D structuring and less stringent production requirements. Micro-structuring using excimer laser, best known as laser ablation, is a non-contact micro- and nano-machining based on the projection of high-energy pulsed UV masked beam on to a material of interest such that pattern(s) on the mask is transferred to the substrate, often at a demagnified dimension with high resolution and precision. The use of mask with desired patterns and beam delivery system makes the fabrication in this case accurate, precise and easily controllable. The first part of this chapter introduces the fundamentals of laser technology and material processing. In the second part, optical interconnects as a solution to 'bottlenecked' conventional copper interconnections is introduced with emphasis on excimer laser ablation of polymer waveguides and integrated mirrors. Key research findings in the area of optical circuit boards using other techniques are also briefly covered.

2. Introduction to laser technology

The word 'laser' has been part of the lexis of the English language since its invention in 1960 and subsequent commercialisation few years later. It is an acronym that stands for Light Amplification by Stimulated Emission of Radiation, which is considered a modified version of its predecessor - 'maser' (Microwave Amplification by Stimulated Emission of Radiation); in other words, laser is an optical maser. The first laser, ruby, emitted red-coloured light at λ = 694.3 nm. Just over five decades later, laser (and laser technology) controls a remarkable market share in various applications ranging from research and medicine, to manufacturing and domestic applications. One of the sectors that have seen dramatic advancement with the advent of lasers is medical surgery (e.g. ophthalmology, cosmetic surgery and dentistry).

Laser generation has been extensively covered in the literature, but essentially, but essentially there are three principles that must first take place: (i) stimulated emission to defeat spontaneous emission and absorption, (ii) population inversion to temporarily disturb normal distribution - these two processes require movement of species from a lower energy level to a higher one, and (iii) a feedback system to amplify the photon population.

2.1 Laser micromachining (or material processing)

Laser material processing is generally, though not technically, referred to as laser micromachining of engineering materials e.g. polymer, metals, glass and ceramics. This definition thus excludes applications of lasers to, for example, human tissues even though the mechanism is similar. The possible reason for this exclusive usage might be because early laser candidates found application in engineering sectors such as drilling and cutting of materials where high energies are needed. For laser micromachining, there are four key processes of importance (Figure 1).

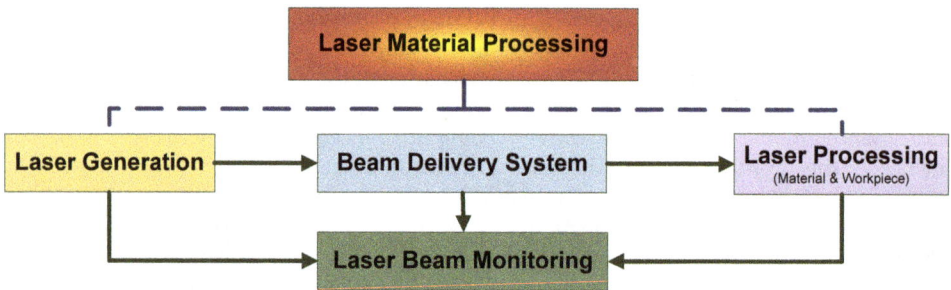

Fig. 1. Schematic diagram showing key stages of a typical laser material processing.

Beam generation
This is the first stage and the backbone of any material processing; its output determines the components of the remaining stages. For example, if a ceramic material is to be processed then the output at this stage should be a high-powered laser. Furthermore, if the ceramic is to be processed with minimum thermal damage then the output beam should, for example, be a pulsed laser with short pulse duration to provide a minimum time interaction between the beam and the material.

Beam delivery or propagation
This involves transporting the output beam to the site of processing or workpiece. What constitutes the beam delivery system depends on the application. In general, the elements of the stage, whose number and arrangement varies, include various optical devices such as mirrors, lenses and attenuator among others. It is therefore imperative that careful combination is made to achieve optimum result without losing much power as a small fraction of beam energy is lost per element. Also to be considered is the length of the path between the laser chamber output window and the workpiece. This needs to be kept to a minimum in order to avoid beam profile distortion and divergence. Excimer laser usually has the longest beam path with the highest number of optical components while a CO_2 laser employs the least.

Laser beam monitoring

Many of the laser beam properties are essential for an optimum process. However, three of these - energy, beam diameter and beam profile – are highly important in micromachining. There are two methods of obtaining the beam energy. In the first approach, the beam is sampled during the processing; this provides an accurate account of beam energy utilised during a particular process. It is pertinent to note that this task is in some way difficult and risky. Three methods of beam sampling: static beam splitter, rotating chopper mirror and leaky resonator mirror are discussed in [Crafer & Oakley, 1993]. The second approach is by total beam measurement; this approach involves measuring the energy at the workpiece using a power meter. Although the method might not totally account for what happens during a process, it is easier than the sampling method [Crafer and Oakley, 1993]. A common way of examining both the beam diameter and profile is by using low energy to irradiate a suitable material; the etched sample is then analysed to measure the diameter and observe the profile. This is an indicative method especially when the process is thermal. Alternatively, beam profile and homogeneity is monitored using a beam profiler which shows the shape of the beam, in real-time, during a process.

Laser-matter interaction (Laser processing)

The wave-particle duality concept is quite useful in treating laser-matter interaction. For example, laser generation is better described using the quantum (or particle) approach while propagation and delivery is suitably described using the wave concept. For laser-matter interaction, it is appropriate to use quantum physics. Thus viewing the beam as a packet of photons hitting the matter with which it is interacting. When the laser beam strikes the material, the photon energy is transferred to the material and subsequently converted to other forms of energy depending on the material. With metals, this is transferred to the mobile electrons which results in the heat energy that can cause vaporisation and disintegration of the metal. However, with non-metals, the energy can either be converted to chemical energy required for bond-breaking or heat energy for vaporization. These two possibilities depend on the type of material, its bond energy and the wavelength of the laser or more precisely the photon energy. Essentially, there are two common mechanisms for laser material interactions, which can occur at varying degrees while processing a material.

- Thermal (photothermal or pyrolytic): This is an electronic absorption in which the photon energy is used to heat up the material to be processed and thus part of the material is removed as a result of molecule vaporization, such as in CO_2 laser cutting. This type of process is broadly referred to as laser micromachining.

- Athermal (photochemical or photolytic): This is a photochemical process whereby the material is ablated by direct breaking of molecular bonds when hit by photons (energy) of the incident beam. In principle, this is only possible if the photon energy is equal or greater than the bond energy of the molecules of the material to be processed. During this process, a particular area of the surface of the material is removed with minimum (or without any, theoretically) thermal damage to the surrounding material. This process is generally called ablation, though photothermal processes are also referred to as ablation. Ablation is generally used in reference to polymer and/or soft materials, but laser ablation is also possible with other materials such as ceramic and glass. However higher fluencies are required in their case.

The etch rate – the amount of material removed per pulse – is mainly a function of the photon energy and the material being processed. However, it is impractical to model laser-matter interactions based on the aforementioned two quantities as the mechanism is also influenced by numerous other factors (e.g. thermal diffusion, absorption saturation, surrounding medium, etc.) such that the measured ablation depths seldom agree with these predictions; this necessitates more complex 'models' often based on these two quantities [Tseng, et al., 2007]. Equations 1 & 2 provide two often referenced mathematical representations: Beer's law and the Srinivasan-Smrtic-Babu (SSB) model [Shin, et al., 2007], which are based on pure photochemical and combination of photochemical and photothermal mechanisms respectively. The two formulae are similar except that SSB's adds a photothermal part to Beer's model where L, β, f and f_{th} are the etching depth per laser pulse, coefficient of absorption (cm⁻¹), laser fluence per pulse (J/cm²) and threshold fluence (J/cm²) respectively.

$$L = \frac{1}{\beta} \ln \left(\frac{f}{f_{th}} \right) \text{ for } f > f_{th} \tag{1}$$

$$L = \frac{1}{\beta} \ln \left(\frac{f}{f_{th}} \right) + \text{photothermal for } f > f_{th} \tag{2}$$

2.1.1 Beam profile

The most common laser beam profile is the Gaussian beam (TEM$_{00}$ or fundamental mode) schematically shown in Figure 2a. Its beam intensity variation can be described according to equation 3, where $I_0 = I_{max}$ = intensity at the centre of the profile, I is the intensity at any other point, and r is the radius of the beam taken at a point where the beam axis intensity has fallen to $1/e^2$ of its maximum. Although this Gaussian profile is better than and preferred to higher order modes, its intensity variation is still a source of concern in laser material processing and particularly in laser ablation. For this reason, a modified version - which is thought to improve the tapering of the beam profile - is generated with uniform intensity across the entire profile similar, in principle, to that shown in Figure 2b. This is described as a 'top-hat' (or 'flat-top') profile perhaps due to the 'flatness' of the top of the profile. As shown in Figure 2c, a top-hat profile is obtained from its Gaussian counterpart by taking the energy from the weak intensity region, where beam intensity distribution is lower than $1/e^2$ (i.e. 13.5 %) of the centre and folding it back into the region within the beam waist. A point should be made here: saying that a laser operates in a single mode e.g. TEM$_{00}$, simply means that this is the dominant mode of operation just like a given wavelength implies the fundamental (i.e. dominant) wavelength of operation.

$$I = I_0 e^{-(r/r_0)^2} \tag{3}$$

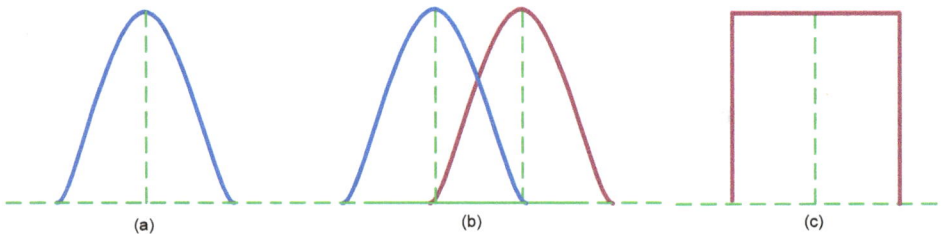

Fig. 2. Typical laser bean profile (a) Gaussian beam profile, (b) overlapping of Gaussian profile to generate 'top-hat', and (c) 'Top-hat' beam profile.

2.1.2 Ablation threshold

The ablation threshold is the point at which the applied energy density is enough to cause ablation either photolytic or pyrolytic. The value of this varies from polymer to polymer depending on the nature and strength of the bonds in the polymer and also on laser wavelength (Tables 1 & 2). An ablation threshold can be obtained from a plot of etch rate against a logarithmic scale of fluence at zero ablation rate [Jackson, et al., 1995; Tseng, et al., 2007]. Zakariyah (2010) obtained a threshold as the x-intercept value on a graph of ablation rate against incident fluence. Irrespective of the base of the logarithmic scale taken, the two approaches are found to produce the same value. Table 2 shows a list of common bonds in polymers with their respective bond energies, which need to be overcome during any laser ablation regardless of the nature of the mechanism. For photochemical ablation, the laser wavelength has to be carefully chosen such that the photon energy obtained from the laser is equal or greater than the bond energy of the polymer to be processed. When working below this threshold, no ablation is expected to occur, however, the chemical properties of the materials are subject to certain changes. Furthermore, operating at well above the threshold can cause or increase the heat-affected zone (HAZ) and debris deposition. The former is due to high energy while the latter is as a result of bombarding the ejected materials. It should be noted that intense bombardment of ejected particles above the ablation zone can retard the ablation rate. This is because the ejected materials might absorb fractions of the incoming beam thus reducing the effective fluence at the ablation zone. Wavelength is one of the factors that determine the thresholds of ablation. For example, the ablation threshold for PMMA (PolyMethyl MethAcrylate) is ~150 mJ/cm² at 193 nm and ~500 mJ/cm² at 248 nm – this is a 3-time increase in value between the two wavelengths. The rule-of-thumb for laser ablation of polymers is to have lower threshold fluences for ablation at shorter wavelengths [Pfleging, 2006].

Material	Fluence (mJ/cm²)	λ (nm)	Material	Fluence (mJ/cm²)	λ (nm)
PS	15.3	193	PMMA	150[1]	193
PET	18.4	193	Silicon nitride	195	-
Truemode™ acrylate polymer	20	248	SiO₂	350	-
PC	21.5	193	PMMA	500	248
PI	25.1	193	Nd:glass	500	193
Photo resist	30	-	Nd:YAG	800	193
PC	40	-	Glass, metal oxide	700-1200	-
PI	~ 40	248	Nd:YAG	1200	248
PI	50	308	Nd:glass	1600	248
PI	100	355			

Table 1. Ablation threshold fluence for some selected material [Chen, Y.-T., et al., 2005; Jackson, et al., 1995; Meijer, 2004; Pfleging, 2006; Yung, et al., 2000; Zakariyah, 2010; Zeng, et al., 2003].

---

[1] A threshold of 33.8 mJ/cm² is reported for PMMA at 193 nm by Chen, Y-T, et al. (2005)

Group	Bond Energy (eV)	Group	Bond Energy (eV)
C = C	7.0	O-H	4.5
C = O	6.7, 4.2	H-H	4.6
Si-Si, Cl-Cl	1.8 – 3	O-O	5.1
C-H	3.5	C-C	6.2
C-N, C-C	3 – 3.5	C-O	11.2
-N = N	3.5, >4.8	Benzene Ring	4.9, 6.2, 7.75

Table 2. Table showing typical bonds in photopolymers and their respective bond energies [Basting, 2005; Crafer & Oakley, 1993; Meijer, 2004; Tseng, et al., 2007].

2.2 Industrial laser – Excimer

Lasers can be classified based on a number of factors e.g. active medium (solid, liquid and gas), output power (low, medium and high power lasers), excitation method (electrical, optical and chemical), operating mode (continuous wave, pulsed mode and Q-switched output mode), efficiency and applications. CO_2, Nd:YAG and excimer lasers, with Ti-Saphire following suit, are the key lasers in material processing due to their relatively high power. These three form a complete laser assembly in PCB (printed circuit board) manufacturing processes. Excimer laser is described here as it is the prominent laser candidate for polymer waveguide fabrication; however, a UV Nd:YAG has recently been reported [Zaakriyah, et al., 2011] as a competitive alternative.

An excimer laser - a commonly used gas laser and the halide of noble gases – obtained its name from the contraction of the term 'EXCIted diMER'. Because a dimer strictly refers to a molecule composed of two similar subunits (ions, monomers, etc.), it is therefore more technical to refer to excimer as 'exciplex' meaning EXCIted comPLEX. The wavelengths of excimer lasers vary from about 190 nm (deep UV) to 350 nm (near UV)[2] (Figure 3) but ArF, KrF and XeCl are the most commonly used. F_2 (λ = 157 nm) laser is sometimes classified as a gas laser and sometimes as an excimer laser as implied in [Basting, et al., 2002; Tseng, et

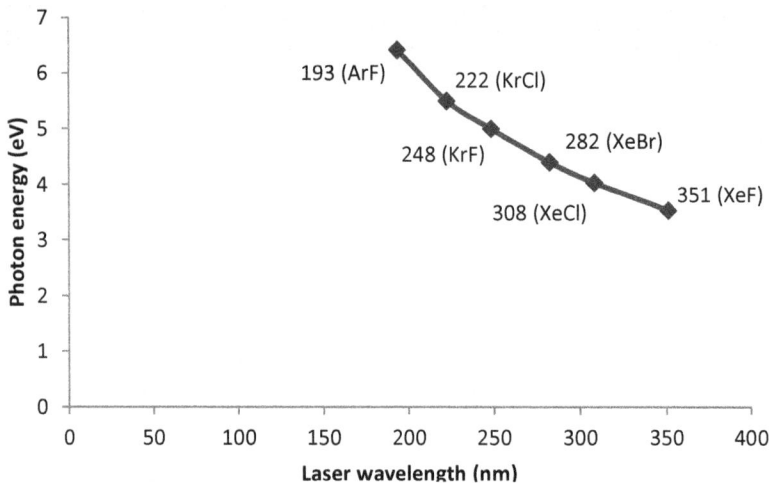

Fig. 3. A graph of photon energy (eV) against excimer laser wavelengths.

[2] Basting, et al., (2002) put the range between 126 nm and 660 nm (visible region).

al., 2007]. The pulse duration and repetition rate are in the ranges of 5 – 50 ns and 1 – 100 Hz respectively.

Since its discovery and introduction into the market in 1970 and 1977 respectively, the excimer laser has turned out to be a multi-purpose, multi-featured laser with increasing market shares in industrial and medical applications. Its first commercially available product from Lamda Physik is called EMG 500 [Basting, et al., 2002]. Although other lasers such as YAG and CO_2 lasers are also extensively used in High Density Interconnection (HDI) technology, the excimer laser ablation is indispensable when it comes to 'fine' finish micro- and nano-fabrications. This is particularly true for hard and delicate materials. This is largely due to its wavelength, pulse duration, and of course its pulse energy allowing for what is generally termed as a 'cold ablation' process. The excimer laser also excels others in its ability to 'mask-project' patterns, using stencil or metal-on quartz masks [Tseng, et al., 2007], on to a sample with a minimal HAZ. The minimal HAZ is argued to be due to the short interaction between the laser beam and the material. In addition, the short pulse duration of the excimer is also a contributing factor. Nevertheless, picosecond and femtosecond lasers are now available today. These classes of lasers are designed to further reduce the HAZ. They are also characterized by higher etch rate, strong absorption by the material, improved surface roughness and lower ablation thresholds [Li, L., et al., 2011; Sugioka, et al., 2003].

These aforementioned features of the excimer laser have attracted and favoured its use not only for polymers [Wei & Yang, 2003] but also with other materials such as ceramics [Ihlemann, 1996], glasses [Tseng, et al., 2007] and silicon [Li, J. & Ananthasuresh, 2001] which are often hard to machine. Besides, excimer lasers are now used for surface modification of various materials. Pfleging, et al. (2006) have used excimer at fluences below the ablation threshold to fabricate single mode optical waveguides in PMMA similar to that employed using CO_2 laser in [Ozcan, 200 8]. Thomas, et al. (1992) also used an excimer laser to effect changes to the chemical structures of materials (polymer and ceramic) with potential application in enhanced material adhesion and surface wettability among others.

3. Polymer waveguide fabrication for optical interconnect on PCB

3.1 Optical Interconnects (OI)

The miniaturisation in consumer electronics, dictated by the rise in demand for more features and the change in the manufacturing technology, has caused an increase in the data rate on the micro-levels such as backplane, board-to-board, and chip-to-chip. The bottleneck for copper transmission in PCB with high interconnection density and high-frequency is more pronounced at the 10 Gb/s limit where problems such as crosstalk, electromagnetic interference (EMI) and power dissipation, inter alia, cannot be tolerated [Holden, 2003; Offrein, 2008; Shioda, 2007]. To overcome this barrier, optical interconnect – as it has been successfully used for long haul communication - is being considered. The deployment suggested here is not to overhaul traditional copper technology but to create a hybrid electric-optical interconnect.

To address the bottleneck caused by the inherent problems in the copper transmission used in backplanes and boards, the last two decades have witnessed vigorous research input and output from researchers around the world to deploy OI on PCB. Japan, the EU and Asia-Pacific/North America, who led in the microvia technology, are also key figures in the OI deployment [Holden, 2003; Lau, 2000; Shioda, 2007]. Undoubtedly, the cost-effectiveness of OI is a major consideration if it is to be implemented [Huang, et al. 2003]. Hopkins & Pitwon (2007) asserted that at higher bandwidth for current and near future requirements for telecom and datacom systems, the application of OI at the backplane is unavoidable. It was argued that the cost of solving the bottleneck of copper transmission will surpass that of implementing OI at ~ 6.25 Gb/s (Fig. 4). Furthermore, the total power loss, commonly referred to as power budget, is also a consideration and is currently being investigated. It is written in [Uhlig & Robertson, 2005] that a ~20 dB would be an acceptable total loss for an optic link at the backplane; Dangel, et al. (2006) put this at 12 – 15 dB for board-to-board optical link of 30 – 100 cm. Uhlig and Robertson (2005, 2006) argued that at some point along the transmission, optical amplification would be needed for a realistic OI on PCB to be implemented. While optical loss is important, reliability (thermal cycling, athermal aging, high temperature reflow, environment, humidity tests, etc.) is another key characteristic and requirement for the deployment of the polymer waveguide [Dangel, 2006; Hwang, et al., 2010].

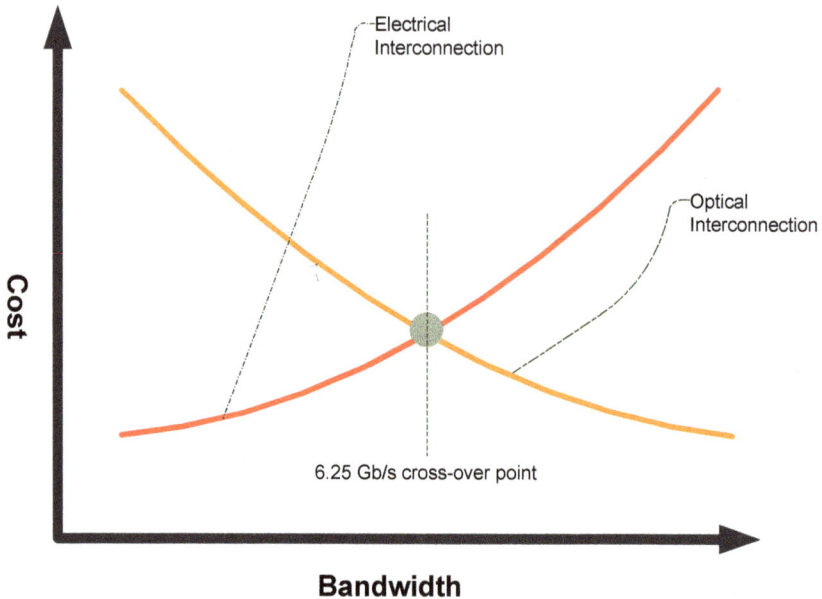

Fig. 4. Relative cost of copper technologies as compared to optical technologies on PCB [Adapted from Hopkins & Pitwon, 2006].

The two OI approaches under consideration are either unguided or guided; both having their pros and cons. The latter can be further divided into fibre- and polymer-based technologies with silicon-based waveguides also gaining momentum (Figure 5). Current literature reports suggest that a polymer-waveguide is the favoured candidate. This is because: (i) polymers are relatively cheap, (ii) low acceptable loss is achievable with polymer, (iii) they are easily available, and (iv) most importantly, polymer waveguide fabrication which is being considered, is compatible with the standard processes employed in PCB manufacturing such as soldering temperature, Coefficient of Thermal Expansion (CTE) matching, thermal stability and stress during lamination [Tooley, et al., 2001].

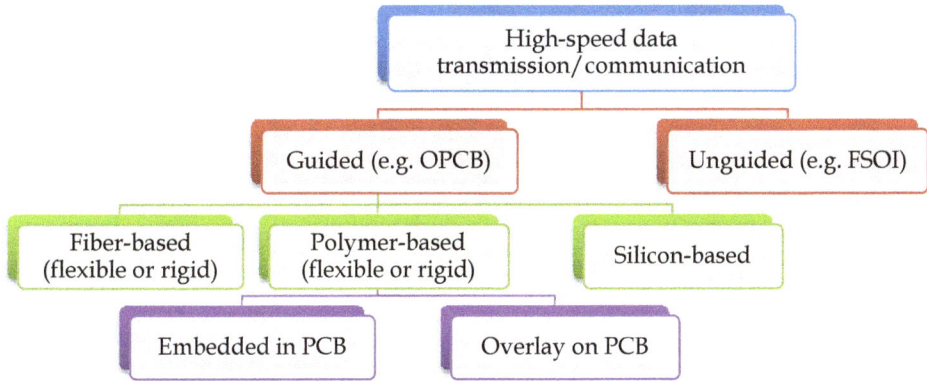

Fig. 5. Hierarchical classification of optical data communication system based on medium of transmission.

3.2 Deposition of optical polymer

The stages involved in laser ablation of a polymer waveguide are typified in Figures 6 and 7. In the first stage, liquid optical polymer is spun on FR4 substrate and subsequently UV cured to form both the lower cladding and the core layers. The samples were then dried in an oven (at 80 °C – 100 °C for about for about 60 minutes for Truemode™ acrylate polymer, $\Delta n \approx 0.03$ variable @ 850 nm) to ensure they were moisture-free. Laser ablation is carried out in the second stage to machine channels such that a ridge of polymer is left in-between the channels to form the waveguide. For one or more adjacent waveguides, the number of grooves required is equal to $(n+1)$, where n is the number of adjacent waveguides. Finally, a layer of upper cladding is deposited using spin coating (or any other suitable coating technique) and then UV cured.

A single layer of waveguide fabrication is common as this is currently enough to provide the data rate requirements for OI, but a multilayer waveguide has also been demonstrated [Hendrickx, et al., 2007a, 2007b; Matsuoka, et al., 2010]. Multimode waveguides are also common; dimensions such as 20 μm × 20 μm, 30 μm × 30 μm, 35 μm × 35 μm, 45 μm × 45 μm, 50 μm × 50 μm, 50 μm × 20 μm, 70 μm × 70 μm, 75 μm × 75 μm, 85 μm × 100 μm have already been reported [Albrecht, et al., 2005; Bamiedakis, et al., 2007; Dangel, et al., 2004; Immonen, et al., 2005, 2007; Liang, et al., 2008; Tooley, et al., 2001; Van Steenberge, et al., 2004; Zakariyah, 2009, Zakariyah, et al., 2011]. Two or more adjacent waveguides with a pitch of 250 μm [Albrecht, et al., 2005; Horst, 2009; Hwang, et al., 2010; Kim, et al., 2007; Van Steenberge, et al., 2004] is preferred as it is the pitch used for Vertical Cavity Surface Emitting Lasers (VCSEL) and photodector arrays, but other pitch sizes such as 80 μm [Dangel, et al., 2007], 100 μm [Dangel, et al., 2004] and 125 μm [Matsuoka, et al., 2010; Van Steenberge, et al., 2006] have also been used. Since the optical link required for OI is relatively short, loss due to multimode is acceptable and that alignment between various optical components would be relaxed. However, single mode waveguides is much suitable with silicon-based waveguides due to their high refractive indices, though they still pose alignment challenges [Horst, 2009]. Papakonstantinou, et al. (2008) reported a low cost method of achieving high alignment accuracy.

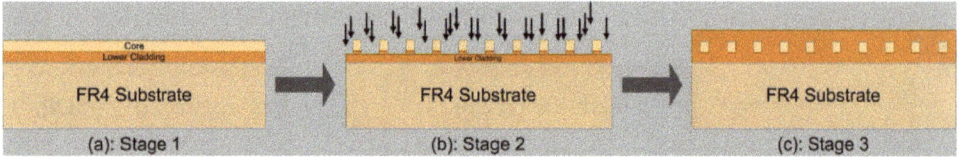

Fig. 6. Schematic diagram (side view) of the three major stages in the fabrication of optical waveguides by laser ablation.

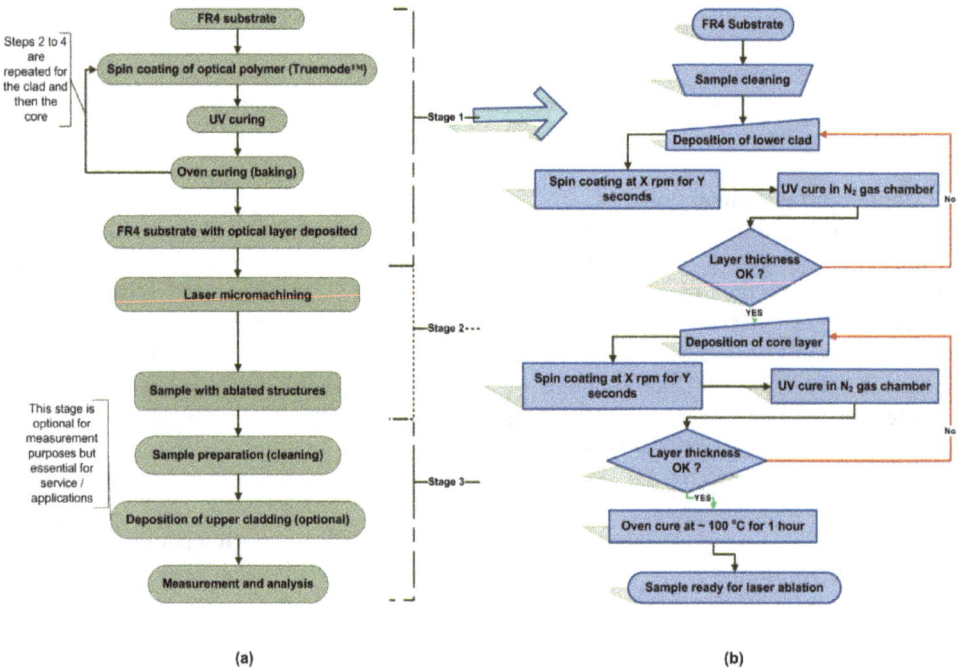

Fig. 7. (a) Flow diagram of the processes involved in patterning optical polymer waveguides using laser ablation, and (b) Schematic flow diagram showing procedure for depositing optical polymer on an FR4 substrate.

3.3 Laser ablation of polymer waveguides

Polymer waveguide fabrication for optical-PCB applications has been reported using a number of techniques, and more methods are still emerging. Selviah, et al. (2010) reported the use of four techniques - photolithography, laser direct writing, inkjet printing and laser ablation - in a flagship entitled '*Integrated Optical and Electronic Interconnect PCB Manufacture - OPCB*'. However, excimer laser ablation of optical waveguides is an emerging and competitive approach as it involves fewer steps when compared to others with great flexibility in pattern design. Furthermore, laser micromachining is currently being used for the drilling of vias for blind, buried and through holes in PCB manufacturing making it a more suitable choice when compatibility issues are taken into consideration. The key feature of this class of laser i.e. excimer is its wavelength and pulse duration. The latter reduces the degree of thermal diffusivity while the former is a key to high-energy intensity, high resolution and absorptivity of the laser beam not only in the polymer but also in tough materials such as glass [Tseng, et al., 2007]. The pulse duration of excimer laser is of significance when it comes to quality because shorter pulse width lasers give better machined quality though it is a costly task quality though it is a costly task [Chen, X. & Liu, 1999]; it also helps in reducing the ablation threshold [Ihlemann, 1996]. In fact most of the close competing lasers, for example YAGs and Ti-Sapphire, are found to operate in the UV regions and/or with very short pulse duration, thus intensifying competition.

The suitability of a UV laser (e.g. excimer) for a photochemical ablation over any other laser operating in the IR (or visible) region of wavelengths, such as CO_2, could be demonstrated as follows. The photon energy is given by $E = h\nu$, which is inversely proportional to its wavelength, thus a CO_2 laser operating at 10.6 μm will produce an energy more than 40 times less than that produced by a 248 nm KrF laser. Obviously, this is not in the order of magnitude of the energies for chemical bond scission of typical polymers, usually between 3 – 8 eV [Tseng, et al., 2007]. Increasing the number of pulses to match the required bond energy will merely result in a cumulative heat effect on the polymer surface. It is thus clear that excimer lasers have the right order of photon energy to athermally ablate polymers, while on the other hand, IR laser sources have photon energies much lower than 3 eV causing the dominance of a thermal mechanism. Therefore, in principle using the aforementioned assertion, laser of a maximum wavelength of 414 nm is required in order to photochemically ablate a polymer material with a bond energy of 3 eV. There would be a shift in the dominance of the mechanism by changing the wavelength of the laser source. For example, a shorter wavelength e.g. 355 nm would guarantee or increase the dominance of a photochemical process. On the other hand, a longer wavelength e.g. 1064 nm in the IR would not only reduce the dominance of photochemical but also initiate thermal process for the same polymer.

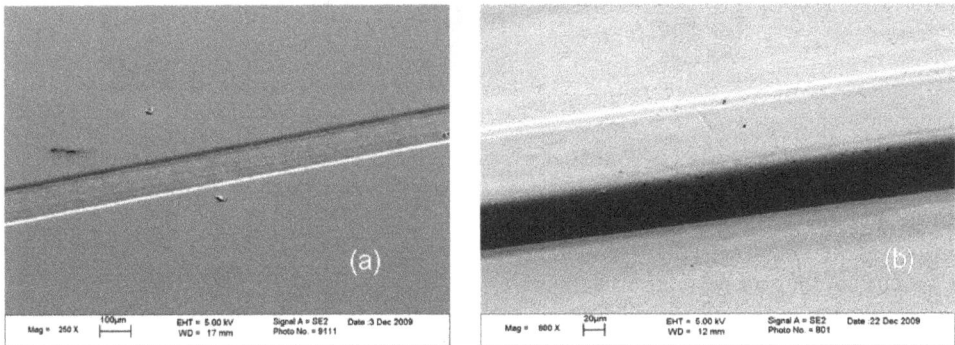

Fig. 8. Samples machined at 30 Hz, 50 shots per point and 3.6 mm/min with different fluences of 80 mJ/cm².

In Figure 8a above, a straight, shallow track is machined in an acrylate-based photopolymer while in Figure 8b above, two parallel tracks were etched leaving a ridge that constitutes a waveguide. In this case no upper cladding (as per stage 2, Figure 6) is applied. Sometimes, the ridge or waveguide may not be continuous. To examine this, light can be passed into one end of the guide for possible detection at the other end i.e. backlighting, as shown in Figure 9 where a single multimode waveguide of 50 μm × 35 μm and 60 mm long was illuminated from behind using a Flash™200 optical measuring device. The structure was made by ablating ~200 μm wide grooves in Truemode™ polyacrylate. Furthermore, waveguides can be 'crossed' at 90 degree (Figure 10) or other shapes may be desired. While excimer laser ablated waveguides is favoured, UV Nd:YAG (λ = 355 nm) [Van Steenberge, et al., 2004; Zakariyah, et al., 2011] and 10.6 μm CO_2 [Zakariyah, 2010] have been demonstrated as promising candidates especially for mass production at a low-cost.

Fig. 9. Excimer laser ablation of optical waveguide showing cross-section of a 50 μm x 35 μm multimode waveguide in Truemode™.

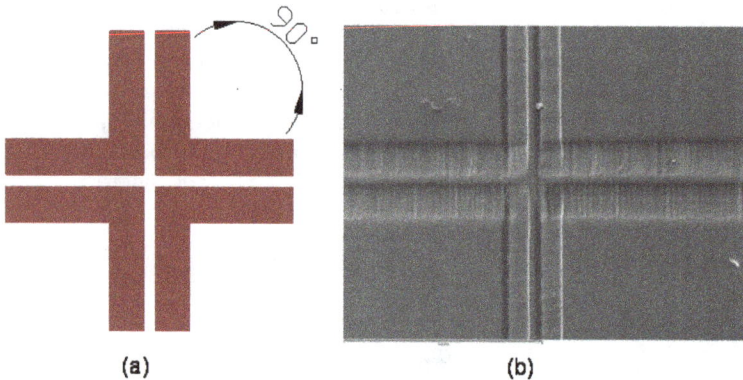

(a)   (b)

Fig. 10. Waveguides crossed over at 90 degree to each other machined at 100 mJ/cm2, 45 shots per point, 3.3 mm/min, 25 Hz and a single pass showing (a) a schematic diagram, and (b) an SEM image of an initial trial.

3.4 Integrated mirror fabrication

Optical signals on PCBs need to be routed to different parts of a device, such as between the boards of a backplane, if OI is to be fully utilised. Various proposals have been made on how to direct signals out of the plane of the board. These include 45-degree ended optical connection rods, microlens, 90⁰-bent fibre connectors, 45⁰-ended blocks, 45⁰-ended I-shape waveguides, optical coupler and microprism. These aforementioned concepts of out-of-plane coupling utilises blade cutting, laser ablation, moulding, dicing and RIE among others with each having its benefits and limitations [Byung, et al., 2004; Cho, et al., 2005; Cho, 2005; Teck, et al., 2009; Van Steenberge, et al., 2006]. To improve the coupling efficiency, Glebov, et al.(2005) proposed a curved micro-mirror instead of the flat 45-degree commonly employed.

Coupling light in and out of the polymer waveguides could be achieved by relying on the air/vacuum refractive index which is capable of causing total internal reflection (TIR) (Figure 11a) at this interface as used in [Teck, et al., 2009], but this can be difficult in real application because: (i) a vacuum is not guaranteed in a typical electronics assembly, (ii) air content and temperature are subjects of the environmental conditions, and (iii) even if air refractive index is guaranteed to be constant, air reflectivity is not efficient for coupling. For these reasons, end facets of mirrors are coated with a metal to improve its reflectivity and for a good surface finish. The chosen deposition technique depends largely on the sample to be coated and adhesion adhesion inter alia. For example, the authors of [Glebov, et al., 2005] used sputtering to deposit a thin layer of gold on the mirror surface before filling the trench with upper cladding; similar process was used for laser ablated mirror [Van Steenberge, et al., 2006]. It should be noted that there is a potential of light scattering or reflection at the clad-core interface [Hendrickx, et al., 2007a, 2007b]. Furthermore, the inaccuracy of the fabricated mirror angle can cause a significant reduction in the amount of light emanating from the core-clad exit of the waveguide to that reaching the metallised mirror surface thus affecting the coupling efficiency; a short path with a minimum angle deviation can mitigate this challenge.

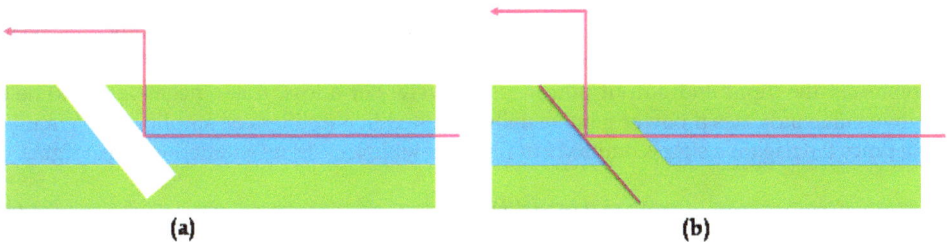

(a)　　　　　　　　　　　**(b)**

Fig. 11. Mirror fabrication schemes (a) TIR is used to deflect incoming signal out of the waveguide at the waveguide-air interface, and (b) light is coupled from a metal deposited at the surface of mirror trench which is the trench filled with cladding material.

While the out-of-plane coupling scheme is gaining impetus, there is no doubt that in-plane lateral routing of optical signals is also needed. A typical system architecture would require routing of signals not only from one layer to the other, but also within a layer; the latter would be extremely important if OI is extended to the board (and even chip) level as various roadmaps have laid down this possibility. Figure 12 is a schematic diagram of the in-plane mirror fabrication, which can be used to couple light between multiple components in the same layer. With this design, an effective turning angle of zero, 90-degree and multiples of 90-degrees are possible; a scheme demonstrated in [Glebov & Lee, M-G., 2006; Lamprecht, et al. 2009; Zakariyah, 2010]. Glebov & Lee, M-G. (2006) placed a vertical terminator at the end of the waveguide to form the mirror with a loss of 0.5 – 1.0 dB recorded for this approach; however, Zakariyah (2010) employed excimer laser ablation to manufacture the 45⁰ lateral mirrors. It is argued [Zakariyah, 2010] that laser ablation is a more suitable fabrication technique as it allows for both the waveguide and the mirrors to be manufactured using a single process on the same system. The laser ablation approach was also used for out-of plane mirror coupling such as in [Teck, et al., 2009] for 3D out-of-plane coupling.

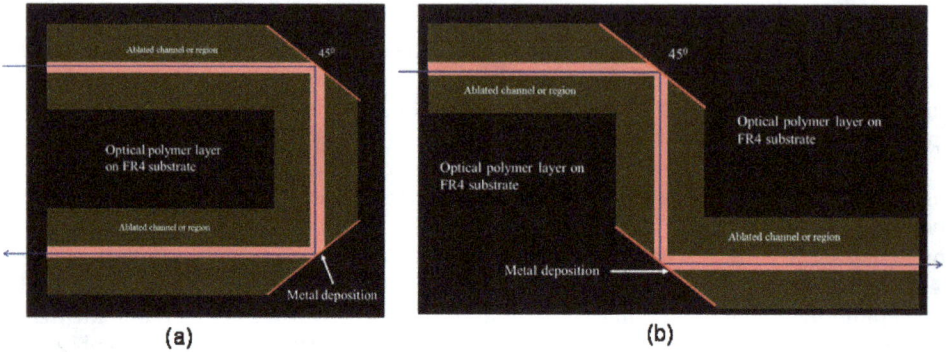

(a)   (b)

Fig. 12. Proposed 2D in-plane scheme showing (a) 45-degree in-plane coupling mirror design with 180-degree effective turning angle, and (b) 45-degree in-plane coupling mirror design with zero-degree effective turning angle.

3.5 Loss measurement

Signals launched at one end of an optical waveguide are not ideally identical in many cases to those arriving at the receiving end, either due to attenuation (change in amplitude) or distortion (change in waveform). These losses (propagation, coupling, angular misalignment, etc.) are quantified using a logarithmic unit called decibels (dB) using equation 4, where P_1 and P_2 represent the input and output power respectively. For a loss, it is a negative dB while a positive value indicates a power gain, usually obtained in amplifiers or amplification circuits. Sometimes the negative sign is omitted but replaced with 'loss' to mean attenuation in signal.

$$Loss\ (dB) = 10\log_{10}\frac{p_2}{p_1} \tag{4}$$

Reports have shown different values for the waveguide propagation loss depending mainly on the materials and the fabrication process used; Teck, et al. (2009) put the loss values in the range of 0.05 – 0.6 dB/cm, and the loss at a datacom (λ = 840 nm) in the range of 0.01 dB/cm – 0.8 dB/cm was given in [Holden, 2003]. Propagation loss of 0.24 dB/cm was recorded for a single mode waveguide in polyetherimide at 830 nm using laser ablation [Eldada, 2002]. A polymer waveguide manufactured by excimer laser ablation produced a propagation loss of 2 dB /cm at 1550 nm [Jiang, et al., 2004]; this high loss was attributed to the sidewall roughness of the guides. At 850 nm, propagation loss between 0.04 dB/cm and 0.2 dB/cm and 0.04 dB/cm and 0.18 dB/cm are reported for flexible and rigid waveguides respectively measured for different polymers [Shioda, 2007]. Table 3 is a list, though not exhaustive, of recent optical waveguide reports. While propagation loss is dependent on the waveguide characteristics, it is possible to reduce the insertion loss by reducing the coupling efficiency. One way of achieving this is through a good alignment between the coupling device and the waveguide. Jiang, et al. (2004) proposed an excimer laser ablation of the end facets for efficient coupling of light which in turn can reduce the loss.

Material	Process	Waveguide Dimension	Loss		Reference
			Waveguide	Mirror	
Custom multifunctional acrylate based photo-polymer	UV Laser Direct Writing (He: Cd, 325 nm and 3 mW)	50 μm × 50 μm multimode	< 0.17 dB/cm @ 850 nm & < 0.5 dB/cm @ 1300 nm	-	Tooley, et al., 2001; Walker, et al., 2008
SU-8-50 epoxy (core) & MR-L6100XP (cladding)	UV Lithography	85 μm × 100 μm	0.60 ± 0.03 dB/cm at λ = 850 nm	1.8 – 2.3 dB (estimated)	Immonen, et al., 2005
Perfluorocyclobutane (PFCB)	Rubber molding	47 μm × 41μm	0.4 dB/cm (1300 nm) & 0.7 dB/cm (1550 nm)	-	Lee, B-T., et al., 2000
Photosensitive polymer	UV photolithography	30 μm × 30 μm	0.06 dB/cm (850 nm) & ~ 0.25 dB/cm (1310 nm)	-	Matsuoka, et al., 2010
-	Imprinting	50 μm × 50μm	0.035 dB/cm (850 nm)	0.5 dB per each facet	Hwang, et al., 2010
Truemode™ acrylate-based photopolymer	Excimer Laser Ablation (3 ± 0.5 J/cm², 200 Hz & 240 μm/s ablation speed)	50 μm × 50 μm	0.13 dB/cm at 850 nm	-	Steenberge, et al., 2006
Polycarbonate (cladding) epoxy resin (core)	Hot-embossing	-	0.5 dB at 850 nm	-	Kim, et at., 2007
Polysiloxane-based polymer	Photolithography and dry etching	8 μm × 8 μm single mode	0.17 dB/cm at 1310 nm & 0.43 dB/cm at 1550 nm	-	Usui, et al., 1996
Truemode™ & ORMOCER	Photolithography and Excimer Laser Ablation	50 μm × 50 μm two layers	0.12 dB/cm at 850 nm	-	Hendrickx, et al., 2007a, 2007b; Steenberge, et al., 2006
Proprietary to Mistui Chemicals Inc., Tokyo, Japan	Excimer laser ablation (mirror)	70 μm × 50 μm	0.1 – 0.3 dB/cm at 850 nm	< 4 dB loss for two 45⁰ 82 mm long mirrors	Teck, et al., 2009

Material	Process	Waveguide Dimension	Loss		Reference
			Waveguide	Mirror	
UV curable resins (core)	Hot-embossing	60 μm × 60 μm	~ 0.1 dB/cm at 850 nm	-	Yoon, et al., 2004
Photopatternable polymer	Photolithography (WGs) & Microdicing (mirrors)	30 μm × 30 μm	0.05 dB/cm at 850 nm	0.5 – 0.8 dB at 850 nm	Glebov, et al., 2005, 2007.
Photosensitive acrylate polymer	Photolithography	50 μm × 50 μm (250 μm pitch) & 35 μm × 35 (100 μm pitch)	0.035 – 0.05 dB/cm at 850 nm & 0.12 dB/cm at 990 nm	-	Dangel, et al., 2004
ORMOCER	-	≤ 50 μm × 10 μm multimode	-	-	Uhlig, et al., 2006
Truemode™	UV Nd:YAG Laser Ablation	45 μm × 45 μm	1.4 ± 0.5 dB/cm at 850 nm	-	Zakariyah, et al., 2011
Polysiloxane	Casting + Doctor blade	-	0.05 dB/cm at 850 nm	-	Kopetz, et al., 2004.
Fluorinated acrylate polymer	Soft molding (core) & spin-coating (cladding)	70 μm × 70 μm	-	-	Liang, et al., 2008
Epoxy resin	Spin-coating	50 μm × 50 μm	0.15 dB/cm at 850 nm	-	Albrecht, et al., 2005
Siloxane polymer	Photolithography	50 μm × 20 μm	0.03 – 0.05 dB/cm at 850 nm	-	Bamiedakis, et al., 2007
SU-8 (NANOTM SU-8-50)	Photolithography	50 μm × 50 μm	-	-	Chen, Y-M, et al., 2005
Deuterated PMMA (core) & UV-cured epoxy resin (cladding)	Spin coating, photolithography & RIE	40 μm × 40 μm	< 0.02 dB/cm at 830 nm	0.3 – 0.7 dB	Hikita, et al., 1998

Table 3. Optical polymer waveguide fabrication techniques

4. Conclusion

In this chapter, the author presented the need for OI for both intra- and inter-board applications due to prevailing limitations with electrical interconnection on the PCB despite the various rectifying measures being considered. For successful implementation of OI, the following are needed: materials that would be compatible with PCB manufacturing procedures; fabrication techniques that would be easy, cost effective and efficient from the production point of view; and finally materials / waveguides that would satisfy the optical power budget requirement. A polymer-based waveguide is favoured for this technology primarily due to its low cost and compatibility. Multimode polymer waveguides with typical dimensions 50 ± 20 μm square are common as it relaxes alignment constrain thus lowering coupling. While various fabrication techniques have been reported with new still emerging procedures, laser ablation is a preferred approach since it is the technique currently being used for the drilling of μvias, which makes it a much compatible candidate. Furthermore, for the fabrication of integrated mirrors, either in-plane or out-of-plane, laser ablation using an excimer laser for example, is a much suitable option for this due to its excellent laser matter interaction, resulting in clean removal at micro-level scales. In addition, the mask projection available with excimer laser makes it possible for complex features to be easily defined. Although the cost and speed of excimer laser could be an issue from the production point of view at this stage of the deployment, other lasers such as UV Nd:YAG and CO_2 can offer both prototyping and mass production opportunities as it has been demonstrated, thus making laser ablation an all-encompassing technique meeting required production speed, cost, efficiency and quantity. In light of this, the chapter also provides an overview of laser technology for material processing and in particular for polymer waveguide fabrication.

5. Acknowledgment

The authors wish to thank Khadijah Olaniyan, Abdul Lateef Balogun, Mayowa Kassim Aregbesola and Witold Kandulski for helpful discussions.

6. References

Albrecht, H., Beier, A., Demmer, P., Franke, M., Modinger, R., Pfeiffer, K., Beil, P., Kostelnik, J., Bauer, J., Ebling, F., Schroder, H. & Griese, E. (2005). New-generation interconnect, *Information Photonics*, 2005. IP 2005. OSA Topical Meeting, pp 1-3.

Bamiedakis, N., Beals, J., Penty, R.V., White, I.H., Degroot, J.V. & Clapp, T.V. (2007). Low Loss and Low Crosstalk Multimode Polymer Waveguide Crossings for High-Speed Optical Interconnects, *Proceedings of Lasers and Electro-Optics*, 2007. CLEO 2007, pp. 1-2.

Basting, D., Pippert, K. & Stamm (2002). History and future prospects of excimer laser technology, *Prooceedings of the 2nd International Symposium on Laser Precision Microfabrication*, *RIKEN Review*, No.43, pp. 14-22.

Basting, D. & Marowsky, G. (1st Edition). 2005. *Excimer Laser Technology*, Springer, ISBN-10: 3540200568, Berlin Heidelberg, New York.

Byung, S.R., Kang, S., Han, S.C., Park, H-H., Ha, S-W. & Rhee, B-H. (2004). PCB-compatible optical interconnection using 45 deg -ended connection rods and via-holed waveguides, *IEEE Journal of Lightwave Technology*, Vo.22, No.9, pp. 2128-34.

Chen, X. & Liu, X. (1999). Short pulsed laser machining: How short is short enough?, *Journal of laser applications*, Vol.11, No.6, pp. 268-72.

Chen, Y-T, Naessens, K., Baets, R., Liao, Y. & Tseng, A. (2005). Ablation of transparent materials using excimer lasers for photonic applications, *Optical Review*, Vo.12, No.6, pp. 427-441.

Chen, Y-M., Yang, C-L., Cheng, Y-L., Chen, H-H., Chen, Y-C., Chu, Y. & Hsieh, T-E. (2005). 10Gbps multi-mode waveguide for optical interconnect, *Proceedings of the 55th Electronic Components and Technology Conference*, 2005, Vo.2, pp. 1739-1743.

Cho, M.H. (2005). High-coupling-efficiency optical interconnection using a 90-bent fiber array connector in optical printed circuit boards, *IEEE Photonics Technology Letters*, Vol.17, No.3, pp. 690-692.

Crafer, R. & Oakley, P.J. (1993), *Laser processing in manufacturing*, Chapman and Hall, ISBN: 0412415208, London.

Dangel, R., Bapst, U., Berger, C., Beyeler, R., Dellmann, L., Horst, F., Offrein, B. & Bona, G-L. (2004). Development of a low-cost low-loss polymer waveguide technology for parallel optical interconnect applications, *Biophotonics/Optical Interconnects and VLSI Photonics/WBM Microcavities*, 2004 Digest of the LEOS Summer Topical Meetings, 2004, pp. 29 - 30.

Dangel, R., Beyeler, R., Horst, F., Offrein, B.J., Sicard, B., Moynihan, M., Knudsen, P. & Anzures, E. (2007). Waveguide Technology Development based on Temperature- and Humidity-Resistant Low-Loss silsesquioxane Polymer for Optical Interconnects, *Proceedings of Optical Fiber Communication and the National Fiber Optic Engineers Conference*, 2007. OFC/NFOEC 2007, pp 1-3.

Do-Won Kim, In-Kui Cho, Seung Ho Ahn & Hyo-Hoon Park (2007). 5-Gb/s Chip-to-chip Optical Interconnection Using Polymeric Waveguides, *Proceedings of Lasers and Electro-Optics - Pacific Rim*, 2007. CLEO/Pacific Rim 2007, pp. 1-2.

Eldada, L. (2002). Polymer integrated optics: Promise vs. practicality, *Proceedings of SPIE - The International Society for Optical Engineering*, Vol. 4642, pp. 11-22.

Glebov, A. L., Roman, J., Lee, M.G. & Yokouchi, K.(2005). Optical interconnect modules with fully integrated reflector mirrors, *IEEE Photonics Technology Letters*, vol. 17, pp. 1540-1542, Jul. 2005

Glebov, A.L. & Lee, M.G. (2006). 3D Routing on Optical Boards, *Proceedings of IEEE 19th Annual Meeting on Lasers and Electro-Optics Society*, 2006. LEOS 2006. 2006, pp. 22-23.

Glebov, A.L., Lee, M.G. & Yokouchi, K. (2007). Integration technologies for pluggable backplane optical interconnect systems, *Optical engineering: the journal of the Society of Photo-optical Instrumentation Engineers*, vol. 46(1), pp. 15403-1 - 15403-10

Hendrickx, N., Van Erps, J., Van Steenberge, G., Thienpont, H. & Van Daele, P. (2007a). Tolerance analysis for multilayer optical interconnections integrated on a printed circuit board, *IEEE Journal of Lightwave Technology*, Vol. 25, No. 9, pp. 2395-2401.

Hendrickx, N., Van Steenberge, G., Geerinck, P. & Van Daele, P. (2007b). Laser ablation as enabling technology for the structuring of optical multilayer structures, *Journal of Physics*: Conference Series, Vol. 59, No.1, pp. 118-21.

Hikita, M., Yoshimura, R., Usui, M., Tomaru, S. & Imamura, S. (1998). Polymeric optical waveguides for optical interconnections, *Thin Solid Films*, Vol.331, No.1-2, pp. 303-308.

Holden, H.T. (2003). The developing technologies of integrated optical waveguides in printed circuits, *Circuit World*, Vol.29, No. 4, pp. 42-50.

Huang, D., Sze, T., Landin, A., Lytel, R. & Davidson, H.L. (2003). Optical interconnects: out of the box forever?, *IEEE Journal of Selected Topics in Quantum Electronics*, Vol.9, No.2, pp. 614-623.

Hwang, S.H., Lee, W.J., Lim, J.W. & Rho, B. S. (2010). Fabrication and reliability test of rigid-flexible optical printed circuit boards for mobile devices, *Proceedings of 15th OptoeElectronics and Communications Conference (OECC)*, 2010, pp. 256-257.

Ihlemann, J. & Wolff-Rottke, B. (1996). Excimer laser micro machining of inorganic dielectrics, *Applied Surface Science*, Vol.106, pp. 282-286.

Immonen, M., Karppinen, M. & Kivilahti, J.K. (2005). Fabrication and characterization of polymer optical waveguides with integrated micromirrors for three-dimensional board-level optical interconnects, *IEEE Transactions on Electronics Packaging Manufacturing*, Vol. 28, No.4 pp. 304-311

Immonen, M.P., Karppinen, M. & Kivilahti, J.K. (2007). Investigation of environmental reliability of optical polymer waveguides embedded on printed circuit boards. *Microelectronics Reliability*, Vol.47, No.2, pp. 363-371.

Jackson, S.R., Matheringham, P.E. & Dyer, P.E. (1995). Eximer laser ablation of Nd:YAG and Nd:glass, *Applied Surface Science*, Vol.86, pp.223-227.

Jiang, J.. Callender, C.L Noad, J.P. Walker, R.B. Mihailov, S.J. Ding J. & Day, M. (2004). All-polymer photonic devices using excimer laser micromachining, *IEEE Photonics Technology Letters*, Vol.16, No.2, pp. 509-511.

Kopetz, S., Rabe, E., Kang, W. & Neyer, A., (2004). Polysiloxane optical waveguide layer integrated in printed circuit board, *Electronics Letters*, Vol.40, No.11, pp. 668-669.

Lamprecht, T., Beyeler, R., Dangel, R., Horst, F., Jubin, D., Meier, N., Weiss, J. & Offrein, B.J. (2009). Integrated micro-mirrors for compact routing of optical polymer waveguides, *Proceedings of LEOS Annual Meeting Conference Proceedings, 2009. IEEE 2009, pp. 20-21*.

Lau, J.H. (2000). An overview of microvia technology, *Circuit World*, Vol.26, No.2, pp. 22 - 32.

Lee, B-T., Kwon, M-S., Yoon, J-B. & Shin, S-Y. (2000). Fabrication of polymeric large-core waveguides for optical interconnects using a rubber molding process, *IEEE Photonics Technology Letters*, Vol.12, No.1, pp. 62-64.

Li, J. & Ananthasuresh, G.K. (2001). A quality study on the excimer laser micromachining of electro-thermal-compliant micro devices, *Journal of micromechanics and microengineering : structures, devices, and systems*, Vol.11, pp. 38-47.

Li, L., Hong, M., Schmidt, M., Zhong, M., Malshe, A., Huis In'tveld, B. & Kovalenko, V. (2011). Laser nano-manufacturing – State of the art and challenges, *CIRP Annals - Manufacturing Technology*, In Press, Corrected Proof.

Liang, C.T.W., Yee, H.L., Shiah, L.L., Wei, T.C., Yoon, J.C.Y., Jie, Y.G., Guan, L.T., Ramana, P.V., Lau, J.H., Chang, R., Tang, T., Chiang, S., Cheng, D. & Tseng, T.J. (2008). Fabrication and Optimization of the 45Â° Micro-mirrors for 3-D Optical Interconnections, *Proceedings of the 10th Electronics Packaging Technology Conference*, 2008. EPTC 2008, pp. 1121-1125.

Matsuoka, Y., Kawamura, D., Ban, T., Mita, R., Lee, Y., Adachi, K., Sugawara, T., Hamamura, S., Matsushima, N., Cyujyo, N., Shibata, T., Masuda, H. & Takahashi, A. (2010). Optical printed circuit board with an efficient optical interface for 480-Gbps/cm/sup 2/ (20 Gbps X 12 ch X 2 layers) high-density optical interconnections, *Optical Fiber Communication (OFC), collocated National Fiber Optic Engineers Conference, 2010 Conference on (OFC/NFOEC)*, pp. 1-3.

Meijer, J. (2004). Laser beam machining (LBM), state of the art and new opportunities, *Journal of Materials Processing Technology*, Vol.149, No.1-3, pp. 2-17.

Offrein, B.J. (2008). Optical interconnects and nanophotonics, *Proceedings of IEEE/LEOS Internationall Conference Optical MEMs and Nanophotonics*, 2008, pp. 21-22.

Ozcan, L.C. (2008). Fabrication of buried waveguides in planar silica films using a direct CW laser writing technique, *Journal of non-crystalline solids*, Vol.354, No.42, pp. 4833.

Papakonstantinou, I., Selviah, D.R., Pitwon, R. & Milward, D. (2008). Low-Cost, Precision, Self-Alignment Technique for Coupling Laser and Photodiode Arrays to Polymer Waveguide Arrays on Multilayer PCBs, *IEEE Transactions on Advanced Packaging*, Vol.31, No.3, pp. 502-511.

Pfleging, W. (2006). Excimer laser material processing – state of the art and new approaches in microsystem technology, *Proceedings of SPIE - The International Society for Optical Engineering*, Vol. 6107.

Pitwon, R., Hopkins, K. & Milward, D. (2006). An optical backplane connection system with pluggable active board interfaces, *Proceedings of the Sixth IASTED International Multi-Conference on Wireless and Optical Communications*, pp. 297- 302.

Hopkins, K. & Pitwon, R. (2007). Pluaggable Optical Backplane Technology, In: *White Paper, Xyratex Technology*. Accessed : November 2011, Avaialble at : http://www.xyratex.com/pdfs/whitepapers/Xyratex_white_paper_Pluggable_Optical_Backplane_2-0.pdf.

Selviah, D.R., Walker, A.C., Hutt, D.A., Wang, K., Mccarthy, A., Fernandez, F.A., Papakonstantinou, I., Baghsiahi, H., Suyal, H., Taghizadeh, M., Conway, P., Chappell, J., Zakariyah, S., S., Milward, D., Pitwon, R., Hopkins, K., Muggeridge, M., Rygate, J., Calver, J., Kandulski, W., Deshazer, D.J., Hueston, K., Ives, D.J., Ferguson, R., Harris, S., Hinde, G., Cole, M., White, H., Suyal, N., Rehman, H.U. & Bryson, C. (2010). Integrated optical and electronic interconnect PCB manufacturing research. *Circuit World*, Vol.36, No.2, pp. 5-19.

Shin, B.S., Oh, J.Y. & Sohn, H. (2007). Theoretical and experimental investigations into laser ablation of polyimide and copper films with 355-nm Nd:YVO4 laser, *Journal of Materials Processing Technology*, Vol.187-188, pp. 260-263.

Shioda, T. (2007). Recent Progress and Potential Markets for Optical Circuit Boards, *Proceedings of the 6th International Conference on Polymers and Adhesives in Microelectronics and Photonics*, 2007, Polytronic 2007, pp. 167-169.

Sugioka, K., Obata, K., Midorikawa, K., Hong, M.H., Wu, D.J., Wong, L.L., Lu, Y.F. & Chong, T.C. (2003). Advanced materials processing based on interaction of laser beam and a medium, *Journal of Photochemistry and Photobiology A: Chemistry*, Vol.158, No.2-3, pp. 171-178.

Teck, G.L., Ramana, P.V., Lee, B.S.P., Shioda, T., Kuruveettil, H., Li, J., Suzuki, K., Fujita, K., Yamada, K., Pinjala, D. & Shing, J.L.H. (2009). Demonstration of direct coupled optical/electrical circuit board, *IEEE Transactions on Advanced Packaging*, Vol.32, No.2, pp. 509-16.

Thomas, D.W. (1992). Surface modification of polymers and ceramics induced by excimer laser-radiation, *Laser ablation of electronic materials*, Vol.4, pp. 221-228.

Tooley, F., Suyal, N., Bresson, F., Fritze, A., Gourlay, J., Walker, A. & Emmery, M. (2001). Optically written polymers used as optical interconnects and for hybridization, *Optical Materials*, Vol.17, No.1-2, pp. 235-241.

Tseng, A.A., Chen, Y., Chao, C., Ma, K. & Chen, T.P. (2007). Recent developments on microablation of glass materials using excimer lasers, *Optics and Lasers in Engineering*, Vol.45, No.10, pp. 975-992.

Uhlig, S., Frohlich, L., Chen, M., Arndt-Staufenbiel, N., Lang, G., Schroder, H., Houbertz, R., Popall, M. and Robertsson, M. (2006). Polymer Optical Interconnects -- A Scalable Large-Area Panel Processing Approach, *IEEE transactions on advanced packaging* : a publication of the IEEE Components, Packaging, and Manufacturing Technology Society and the Lasers and Electro Optics Society, Vol. 29, No.1, pp. 158 - 170.

Uhlig, S. & Robertsson, M. (2005). Flip chip mountable optical waveguide amplifier for optical backplane systems, *Proceedings of the 55th Electronic Components and Technology Conference*, 2005, Vol. 2, pp. 1880-1887.

Usui, M., Hikita, M., Watanabe, T., Amano, M., Sugawara, S., Hayashida, S. & Imamura, S. (1996). Low-Loss Passive Polymer Optical Waveguides with High Environmental Stability, *IEEE Journal of Lightwave Technology* : a joint IEEE/OSA publication., Vol.14, No.10, pp. 2338 -2343.

Van Steenberge, G., Geerinck, P., Van Put, S., Van Koetsem, J., Ottevaere, H., Morlion, D., Thienpont, H. & Van Daele, P. (2004). MT-compatible laser-ablated interconnections for optical printed circuit boards, *IEEE Journal of Lightwave Technology*, Vol.22, No.9, pp. 2083-2090.

Van Steenberge, G., Hendrickx, N., Bosman, E., Van Erps, J., Thienpont, H. & Van Daele, P. (2006). Laser ablation of parallel optical interconnect waveguides, *IEEE Photonics Technology Letters*, Vol.18, No.9, pp. 1106-1108.

Walker, A.C. ; Suyal, H. ; McCarthy, A. (2008). Direct laser-writing of polymer structures for optical interconnects on backplane printed circuit boards, *Proceedings of 2nd Electronics Systemintegration Technology Conference, ESTC*, pp. 977-979.

Wei, M. & Yang, H. (2003). Cumulative Heat Effect in Excimer Laser Ablation of Polymer PC and ABS, *International Journal of Advanced Manufacturing Technology*, Vol.21, No.1, pp. 1029-1034.

Yoon, K.B., Park, H., Jeong, M.Y., Rhee, B., Heo, Y.U., Rho, B.S., Cho, I., Lee, D.J. & Ahn, S.H. (2004). Optical backplane system using waveguide-embedded PCBs and optical slots, *IEEE Journal of Lightwave Technology*, Vol.22, No.9 pp. 2119-2127.

Yung, W.K.C., Liu, J.S., Man, H.C. & Yue, T.M. (2000). 355 nm Nd:YAG laser ablation of polyimide and its thermal effect, *Journal of Materials Processing Technology*, Vol.101, No.1-3, pp. 306-311.

Zakariyah, S.S., Conway, P.P., Hutt, D.A., Selviah, D.R., Wang, K., Baghsiahi, H., Rygate, J., Calver, J. & Kandulski, W. (2009). Polymer optical waveguide fabrication using laser ablation, *Proceedings of the 11th Electronics Packaging Technology Conference, 2009. EPTC '09*, pp. 936-941.

Zakariyah, S.S. (2010). Laser Ablation of Polymer Waveguide and Embedded Mirror for Optically-Enabled Printed Circuit Boards (OEPCB), *PhD Thesis*, Loughborough University, UK.

Zakariyah, S.S., Conway, P.P., Hutt, D. A., Selviah, D.R., Wang, K., Baghsiahi, H., Rygate, J., Calver, J., Kandulski, W. (2011). Fabrication of polymer waveguides by laser ablation using a 355 nm wavelength Nd:YAG laser, *IEEE Journal of Lightwave Technology*, 2011 (DOI : 10.1109/JLT.2011.2171318) (Manuscript No: JLT-13362-2011).

Zeng, D.W., Yung, K.C. & Xie, C.S. (2003). UV Nd:YAG laser ablation of copper: chemical states in both crater and halo studied by XPS, *Applied Surface Science*, Vol. 217, No.1-4, pp. 170-180.

6

Laser Micromachining and Micro-Patterning with a Nanosecond UV Laser

Xianghua Wang[1], Giuseppe Yickhong Mak[2] and Hoi Wai Choi[2]
[1]Key Lab of Special Display Technology, Ministry of Education,
National Engineering Lab of Special Display Technology,
National Key Lab of Advanced Display Technology,
Academy of Photoelectric Technology,
Hefei University of Technology, Hefei,
[2]Department of Electrical and Electronic Engineering,
The University of Hong Kong,
[1]China
[2]Hong Kong

1. Introduction

Laser micromachining specifically refers to drilling and cutting with intensive laser beam usually in the form of pulse trains with energy far exceeding the ablation threshold of the target material. The ablation threshold, however, is dependent on material properties as well as the interacting laser characteristics such as laser wavelength and pulse width. Therefore, laser processes are flexible and can be highly selective. Laser wavelengths have been available in a wide spectrum range from the far infrared to the ultraviolet using a variety of medium material and frequency mixing techniques. Since late 1990s, ultra short laser beams (Gattass and Mazur 2008; Liu, et al. 1997; Molian, et al. 2009; Schaffer, et al. 2001; Stuart, et al. 1996; Varel, et al. 1997) from picosecond to femtosecond are intensively studied as to their interaction mechanism with transparent materials and applications in material processing. Although ultra short pulses within tens of picoseconds can be achieved by advanced techniques like Q-switching and mode-locking, the pulse energy is reduced by several orders of magnitude and the laser system is relatively bulky. Nanosecond lasers with intermediate pulse width are still predominant in industry due to their excellent flexibility and cost effectiveness. This chapter presents the work in laboratory on laser micromachining and micro-patterning of GaN/sapphire LED wafer with nanosecond UV laser for the purpose of chip-shaping and device isolation respectively.

2. Laser micromachining

2.1 Laser-matter interaction

Interaction of material with laser pulses generally involves the following fundamental processes that take place on different time scales after a pulse incidence (Zweig 1991). Firstly, light absorption takes place at a femtosecond time scale accompanied by ejection of excited electrons. Secondly, the excited electrons transfer their energy to the lattice in several tens of picoseconds through electron-phonon collision and induce melted zone in the bulk material. Following this, melted material evaporates on a nanosecond time scale and expands into the air in the form of atomic-sized particles in the ablation plume. Melt expulsion may also occur, termed hydrodynamic ablation, due to the recoil pressure of plume and result in large particles or clusters following the plume.

The specific ablation mechanism in laser micromachining and the resulting morphology of the ablated region such as aspect ratio and sidewall roughness varies with material property, laser wavelength, power intensity as well as pulse duration. The process of multiphoton excitation of electrons and collisional ionization is reported to play vital roles in semiconductor and dielectric laser ablation with a sharp threshold of power intensity observed. This mechanism become dominant when ultra-short laser pulses are employed for ablation, whereby a much increased portion of laser energy is transferred to electrons compared to the portion further transferred to the lattice (Stuart, et al. 1996). Under high intensity picosecond-pulsed laser fluence, the laser irradiated portion of n-type SiC material can be directly transformed into plasma on a few picoseconds time scale trough Coulomb explosion resulting in clean surfaces of the cleaves that outperform the results obtained with thermal ablation (Molian, et al. 2009).

2.2 Optical setup of laser micromachining

The laser micromachining system consists of a UV laser source, beam focusing optics and an x-y motorized translation stage. The laser source used in this experiment is a third harmonic neodymium-doped yttrium lithium fluoride (ND:YLF) diode-pumped solid-state (DPSS) laser manufactured by Spectra Physics. Being an actively Q-switched UV laser at 349 nm, the pulse repetition rate ranges from single pulse to 5 kHz. At the reference diode current of 3.2A, the pulse energy is 120 µJ with a pulse width of around 4 nanoseconds at 1 kHz repetition rate; however, this corresponds to very high intensity up to the order 10^{10} W/cm^2 when focused to a 10-µm-diameter spot.

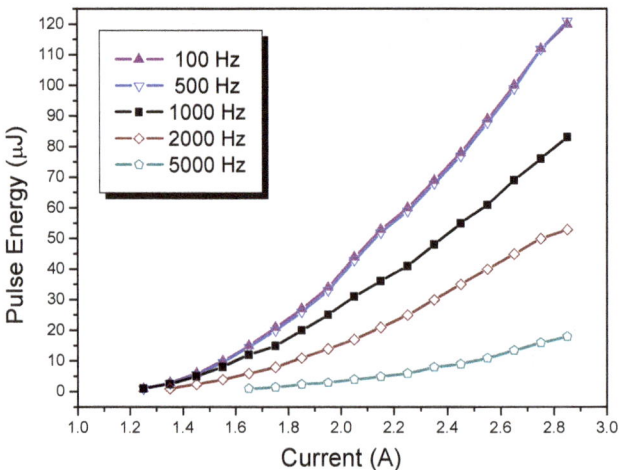

Fig. 1. Pulse energy versus diode current at different pulse repetition rates

The beam intensity distribution is a Gaussian profile. There is a compromise between the pulse energy and the pulse repetition rate. According to the plots of pulse energy versus diode current recorded at a series of repetition frequencies, the pulse energy at lower repetition frequency goes higher until the repetition frequency is lower than 500 Hz as indicated in Figure 1.

As a smaller focus spot, which is diffraction limited, is desired for micro-scale machining, laser of shorter wavelength is preferred. The TEM00 beam in our laser source also allows for tight focusing, offering high spatial resolution. After beam expansion and collimation through a beam expander, the laser beam is reflected 90° by a dielectric laser line mirror and focused onto the horizontal machining plane to a very tiny spot several micrometers in diameter with a focusing triplet. The UV objective lens well suits the Gaussian beam profile of the laser. All optics are made of UV fused silica and anti-reflection coated. The size of the beam at the focal point is not only limited by the capability of the UV objective lens but is also sensitive to the coaxiality of the optics. Schematic diagram of the laser micromachining setup is drawn in Figure 2. A pair of aluminium mirrors is used to adjust the direction of the collimated laser beam to be coaxial with the UV objective lens.

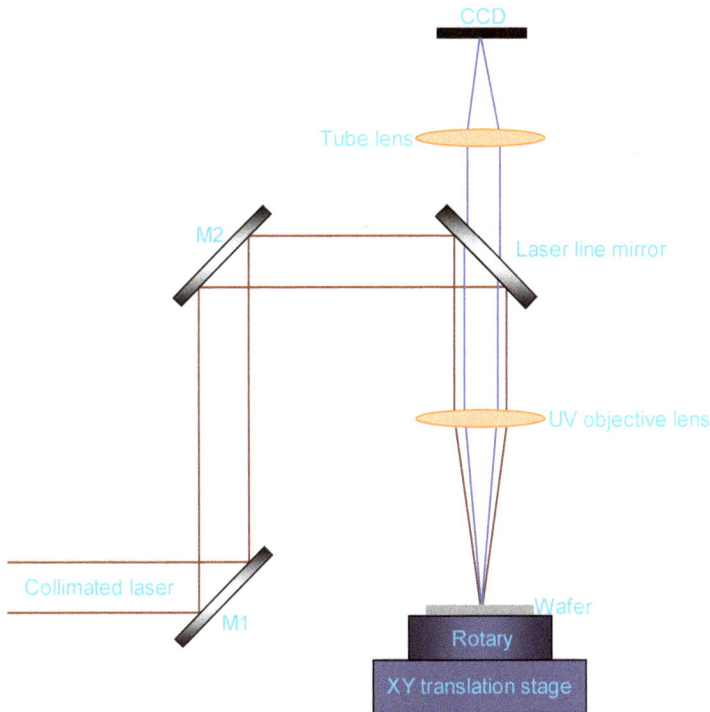

Fig. 2. Schematic of laser micromachining setup

The additional feature of the set-up for tilted cutting, as illustrated in Figure 3, is the insertion of a UV mirror at an oblique angle within the optical path between the focusing optics and the machining plane, which serves to deflect the convergent beam to strike the sample at an oblique angle with respect to the horizontal working plane. With this modified set-up, it is relatively easy to optimize and monitor the beam through the tube lens imaged with a CCD camera. Once the optical setup is optimized before insertion of the tilting mirror, the mirror can be inserted without affecting the coaxiality of the laser beam.

The beam can be effectively applied for micro-sectioning with non-vertical sidewall profiles. The angle of incidence of the deflected laser beam on the wafer is 2θ, where θ as indicated in Figure 3 is the angle between the plane of the mirror and the normal. This angle is readily and precisely controlled by mounting the mirror onto a rotation stage; thus, the incident angle can be varied over a wide range. In this experiment, we have used a UV objective with a focal length of 75 mm, based on two considerations. Firstly, the focal length should be long enough to accommodate the mirror in the optical path. Secondly, an ideal tool for the fabrication of microstructures should have a very long penetration depth and negligible lateral dispersion. Nevertheless, an objective lens with a longer focal length also produces a larger focused beam spot. The diameter of the focused beam spot, d, is determined by the following equation:

$$d = \frac{4\lambda M^2 f}{\pi D} \tag{1}$$

where M^2 is the beam quality factor, λ the wavelength of the laser beam, f the focal length and D the diameter of the incident beam.

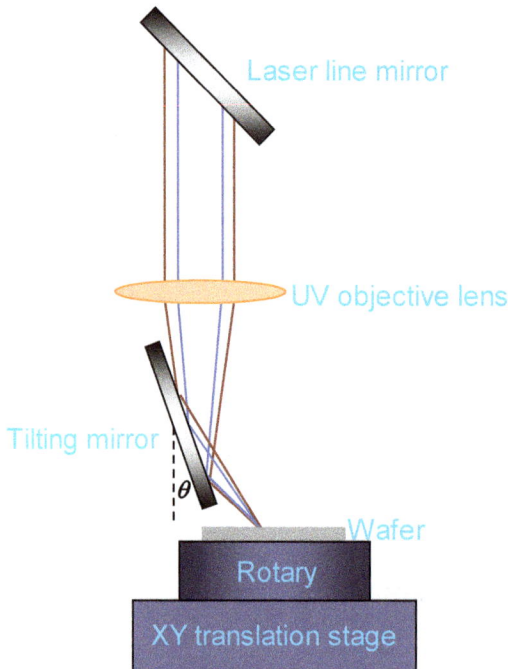

Fig. 3. Schematic of the setup for tilted-cutting with a UV mirror inserted within the optical path

2.3 Laser cutting of sapphire wafer

As our scheme of micromachining is targeted at die separation of GaN-based LEDs, sapphire wafers are used for testing the results, as it is the typical substrate for the metalorganic chemical vapour deposition (MOCVD) growth of GaN. The quality of the cleave can be quantified by the width, depth, linearity and sidewall roughness of the trench formed by the laser beam. Each of these parameters will be investigated. Since the focal length of the focusing lens (f = 75 mm) is much longer than the thickness of the sapphire wafer (t = 420 μm), the depth of the trench mainly depends on the number of micromachining cycles. The number of cycles is controlled by configuring the translation stage to repeat its linear path over a number of times. Since the position repeatability of the stage is better than 5 μm, increasing the number of cycles should not contribute significantly to the width of the feature. Figure 4 shows the cross-sectional optical image of a 420 μm thick sapphire wafer that has been micro-machined with an incident beam inclined at 45°, with scan cycles ranging from 1 to 10. These incisions were carried out by setting the laser pulse energy to 54 μJ at a repetition rate of 2 kHz. The relationship between the inclined cutting depth and the number of passes of the beam are plotted in Figure 5. After the first pass of the beam, a narrow trench with a width of ~20 μm and a depth of ~220 μm was formed. Successive scans of the beam along the trench results in further deepening and widening, but the extent was increasing less. The depth of the trench depends on the effective penetration of the beam. From the second scan onwards, the beam has to pass through the narrow gap before reaching the bottom of the trench for further machining. The energy available at this point is attenuated, partly due to lateral machining of the channel (causing undesirable widening), absorption and diffraction effects. Therefore, the depth of the trench tends to saturate after multiple scans.

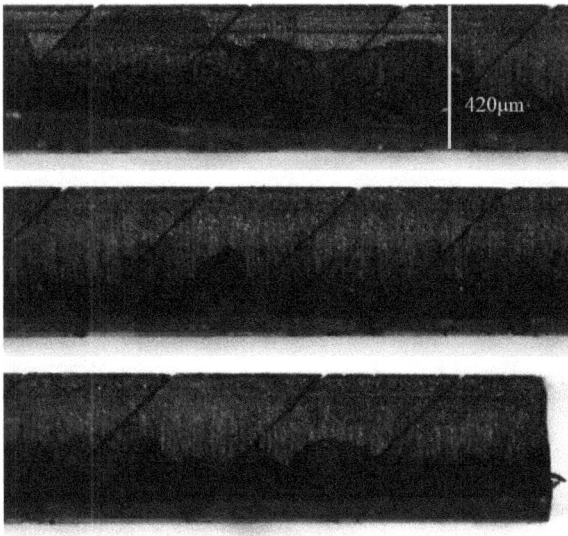

Fig. 4. Cross-sectional optical microphotograph of laser micro-machined micro trenches at an inclination angle of ~45° at a range of scan cycles of between 1 and 10 (left to right then down). (with permission for reproduction from American Institute of Physics)

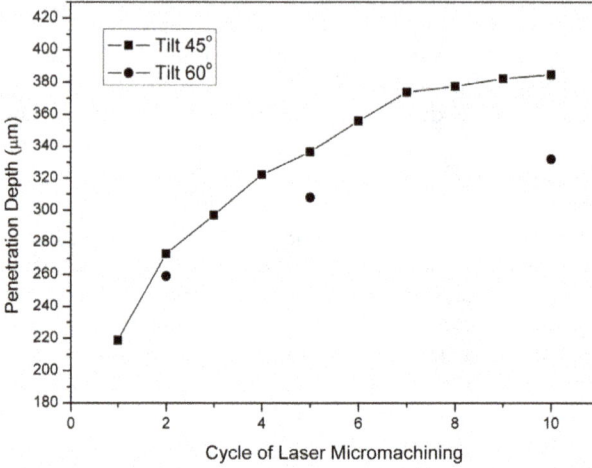

Fig. 5. Depth of tilting micro-trenches as a function of scan cycles. (with permission for reproduction from American Institute of Physics Publishing)

After the chemical treatment, the surface morphologies of the micromachined samples are examined with atomic force microscopy (AFM). 3D images of the AFM scans are shown in Figure 6. The surface topography of the sapphire surface after 2 machining cycles exhibits a uniform roughness with an RMS value of ~150 nm. With more cycles, increasing densities and dimensions of granules are observed on the AFM image and the RMS roughness increased to ~218 nm after 5 cycles. The formation of the larger grains on the surface is a result of uneven aggregation and re-solidification of the melted material. The evacuation rate of the ablated species declines as the beam reaches deeper into the trench, and statistically the density of aggregation is more pronounced at these deeper sites.

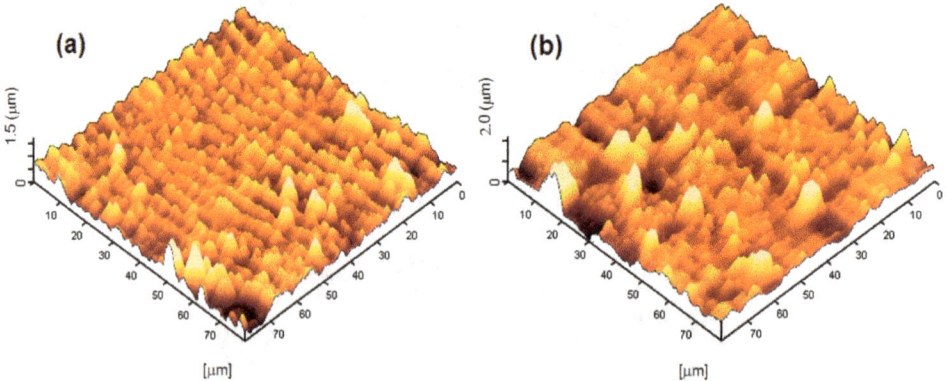

Fig. 6. AFM morphology images of the inclined sapphire surfaces after laser micromachining for (a) 2 cycles and (b) 5 cycles, the corresponding RMS roughness are 150 nm and 218 nm respectively. (with permission for reproduction from American Institute of Physics)

2.4 Front-side Laser micromachining of GaN/sapphire LED wafer

In our laser micromachining setup, using 349 nm wavelength, a front-side machining scheme is employed to avoid damage to the active-layer as well as to achieve higher precision with beam alignment. A tilted incision with 60° (θ=30°) tilting angle on the front side of the GaN/sapphire wafer is ablated after 5 successive scan cycles. The surface of the sidewall is exposed after laser micromachining for FE-SEM examination as shown in Figure 7. With the laser beam tightly focused, the kerf exposed at GaN layer shows a clear brim and the thickness of GaN estimated from the image is 4.5 μm. It is interesting to see a sharp interface between the sapphire substrate and the GaN layer and no heat affected zone (HAZ) is observed in the GaN layer after front-side machining. This finding may be attributed to the relatively low ablation threshold of GaN as it absorbs the 349 nm laser power. According to Figure 7, the sapphire substrate melts on the surface, while no melt is observed for the GaN layer. It is estimated that the surface temperature lies between the melting point of sapphire and GaN, which is in the range from 2040 °C to 2500 °C.

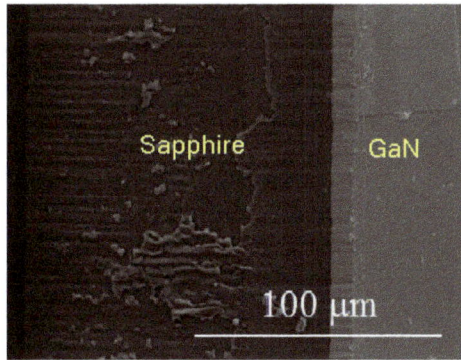

Fig. 7. FE-SEM image of a GaN/sapphire wafer after laser micromachining, the interface of GaN and sapphire and the brim of the 4.5 μm thick GaN layer is clear.

For comparison, surface morphology of backside micromachined LED wafer is illustrated in Figure 8 showing the feature of rugged sidewalls. The two images corresponds to a single scan of the laser beam at 50 μm/sec motion speed, with 30 μJ and 50 μJ pulse energy respectively, repeated at 1 kHz. This feature can be observed at varied pulse energies and scan cycles. With a high ablation threshold and optical transparency at the wavelength of 349 nm, sapphire is ablated with inferior surface quality. A large quantity of clusters is trapped within the groove which blocks light extraction from the sidewalls and also prevents heat dissipation via the sidewall surfaces. Improved quality of sapphire micromachining is possible by using a shorter wavelength or ultrashort pulse duration of the laser to suppress thermal effect during sapphire ablation. Laser ablation of the sapphire substrate with an absorptive wavelength to sapphire also avoids damaging on the epitaxial nitride layers.

Separation of some specially shaped LED such as a circular device after laser micromachining may be difficult if the wafer is cut insufficient in depth. The chips to be separated after machining are subject to uncontrollable fracture and crack whilst applying stress to the incision. In order to shape circular LEDs, the machining has to penetrate through the wafer to ensure separation in good shape. Although the penetration depth depends on laser power, it is important to adjust the focus position in the z direction in order to optimize the process condition.

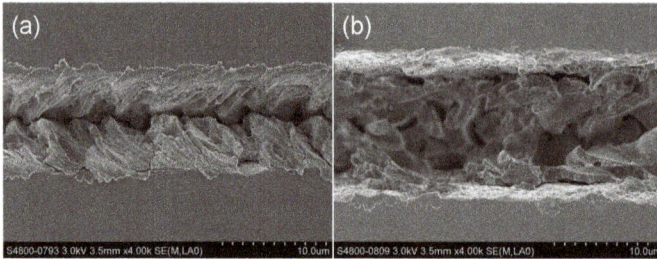

Fig. 8. (a) SEM image of laser scribed lanes on the back-side sapphire substrate with 30 μJ pulses; (b) with 50 μJ pulses

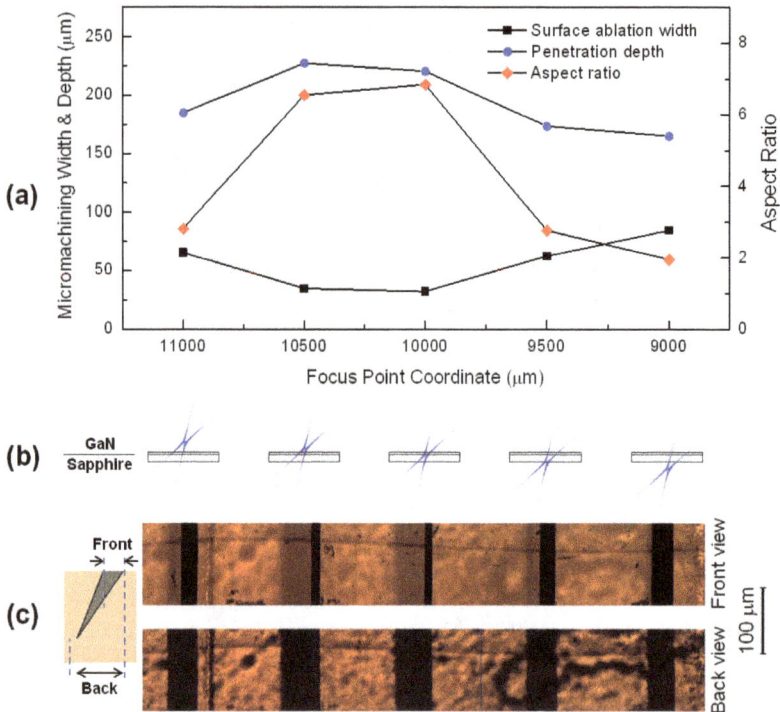

Fig. 9. (a) Trends of surface ablation width, penetration depth and the aspect ratio with changing focusing levels during micromachining; (b) Relative position of the focus point with reference to the wafer; (c) The width of surface damage determined from front view of wafer, and penetration depth estimated from the back view (mirrored) after laser micromachining at 40° tilting angle. (with permission for reproduction from John Wiley and Sons)

Figure 9 shows the surface ablation widths and penetration depths of the incisions machined at 40° tilting angle from the vertical. The pulse energy is about 90 µJ with a pulsed repetition rate of 1 kHz. The beam is scanned over a round trip cycle at a constant scan speed of 50 µm/sec. Figure 9 (a) plots the measured dimensions with relation to a sequence of focal positions. The relative positions of the focal spot with respect to the wafer are depicted in Figure 9 (b). The surface ablation widths and penetration depths are estimated from the front and the back view optical microscopy images as shown in Figure 9 (c). The position where laser focal point coincides with the surface of GaN is recorded at the z coordinate of 9900 µm. The optimized region for micromachining spans over the range of [10000, 10500] as the surface ablation width is at minimum while the penetration depth and aspect ratio are at maximum values. It is also found that when the wafer deviate from the focus position there is a chance of beam deformation and induce additional scribing run parallel to the desired groove. This is observed at the coordinate of 11000 as shown on the leftmost in Figure 9 (c).

2.5 Chip shaping of light-emitting diodes to improve light extraction

Light extraction from GaN-based light-emitting diodes is seriously suppressed by total internal reflections within the semiconductor layers. With a high refractive index around 2.4, light extraction from the top surface is limited within a 23° emission cone as depicted in Figure 10 (c). One effective method to enhance light extraction is employing tilting sidewalls such as those in a truncated pyramid (TP) LED, where the conventionally confined light rays are extracted from the top surface via sidewall reflections that redirect the light ray into the top surface emission cone. Accordingly, top surface emission of a laser fabricated TP LED shown in Figure 10 (b) is particularly stronger compared to the conventional rectangular chip in (a). The overall light extraction can be enhanced by 85%. The improvement is attributed to the additional indirect light extraction from top surface via sidewall reflections.

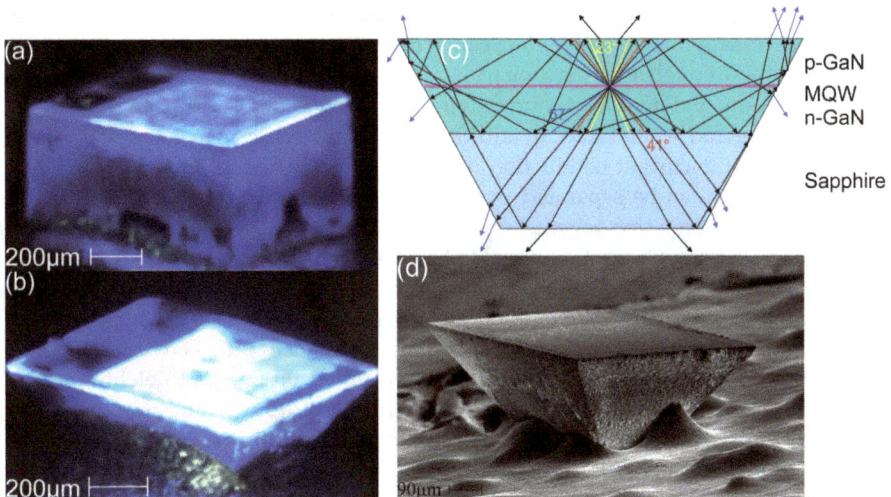

Fig. 10. Optical micrographs of (a) conventional cuboid LED and that of (b)truncated pyramid LED with tilting sidewall,(c)shematic diagram of ehanced top surface light extraction via sidewall reflections. (d) SEM image of the truncated pyramid LED chip shaped by laser micromachining. (adapted from (Fu, et al. 2009) with permission for reproduction from IEEE)

Additional indirect light extraction also exists in a triangular LED, making it unique among polygonal LEDs. However, the mechanism is slightly different with that of a TP LED. In a triangular LED chip, enhanced light extraction is due to indirect light extraction from the sidewall via reflections on neighbouring sidewalls, while in the case of a rectangular chip or other polygons, the indirect extraction is trivial. Actual chip geometry from triangle to heptagon are fabricated with the laser micromachining system and shown in Figure 11.

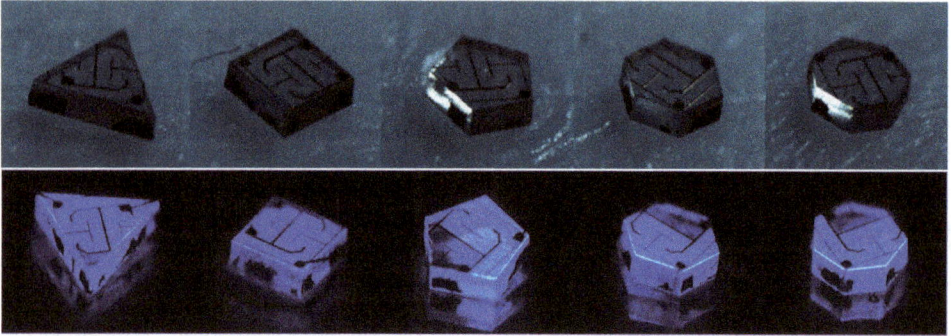

Fig. 11. Optical Micrograph Polygonal LEDs as fabricated by laser beam (upper row) and biased at 2.5 V (lower row). (with permission for reproduction from American Institute of Physics)

3. Device isolation on GaN-on-sapphire wafer via laser micro-patterning

GaN is the major material for the fabrication of state-of-the-art blue light-emitting diodes. It is conventionally grown on sapphire substrates by metalorganic chemical vapour deposition (MOCVD), since sapphire is stable and can withstand the high temperature during the growth process. Although there are many issues involved with sapphire, such as lattice mismatch with GaN and poor heat conductivity, sapphire is still prevalent in the fabrication of low-power blue and white LEDs. In addition, being an electrical insulator, sapphire does not interfere with the current conduction in GaN. By selectively removing certain area of GaN, the GaN layer can be separated into multiple electrically isolated small-area LEDs. These LEDs can be connected together by metal interconnects at a later stage, allowing a variety of integrated optoelectronic circuits to be developed.

As GaN is highly resistant to wet etch, dry etch is the conventional technique for the partial or complete removal of GaN. Reactive ion etching (RIE) (Lee, et al. 1995) using CHF_3/Ar and C_2ClF_5/Ar plasmas, for example, can achieve an etch rate between 60 and 470 angstrom/min (Liann-Be, et al. 2001). Inductively coupled plasma (ICP) etching using Cl_2 and Ar, on the other hand, offers an attractive etch rate of up to 1 μm/min (Smith, et al. 1997). However, dry etch techniques require masking material to cover the regions not to be removed. Typically, with photoresist as an etch mask, the photoresist layer has to be at least as thick as the GaN layer to be etched (Liann-Be, et al. 2001), which is about 3-4 μm. Spin-coating of photoresist layer of this thickness is often cumbersome (for example, edge bead effect may occur (Yang and Chang 2006)), coupled with the fact that thicker photoresists generally offer lower resolutions. Mask thickness can be reduced when hard masks such as SiO_2 are used, but additional lithography and dry etch steps are needed for patterning.

In this section, a maskless direct-write laser micromachining technique for device isolation on GaN-on-sapphire wafer is introduced. Unlike wafer dicing, where GaN and sapphire are to be ablated to complete separation, the laser ablation in our new technique automatically terminates at the GaN/sapphire interface. The principle lies on the large difference between: (1) the ablation thresholds (the minimum laser fluence to achieve ablation), and (2) the optical absorption coefficients at ultraviolet (UV) wavelength of GaN and sapphire, as shown in Table 1.

	GaN	Sapphire
Ablation threshold (J/cm²)	0.25 (Akane, et al. 1999a; Liu, et al. 2002)	4.5 (Li, et al. 2004)
Optical absorption coefficient (cm⁻¹)	100000 – 150000 (Muth, et al. 1997)	0.01 – 1 (Patel and Zaidi 1999)

Table 1. Parameters of GaN and sapphire that facilitate selective laser ablation.

When the laser fluence is controlled between the two ablation thresholds, GaN is ablated while sapphire is left undamaged. A simple way to achieve this is by offsetting the wafer from the best focus plane and adjusting the laser spot size. As shown in Figure 12 (a), the laser energy is concentrated to a small spot in the vicinity of the best focus plane. GaN layer (comprising p-type GaN, InGaN/GaN multi-quantum well (MQW) and n-type GaN) and sapphire layer are cut through, which is the mode for die separation. When the focus offset increases, the laser spot is enlarged and the laser fluence is reduced. At a certain range of focus offset, the laser fluence is just high enough to ablate GaN but not sapphire. By scanning the laser across the wafer, a trench terminating at the GaN/sapphire interface is resulted. This is the desired mode for device isolation (Figure 12 (b)). If the focus offset is increased further, the laser fluence will not be sufficient to ablate GaN completely. Device isolation cannot be achieved (Figure 12 (c)).

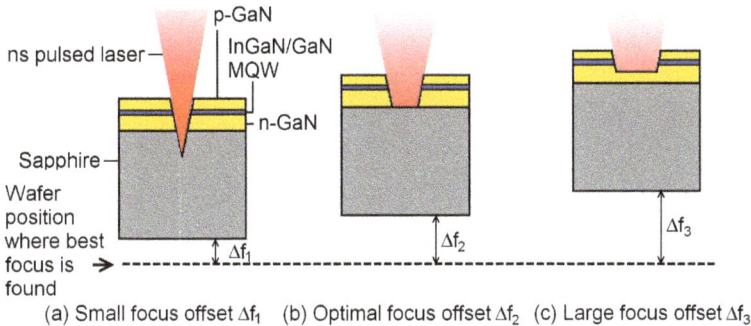

ns pulsed laser — p-GaN / InGaN/GaN MQW / n-GaN

Sapphire —

Wafer position where best focus is found →

(a) Small focus offset Δf₁   (b) Optimal focus offset Δf₂   (c) Large focus offset Δf₃

Fig. 12. Control of laser fluence by focus offset. (with permission for reproduction from American Institute of Physics)

A number of factors affect the quality of trenches. In our study, five laser parameters (focus offset, pulse energy, pulse repetition rate, scan speed and number of scan passes) and two ambient media (air and deionized water) were investigated. By the end of this section, two applications of this laser micromachining technique will also be discussed.

3.1 Trench micromachining in air

The laser micromachining experiment was first performed in ambient air at room temperature by using the setup shown in Figure 13 (schematic diagram shown in Figure 2). The laser source was a third-harmonic neodymium-doped yttrium lithium fluoride (Nd:YLF) diode-pumped solid-state (DPSS) laser, with center wavelength of 349 nm and pulse repetition rate of single pulse to 5 kHz. The full-width-at-half-maximum (FWHM) pulse width was 4 ns, while the pulse energy was varied by changing the diode pumping current. The expanded and collimated beam was guided by several laser mirrors and focused onto a piece of GaN-on-sapphire sample (emission wavelength = 470 nm, thickness of GaN = 3 µm and thickness of sapphire = 300 µm) on an XY motorized stage. The fused-silica focusing triplet lens allowed UV and visible light to pass through and had a focal length of 19 mm. As the stage translated while keeping the laser spot stationary, trenches were scribed onto the sample. The scan speed was controlled by software with a precision up to 25 µm/s. The sample could be shifted away from the focus by manually adjusting the stage height. The accuracy of height adjustment was ±5 µm. A charge-coupled device (CCD) camera was installed confocal to the optical path for real-time observation of the micromachining process. Owing to the high temperature during laser ablation, sedimentary

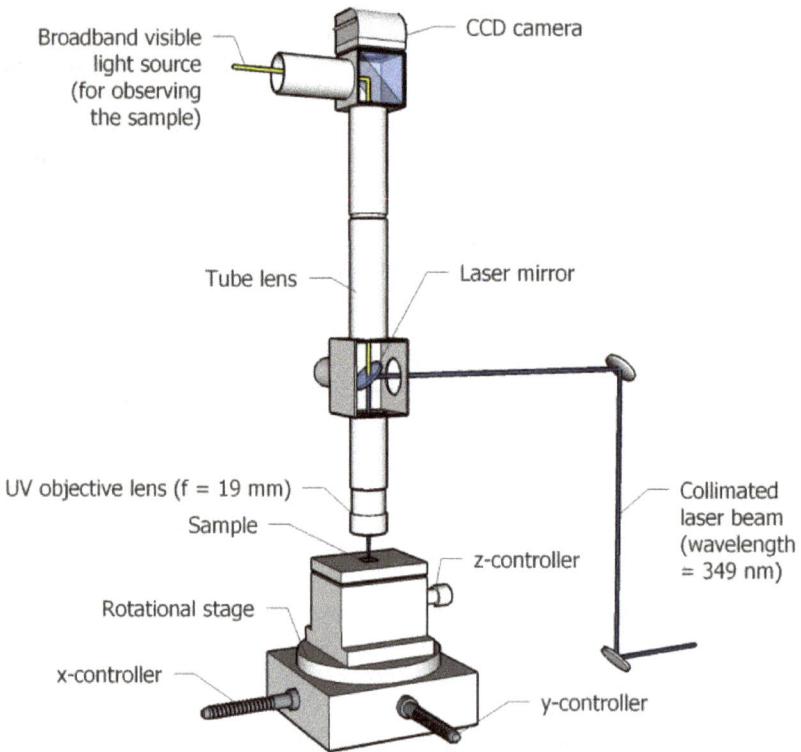

Fig. 13. Experimental setup for laser micromachining in air. (with permission for reproduction from American Institute of Physics)

by-products were formed on the surface of GaN, such as Ga metal (Akane, et al. 1999b; Kelly, et al. 1996) and gallium oxide (Gu, et al. 2006). These substances were effectively removed by sonification of the sample in dilute hydrochloric acid (HCl) (18% by mass) for 15 min. The sample was then rinsed in DI water to remove the remaining acid. The morphology of the resulting trenches were observed by field-emission scanning electron microscopy (FE-SEM), identifying the effect of each laser parameter towards the trench quality.

3.1.1 Focus offset

Figure 14 shows the micromachined trenches at three different focus offset levels while keeping the pulse energy, repetition rate, and scan speed constant. Upward focus offset is taken as positive. The results follow the principle introduced at the beginning of this section. In Figure 14 (a) where the sample is positioned near the best focal plane (300 μm above), the laser beam ablates both the GaN (lighter colour) and sapphire (darker colour). A V-shaped valley is formed in the sapphire layer due to the Gaussian beam shape. Although trenches like these serve the purpose of electrical isolation between adjacent devices, the deep V-shaped valley is not suitable for the conformal deposition of metal interconnect, since the interconnection will become discontinuous at the sharp corners of the valley. At the optimal focal offset plane (450 μm above), as shown in Figure 14 (b), the ablation terminates automatically at the GaN/sapphire interface, exposing a flat and smooth sapphire bottom surface. At a larger focus offset plane of 600 μm, the GaN layer is not completely removed, leaving a shallow and rugged trench on the surface (Figure 14 (c)).

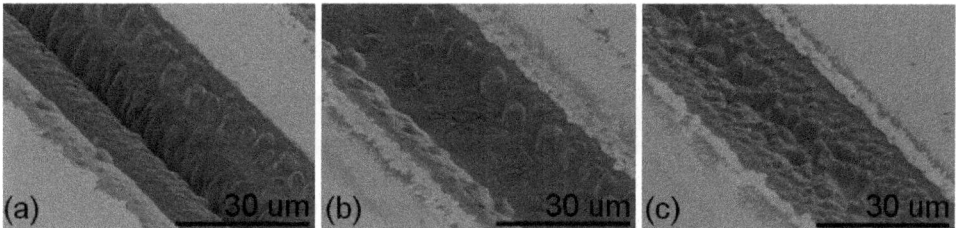

Fig. 14. SEM images of trenches laser micromachined at different focus offset planes: (a) small offset of 300 μm; (b) optimal offset of 450 μm; (c) large offset of 600 μm. The pulse energy, pulse repetition rate, and scan speed were fixed at 23 μJ, 1 kHz, and 25 μm/s, respectively. (with permission for reproduction from American Institute of Physics)

3.1.2 Pulse energy

Pulse energy is another determining factor of trench quality. Figure 15 illustrates micromachined trenches processed at three different pulse energies between 7 and 45 μJ, while keeping all other parameters constant. The focus offset is kept at the optimal value of 450 μm, as determined from the previous set of experiment. When the pulse energy is set too high, the effect is similar to that of having a smaller focus offset, whereby the GaN as well as sapphire are ablated to form a V-shaped trench (Figure 15 (a)). Similar correspondence between low pulse energy and large focus offset can be observed in Figure 15 (c). Notice that the trench width also increases for higher pulse energy. This property will be further explored in laser micromachining in DI water.

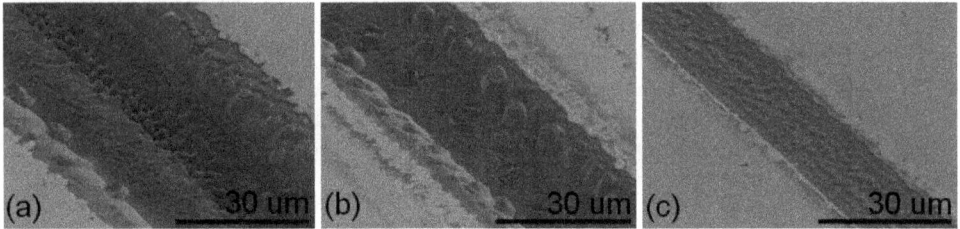

Fig. 15. SEM images of trenches under different pulse energy: (a) higher pulse energy of 45 µJ; (b) optimal pulse energy of 23 µJ; (c) lower pulse energy of 7 µJ. The focus offset level, pulse repetition rate, and scan speed are fixed at 450 µm, 1 kHz, and 25 µm/s, respectively. (with permission for reproduction from American Institute of Physics)

3.1.3 Pulse repetition rate

Trenches that are laser-micromachined under an increasing pulse repetition rate are shown in Figure 16 (a)-(c); all other parameters are kept constant. When the pulse repetition rate increases from 1 to 5 kHz, the trench width remains more or less unchanged, but the sidewall and bottom surfaces become increasingly smooth. This observation can be understood in terms of heat accumulation effects and its consequence to the etch efficiency. As the repetition rate increases, cumulative heating by earlier pulses causes localized melting of the material (Schaffer, et al. 2003). This results in an increase in the average surface temperature and thus the removal rate of the ablated materials, minimizing redeposition of debris over the trench.

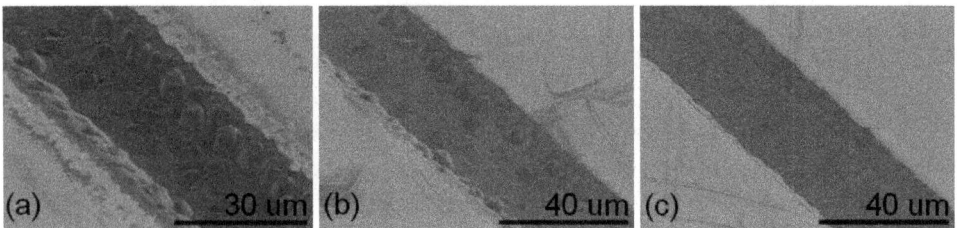

Fig. 16. SEM images of trenches under different pulse repetition rate: (a) 1 kHz; (b) 3 kHz; (c) 5 kHz. The focus offset, pulse energy, and scan speed were fixed at 450 µm, 23 µJ, and 25 µm/s, respectively. (with permission for reproduction from American Institute of Physics)

3.1.4 Scan speed

The rate at which the laser beam scans across the material is also investigated. From Figure 17, a faster translation rate does not result in a change in the trench width. However, it leads to degradation in the trench quality. At a faster translation speed, the exposure time to the laser light at each position becomes shorter. There is no enough time for temperature rise and/or photon-matter interaction. Stalagmite-like structures begin to appear around the sidewalls.

Fig. 17. SEM images of trenches under different scan speeds: (a) 25 µm/s; (b) 75 µm/s; (c) 125 µm/s. The focus offset, pulse energy, and repetition rate were kept at 450 µm, 23 µJ, and 5 kHz, respectively. (with permission for reproduction from American Institute of Physics)

3.1.5 Number of scan passes

The remaining factor to consider is the number of scans. Similar to the effect of increasing scan speed, an increase in the number of scans does not alter the trench width. However, a narrow groove is formed at the center of the trenches for three and five passes, as illustrated in Figure 18 (b) and (c), respectively. There is also no remarkable improvement in the sidewall and bottom surface quality.

Fig. 18. SEM images of trenches under different number of scan passes: (a) single pass; (b) three passes; (c) five passes. The focus offset, pulse energy, repetition rate and scan speed were kept at 450 µm, 23 µJ, 5 kHz, and 25 µm/s, respectively. (with permission for reproduction from American Institute of Physics)

Through the above experiments, we can conclude that an optimal combination of focus offset and pulse energy, higher pulse repetition rate, slower scan speed and single pass of scan are essential for good trench quality. Nevertheless, substantial amount of redeposition and resolidification of ablated material still exists on the trench bottom surface and sidewall when the process is performed in air. This is the result of thermal ablation and photochemical ablation mechanisms of nanosecond lasers. Although the redeposition can be effectively removed by strong acids, this is not feasible when the underlying material also reacts with the acids. In order to reduce the heat load during ablation, laser micromachining with the sample immersed in a liquid is proposed. Criteria for the liquid include good thermal conductivity and high specific heat capacity. In addition, the attenuation of UV and visible light in that liquid should be low, so that laser energy can be transferred efficiently to the substrate and the micromachining process can be monitored concurrently. DI water would be a good choice to match these criteria. The mechanisms involved in the liquid-immersion laser micromachining of GaN will be investigated in the following subsection.

3.2 Trench micromachining in DI water

The characteristics of liquid-immersion laser micromachining for GaN were investigated experimentally using the optical setup shown in Figure 19. The setup was similar to that used for ambient air (Figure 13), except that the GaN-on-sapphire sample was immersed horizontally in a DI water bath with meniscus about 1 mm above the sample surface. Although thicker water layer can improve heat dissipation, attenuation of the laser beam will become more severe. On the other hand, the water layer cannot be too thin since the entire sample will not be wetted under the strong surface tension of water. The water bath was placed on the manual Z translation stage, mounted on the motorized XY translation stage to enable laser scanning. The laser fluence was again adjusted by offsetting the sample surface from the best focal plane. Besides observing the surface morphology by FE-SEM, the trench surface roughness was measured by atomic force microscopy (AFM). Elemental analysis of the trench surface was performed by the energy-dispersive X-ray spectroscopy (EDX) function offered by the FE-SEM.

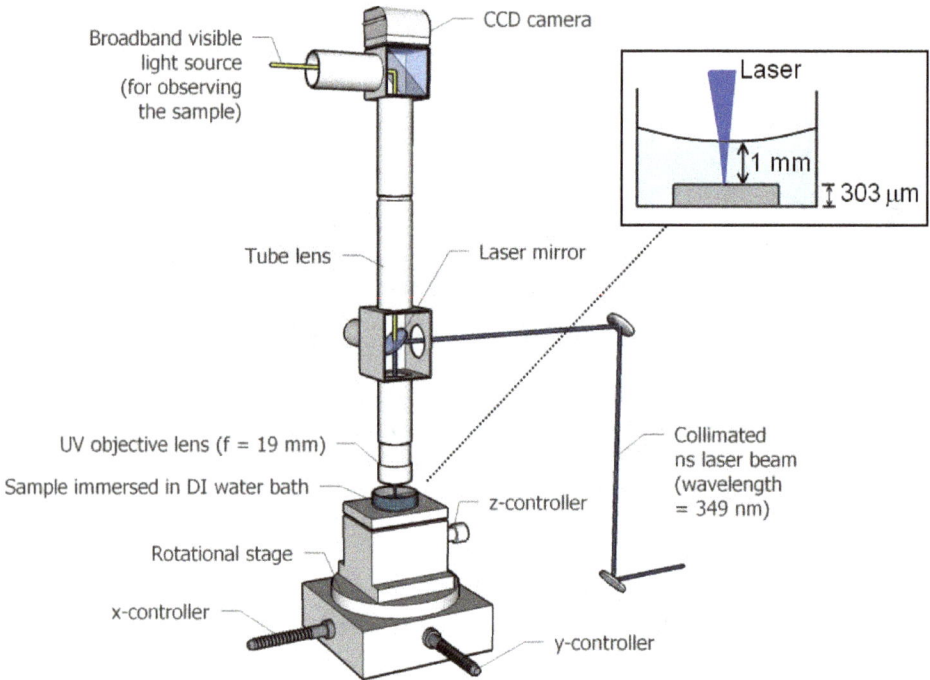

Fig. 19. Experimental setup for laser micromachining in DI water. (with permission for reproduction from Springer)

3.2.1 Trench quality as compared with ambient air

Compared with laser micromachining in ambient air, DI water is capable of producing trenches with substantially smoother sidewalls and bottom surfaces. Figure 20 shows AFM scans of the trenches generated in both ambient media. By measuring the height values y_i

along the trench edge, the sidewall roughness R_a is determined by taking the arithmetic average of the absolute height deviation from the mean height. The trench micromachined in air has an R_a of 312 nm, contrasting sharply with 27.65 nm for the trench produced in DI water. The rms roughness of the bottom surfaces also reveals the superiority of liquid immersion (87.49 nm for air vs. 13.42 nm for DI water). As a comparison with inorganic material whose laser-matter interaction should be similar to that of GaN, we quoted that the R_a value of microchannels micromachined by femtosecond laser in aluminosilicate glass sheet in ambient air is approximately 40 nm (Zheng, et al. 2-006). This indicates that nanosecond laser micromachining in DI water can have comparable performance with femtosecond laser micromachining in air.

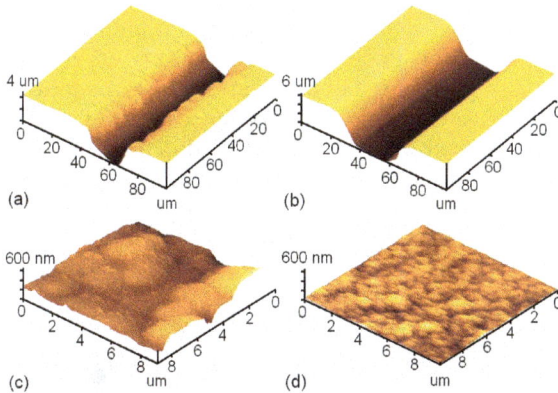

Fig. 20. AFM scans of trenches generated in: (a) and (c) air; (b) and (d) DI water. (c) and (d) are the zoomed-in images of the bottom surfaces. The focus offset, pulse energy, pulse repetition rate, scan speed were fixed at 400 μm, 25 μJ, 1 kHz and 25 μm/s. Only single pass of scan was performed. (with permission for reproduction from Springer)

Liquid-immersion laser micromachining also results in trenches of quality comparable to conventional lithographic and ICP deep etch processes. From the SEM images of sidewalls around ICP-etched regions in the literature (Ladroue, et al. 2010; Qiu, et al. 2011), we see that the smoothness is comparable to that of the laser-micromachined trench sidewalls. Besides, striations are observed over the sidewalls of ICP-etched regions when Ni hard mask is used. This is not observed in liquid-immersion laser micromachining, as shown in Figure 20 (b). Though SiO_2 hard mask can be used to eliminate the striations in ICP, it comes with the price of sidewall steepness reduction.

As a demonstration of reduced redeposition, the EDX results of the trench bottom surface are shown in Table 2. Three elements were found: O, Al and Ga. Al and O are the constituent elements of sapphire (Al_2O_3), whereas Ga is a product from the thermal decomposition of GaN during laser ablation (Ambacher, et al. 1996; Choi, et al. 2002). It is found that there is a lower percentage of Ga over those trenches produced in DI water. It should be noted that while the sample for air had been sonicated in dilute HCl before EDX examination, that for DI water was not subjected the same treatment. The results indicate that liquid immersion is effective in ejecting the molten Ga (m.p. = 29.76°C, b.p. = 2204°C) from the irradiated site and preventing its resolidification around the trench.

	Air	DI water
O	64.67%	65.19%
Al	34.81%	34.65%
Ga	0.52%	0.16%

Table 2. Atomic composition of trench bottom surface.

Besides improved surface quality, liquid-immersion laser micromachining also offers two advantages with respect to process control. The first is increased focus offset tolerance. Figure 21 (a)-(c) show three trenches micromachined in ambient air. When the sample position deviates from the optimum plane (Figure 21 (b)) by 150 μm, either damage to the sapphire layer (Figure 21 (a)) or incomplete trench (Figure 21 (c)) occurs. For micromachining in DI water, the trench quality is not compromised even with a deviation as large as 200 μm, albeit a slight decrease in the trench width (Figure 21 (d)-(f)). The second advantage of liquid immersion is the control over trench width by varying pulse energy. Owing to the focus offset tolerance, the variation of pulse energy does not significantly compromise the trench quality. Figure 22 shows an approximately linear relationship between pulse energy and trench width when performing micromachining in DI water at a fixed focus offset. For micromachining in ambient air, it is difficult to control the trench width just by altering the pulse energy. This is because the optimum focus offset depends strongly on the pulse energy. A new optimum focus offset needs to be found when the pulse energy is altered.

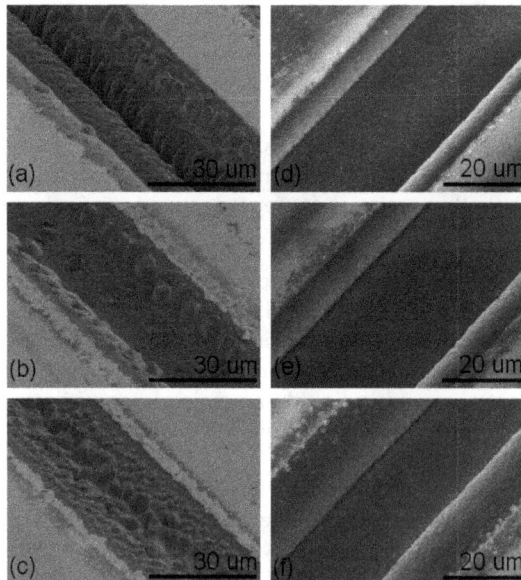

Fig. 21. SEM images of trenches laser-micromachined at different focus offset planes: (a)-(c) in air: (a) 300 μm, (b) 450 μm, (c) 600 μm, where 450 μm is the optimum; (d)-(f): in DI water: (d) 300 μm; (e) 500 μm; (f) 700 μm, where 500 μm is the optimum. The pulse energy, pulse repetition rate and scan speed were fixed at 25 μJ, 1 kHz and 25 μm/s. (with permission for reproduction from Springer)

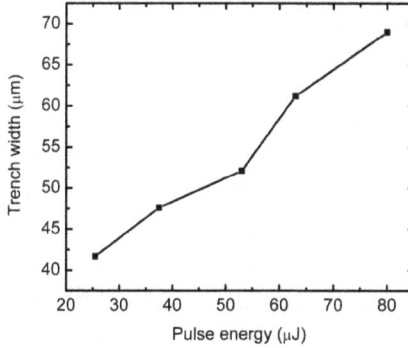

Fig. 22. Trench width control by varying pulse energy. The focus offset, pulse repetition rate and scan speed were fixed at 550 µm, 1 kHz and 25 µm/s. (with permission for reproduction from Springer)

3.3 Theoretical discussions
The improved trench quality can be explained in terms of the following processes during the nanosecond-pulsed laser ablation of GaN: Heat transfer, plasma-induced recoil pressure, plasma shielding effect in water and collapse of cavitation bubbles.

3.3.1 Heat transfer
Referring to Table 3, DI water has a much higher specific heat capacity and thermal conductivity than air. Therefore DI water is expected to carry the excess heat away from the irradiated region faster than air. To verify this, a simulation on the heat conduction process for single-pulse irradiation was performed. The heat equation (in cylindrical coordinates) was solved by finite element method (FEM):

$$\rho c \frac{\partial T}{\partial t} - \nabla \cdot (k \nabla T) = Q \tag{2}$$

where ρ, c and k are as defined in Table 3, T is the temperature distribution (K) and Q represents the heat source (W mm⁻³). Q originates from the Gaussian laser beam, thus both Q and the resulting T are assumed to be radially symmetric. The domain and boundary conditions are defined in Figure 23. Without going into the details, the simulation results are presented as follows. Figure 24 shows the variation of GaN surface temperature during the initial 100 ns at the center of the laser spot $(r = 0, z = 0)$ (solid curves) and near the trench edge $(r = 15 \text{ µm}, z = 0)$ (dotted curves). The GaN surface temperature in air is found to be higher than that in DI water at both positions. In addition, sharper temperature peaks are found when the micromachining is performed in air. The maximum temperature is as high as $1000°C$ even near the trench edge. It is known that GaN begins to decompose into liquid Ga and N_2 gas at a temperature of $900°C$ (Choi, et al. 2002). The rapid heating and cooling cycles in air can result in the increased generation and resolidification of molten Ga within each pulse period. The resolidified Ga droplets deposit around the sidewall, degrading the surface quality. Another consequence of rapid heating and cooling is the increased thermal stress incurred in the crystal structure of GaN. Cracks may be resulted.

	ρ (g cm^{-3})	c (J g^{-1} $^{\circ}$C^{-1})	k (W cm^{-1} $^{\circ}$C^{-1})
GaN	6.15	0.49	1.3
Al$_2$O$_3$	4.025	0.75	0.35
H$_2$O	1.0	4.18 (at 25°C)	0.006
Air	1.184 × 10^{-3}	1.012 (at 23°C)	2.5 × 10^{-4}

Table 3. Density ρ, specific heat capacity c and thermal conductivity k of the substances involved in the laser micromachining process.

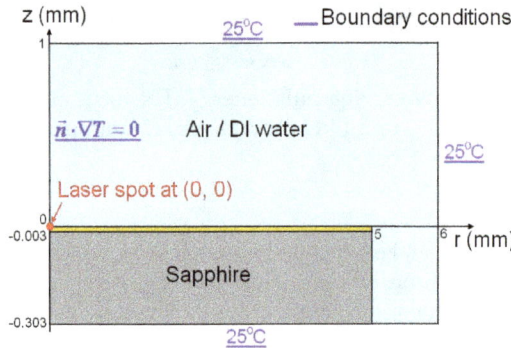

Fig. 23. Simulation domain. (with permission for reproduction from Springer)

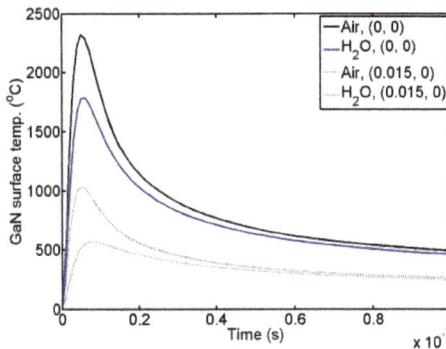

Fig. 24. Temporal variation of GaN surface temperature at two different positions: at the center of laser spot (solid); near the edge of the trench (dotted). (with permission for reproduction from Springer)

The two-dimensional temperature distribution also demonstrates the strong cooling effect of water. As shown in Figure 25, at $t = 1$ µs, the temperature of water immediately above the sample surface is lower than that of air. The slight decrease of water temperature below the room temperature can be attributed to the adiabatic expansion of water vapour as well as the vapour plume of ablated material (Gusarov, et al. 2000). At this instance, the GaN surface temperature in water is in general lower than that in air, indicating efficient heat extraction by water. The heat-affected zone (HAZ) is also seen to be smaller when the sample is immersed in DI water.

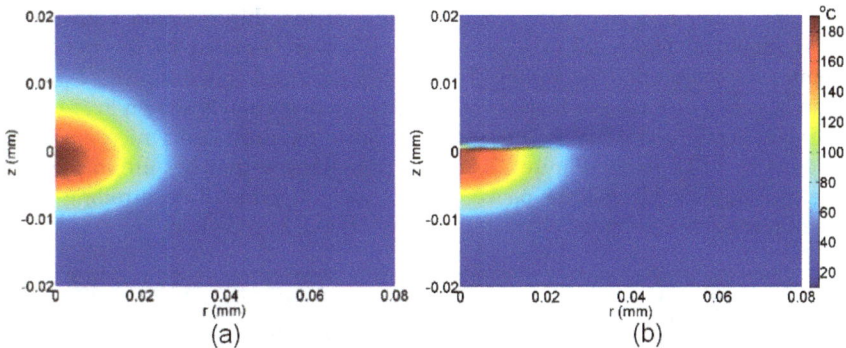

Fig. 25. Temperature distribution at t = 1 μs when the ambient medium is: (a) air; (b) DI water. (with permission for reproduction from Springer)

3.3.2 Plasma-induced recoil pressure
High-energy laser pulses are able to melt, vaporize and ionize the material being irradiated, resulting in the formation of plasma. When the plasma expands, a recoil pressure is exerted on the sample surface, causing material removal. If the ambient medium is water instead of air, the expansion of plasma will be confined by the underwater pressure. This leads to an increase in the plasma-induced recoil pressure and higher removal rate of material.

3.3.3 Plasma shielding effect in water
The plasma generated from laser ablation can absorb part of the incident laser energy and reduce the energy delivered to the sample. This is known as plasma shielding effect. The effect becomes stronger when the plasma has a larger size, longer duration and a starting time that overlaps with the laser pulse. Because of water confinement effects, the plasma size and duration are much reduced in water. The onset of plasma formation is also delayed by 5 ns, reducing the overlap time between the laser pulse and the plasma (Hong, et al. 2002). These factors weaken the plasma shielding effect, causing a stronger coupling of the laser beam with the material. Material removal is thus more efficient in water.

3.3.4 Collapse of cavitation bubbles
Formation of cavitation bubbles is a process that happens solely in liquids. In the experiment, the bubbles may originate from the dissolved gases in DI water, or the N_2 gas from the laser ablation of GaN. When a bubble collapses near the sample surface, a high-speed liquid jet directed towards the surface will be generated. If there is no water between the bubble and sample surface, the liquid jet will produce a strong impulse towards the sample. This impulse can be 5.2-12.4 times that of the laser ablation impact in air (Lu, et al. 2004), which further helps the liquid-phase expulsion of Ga.

3.4 Applications
To close this section, two applications of liquid-immersion laser micromachining are introduced: alternating-current LED (ac-LED) and 5 × 7 dot-matrix microdisplay. Ac-LED is an LED chip that is directly powered by an ac voltage. The basic idea is to make use of the diode property of LEDs to construct an on-chip bridge rectifier. The rectified voltage is then

used to power other on-chip LEDs. The plan view of the chip overlaid with the circuit diagram is shown in Figure 27 (a). The on-chip LEDs are isolated by the trenches from liquid-immersion laser micromachining. This chip is designed to work on 12 V_{rms}, but the design can be easily extended to higher voltages. Its full operation is shown in Figure 26 (b). The other application, dot-matrix microdisplay, also requires laser micromachining to separate rows of LED pixels (Figure 27 (a) and (b)). Narrow trenches are formed by low pulse energy, allowing pixels to be packed more closely together. By controlling the on-off sequence of pixels, alphanumeric characters can be formed, as shown in Figure 27 (c).

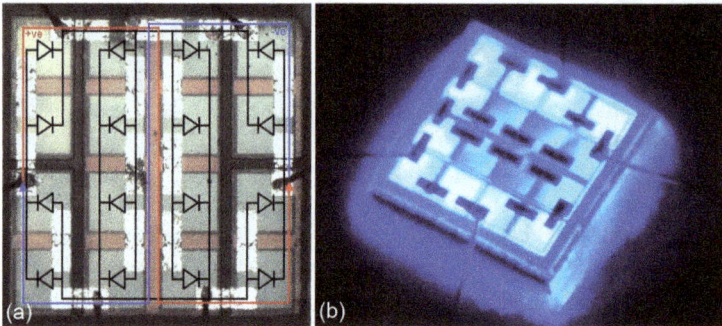

Fig. 26. Microphotograph of ac-LED, with circuit diagram superimposed. The directions of current flow during the positive and negative cycles of ac voltage are shown with red and blue arrows respectively; (b) Microphotograph of ac-LED operating under an ac voltage of 12 V_{rms}. (with permission for reproduction from American Institute of Physics)

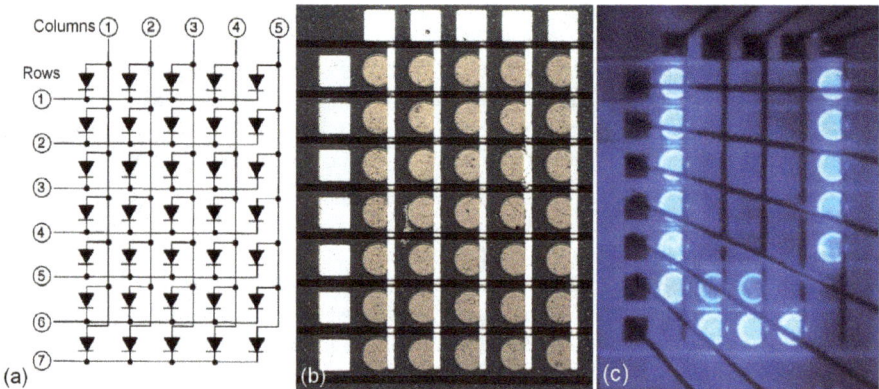

Fig. 27. Circuit diagram of 5 × 7 dot-matrix microdisplay; (b) Microphotograph of the microdisplay chip; (c) Microdisplay in operation, showing the character "U".

4. Conclusion

Laser micromachining involving a nanosecond diode-pumped solid-state laser at 349 nm wavelength is developed for device fabrication of light-emitting diodes on sapphire wafer. The material ablation process can be readily controlled and optimized for varied micromachining purposes such as wafer cutting featuring high aspect ratios and precision etching for device isolation. Highest aspect ratio of the cleave on GaN/sapphire wafer is achieved by front side laser micromachining method with the beam focused at the GaN layer. With laser micromachining process optimized for chip shaping, novel chip geometry such as truncated pyramid and triangular LED is fabricated and the novel LED geometries effectively improve light extraction efficiency. Influence of ambient medium as well as effects of various process parameters are taken into account to develop a reliable laser micro-patterning process. The laser beam, focused with a shorter focal length focusing lens, is tuned to selectively etch conductive GaN layers on the insulating sapphire substrate generating isolation grooves with desired profile and pattern that facilitates further processing.

5. References

Akane, T., K. Sugioka, and K. Midorikawa 1999a High-speed etching of hexagonal GaN by laser ablation and successive chemical treatment. Applied Physics A (Materials Science Processing) A69.

Akane, T., et al. 1999b KrF excimer laser induced ablation-planarization of GaN surface. Applied Surface Science 148(1-2).

Ambacher, O., et al. 1996 Thermal stability and desorption of group III nitrides prepared by metal organic chemical vapor deposition. Journal of Vacuum Science & Technology B (Microelectronics and Nanometer Structures) 14(6).

Choi, H. W., et al. 2002 Surface analysis of GaN decomposition. Semiconductor Science and Technology 17(12).

Fu, W. Y., et al. 2009 Geometrical Shaping of InGaN Light-Emitting Diodes by Laser Micromachining. Ieee Photonics Technology Letters 21(15):1078-1080.

Gattass, Rafael R., and Eric Mazur 2008 Femtosecond laser micromachining in transparent materials. Nature Photonics 2(4):219-225.

Gu, E., et al. 2006 Microfabrication in free-standing gallium nitride using UV laser micromachining. Applied Surface Science 252(13):4897-4901.

Gusarov, A. V., A. G. Gnedovets, and I. Smurov 2000 Two-dimensional gas-dynamic model of laser ablation in an ambient gas. Applied Surface Science 154-155.

Hong, M. H., et al. 2002 Steam-assisted laser ablation and its signal diagnostics. Applied Surface Science 197-198.

Kelly, M. K., et al. 1996 Optical patterning of GaN films. Applied Physics Letters 69(12).

Ladroue, J., et al. 2010 Deep GaN etching by inductively coupled plasma and induced surface defects. Journal of Vacuum Science & Technology A: Vacuum, Surfaces, and Films 28:1226.

Lee, H., D.B. Oberman, and J.S. Harris Jr 1995 Reactive ion etching of GaN using CHF [sub 3]/Ar and C [sub 2] ClF [sub 5]/Ar plasmas. Appl. Phys. Lett 67:1754.

Li, X. X., et al. 2004 Ablation induced by femtosecond laser in sapphire. Applied Surface Science 225(1-4):339-346.

Liann-Be, Chang, Liu Su-Sir, and Jeng Ming-Jer 2001 Etching selectivity and surface profile of GaN in the Ni, SiO 2 and photoresist masks using an inductively coupled plasma. Japanese Journal of Applied Physics, Part 1 (Regular Papers, Short Notes & Review Papers) 40(3A).

Liu, Wei-Min, et al. 2002 Ablation of GaN using a femtosecond laser. Chinese Physics Letters 19(11).

Liu, X., D. Du, and G. Mourou 1997 Laser ablation and micromachining with ultrashort laser pulses. Ieee Journal of Quantum Electronics 33(10):1706-1716.

Lu, J., et al. 2004 Mechanisms of laser drilling of metal plates underwater. Journal of Applied Physics 95(8):3890-3894.

Molian, P., B. Pecholt, and S. Gupta 2009   Picosecond   pulsed   laser   ablation   and micromachining of 4H-SiC wafers. Applied Surface Science 255(8):4515-4520.

Muth, JF, et al. 1997 Absorption coefficient, energy gap, exciton binding energy, and recombination lifetime of GaN obtained from transmission measurements. Applied Physics Letters 71:2572.

Patel, BS, and ZH Zaidi 1999 The suitability of sapphire for laser windows. Measurement Science and Technology 10:146.

Qiu, Rongfu, et al. 2011 Optimization of inductively coupled plasma deep etching of GaN and etching damage analysis. Applied Surface Science 257(7).

Schaffer, C. B., et al. 2001 Micromachining bulk glass by use of femtosecond laser pulses with nanojoule energy. Optics Letters 26(2):93-5.

Schaffer, C. B., J. F. Garcia, and E. Mazur 2003 Bulk heating of transparent materials using a high-repetition-rate femtosecond laser. Applied Physics a-Materials Science & Processing 76(3):351-354.

Smith, SA, et al. 1997 High rate and selective etching of GaN, AlGaN, and AlN using an inductively coupled plasma. Applied Physics Letters 71(25):3631-3633.

Stuart, B. C., et al. 1996 Nanosecond-to-femtosecond laser-induced breakdown in dielectrics. Physical Review B 53(4):1749-1761.

Varel, H., et al. 1997 Micromachining of quartz with ultrashort laser pulses. Applied Physics A (Materials Science Processing) 65(4-5).

Yang, Y.K., and T.C. Chang 2006 Experimental analysis and optimization of a photo resist coating process for photolithography in wafer fabrication. Microelectronics Journal 37(8):746-751.

Zheng, H. Y., et al. 2006 Ultrashort pulse laser micromachined microchannels and their application in an optical switch. International Journal of Advanced Manufacturing Technology 27(9-10):925-929.

Zweig, A. D. 1991 A thermomechanical model for laser ablation. Journal of Applied Physics 70(3):1684-1691.

7

Mechanical Micromachining by Drilling, Milling and Slotting

T. Gietzelt and L. Eichhorn
Karlsruhe Institute of Technology, Campus Nord,
Institute for Micro Process Engineering, Karlsruhe,
Germany

1. Introduction

Micromachining is not only a simple miniaturization of processes using macroscopic tools. As a matter of fact, a lot of specific concerns have to be met for successful fabrication of microstructures. This chapter will be focussed on micromachining using geometrically determined cutting edges, namely on techniques like drilling, milling and slotting. These methods are very flexible. Compared to EDM, ECM or lithographic processes like LIGA, they can be applied to a wide range of materials, like polymers, metals and alloys as well as to some kinds of ceramics, possess a high material removal rate and allow a great degree of freedom concerning design. There are nearly no geometrical limitations and also 3D-structures can be manufactured easily.

2. Micromachining by geometrically determined cutting edges

2.1 Differences as compared to geometrically undefined cutting edges

Micromachining techniques can be divided into two main categories: Processes working with undefined cutting edges e.g. grinding, honing, lapping, and processes are using defined cutting edges like drilling, milling and slotting.

Especially grinding works at high cutting rates. Most of the cutting energy is transferred into heat and absorbed by the work piece [Kön99, Fri08]. The properties of the work piece can be altered or decreased by surface cracks and internal stress due to external forces as well as by microstructural changes due to excessive heat.

Especially for micro grinding using small-diameter tools, extremely high numbers of revolutions are required to achieve a reasonable circumferential speed of up to more than 100 m/s. Compared to processes using defined cutting edges, the energy need is high and the material removal rate is comparably low. Nevertheless, especially for very hard materials like most ceramics where defined cutting edges do not work, grinding is a capable technique. However, since diamonds are used to machine the very hard ceramic materials, machining expenses can be a major cost factor for ceramic parts [War00].

When machining using geometrically determined cutting edges, the cutting energy is mostly used to overcome the cohesion forces of the machined material. The material removal rate is higher than for grinding and most of the heat is transferred to and removed with the chips. A good approximation for the removal of heat is, that 75% are transferred to the chips formed, 18% migrate to the tool and 7% to the work piece [Kön90]. Hence, the work piece and its microstructure are not as affected as in the case of grinding.

When rotating micro-sized tools, attention has to be paid in general to the response on external loads by deformation. The load case and the reaction of the tool are very important as regards the machining result. Especially in the case of rotating tools possessing two chip flutes, the cross section is reduced. Load cases can be distinguished for different machining processes:

In case of micro drilling, only a torsional moment acts on the tool. Depending on the length, bending and buckling may be an issue.

Slotting is an appropriate way if the desired trench width is smaller than that of commercially available end mills or for large aspect ratios where the stability of end mills can be problematic. An additional advantage of slotting is that the tool may not be axially symmetric. Hence, a better stability is accomplished and only bending acts on the tool. Tools with optimized shapes and angles can be made using precision grinding machines e. g. Ewag WS 11 [Ewa_Ws] with worn hard metal end mills made of ultra-fine grain carbides. A disadvantage is the slow feed rate and, hence, a smaller material removal rate than in the case of micro milling.

In the case of micro milling both torsion and bending act on the tool. Predominantly, micro end mills are made of hard metal possessing two chip flutes. However, also tools made of monocrystalline diamonds with only one cutting edge are used.

Hence, dynamic fatigue due to cyclic bending or vibrations and irregular load may be a serious problem especially for two flute micro end mills. Characteristics like appropriate hard metal substrate, manufacturing process affecting roughness and cracks in the surface, coating technology and adapted tool shape will be discussed in Chapter 3 for micro end mills.

2.2 Geometrical limits of tools

Micro drills originate from conductor board manufacturing for contacting through multiple layers. Although the prepregs used consist of a cured resin and very abrasive glass fibres, uncoated hard metal drills are used with good success. Uncoated micro drills are available down to diameters of 20 µm [Ato_Ad, Ham_38]. Fig. 1 shows a 30 µm micro drill bit.

Fig. 1. 30 µm micro drill bit with detail of the cutting edge. Grooves from grinding with jagged edge due to the composite nature of hard metal can be seen.

About five years ago, coating of micro end mills started only above 0.3 mm in diameter due to excessive rounding of the cutting edges by the coating layer. Up to this time, the gain of improved wear resistance due to the coating was less favourable than the increase of the cutting force due to the rounding of the cutting edge. Through improved coating process control, allowing thinner and more uniform layers, the relation was reversed. Today, coated micro mills down to 30 μm in diameter and with aspect ratios of 1.5 are commercially available [Hte_Em], (Fig. 2).

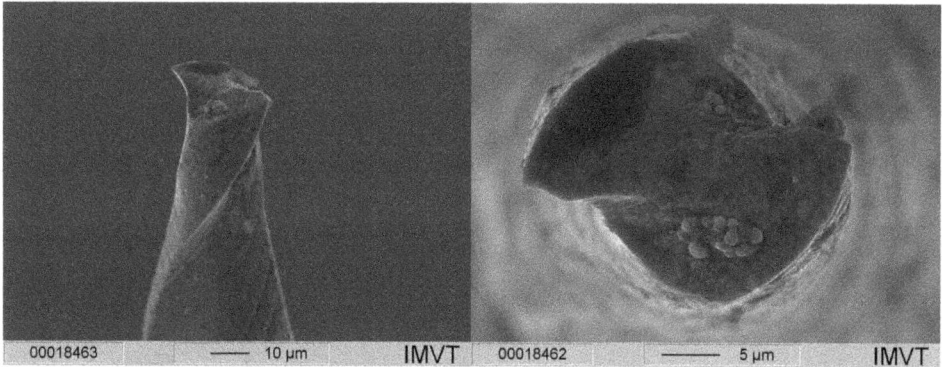

Fig. 2. Left: Coated 30 μm end mill made by Hitachi. Right: Top view.

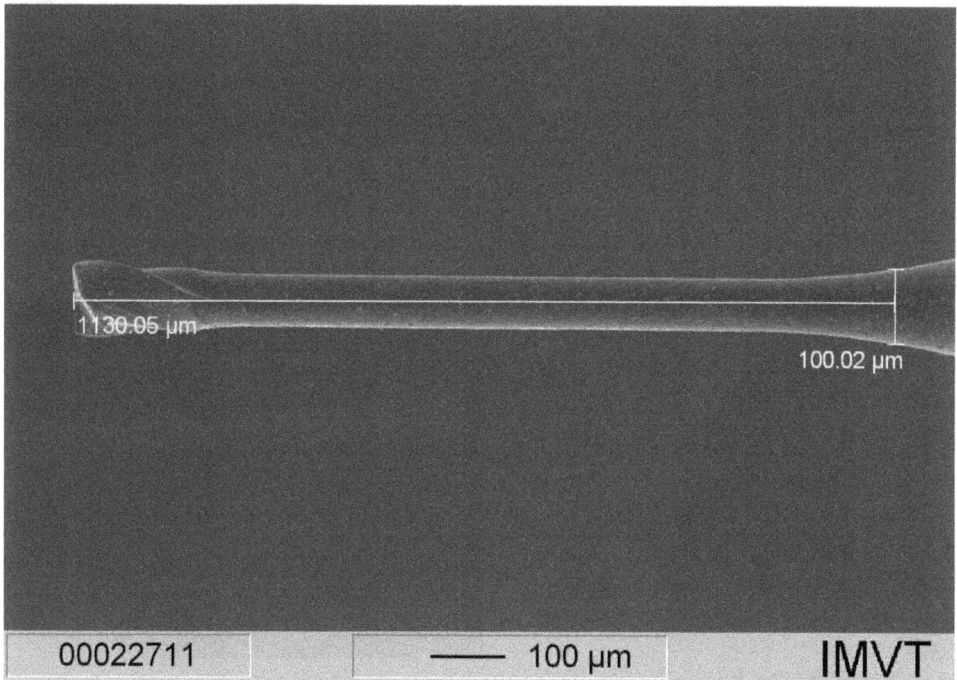

Fig. 3. Micro end mill 100 μm in diameter and 1 mm in length (AR=10).

Starting at a tool diameter of 100 µm, aspect ratios of up to ten are available now, as displayed in Fig. 3 [Nst-Em, Hte_Ep2].

Further miniaturization of micro end mills made of hard metal seems to be useless regarding process yield and tool life. Furthermore, an isotropic mechanical behaviour cannot be achieved since hard metal is a composite material consisting of a hard material and a binder phase with very different mechanical properties.

For manufacturing and stability reasons, micro end mills made of monocrystalline diamond are no less than 50 µm in diameter (Fig. 4). Suppliers are mentioned in [Nst_Di, Med, Möß, Con]. Diamonds are used for very hard and non-iron materials. In contact with iron, the carbon of the diamond would easily diffuse and destroy the tool. An exception occurs in the case of low cutting speeds e. g. during slotting and, hence, low temperatures avoiding diffusion. The advantage of monocrystalline diamond tools is that the cutting edge can be prepared to sharpness nearly at atomic level because diamond is a homogeneous and very hard material. There is much less burr formation on ductile work piece materials than in the case of hard metal tools.

Fig. 4. Left: End mill made of monocrystalline diamond. Right: Detail of the perfect cutting edge.

Micro slotting tools are much more stable than micro end mills because they are not rotationally symmetric and much more rigid. Grinding can be done according to individual requirements (Fig. 5).

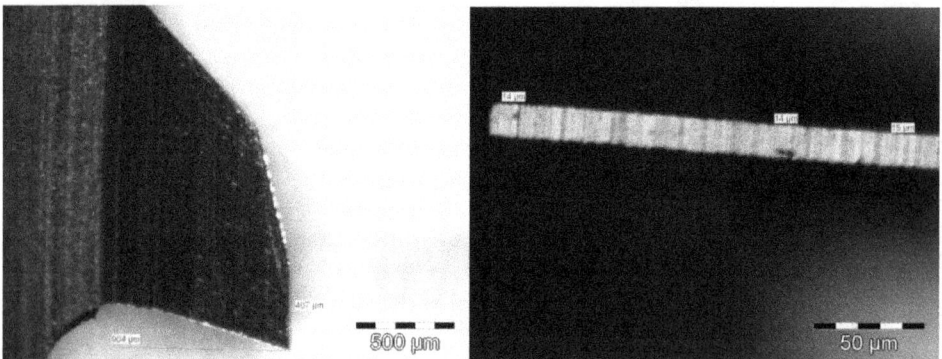

Fig. 5. Left: Side view of a micro slotting tool. Right: Measurement of the cutting width (app. 14 µm).

By micro slotting, minimum sizes of trenches can be reduced to about 15 µm in width at an aspect ratio of about ten (Fig. 6). Such dimensions cannot be achieved by micro milling (Fig. 7).

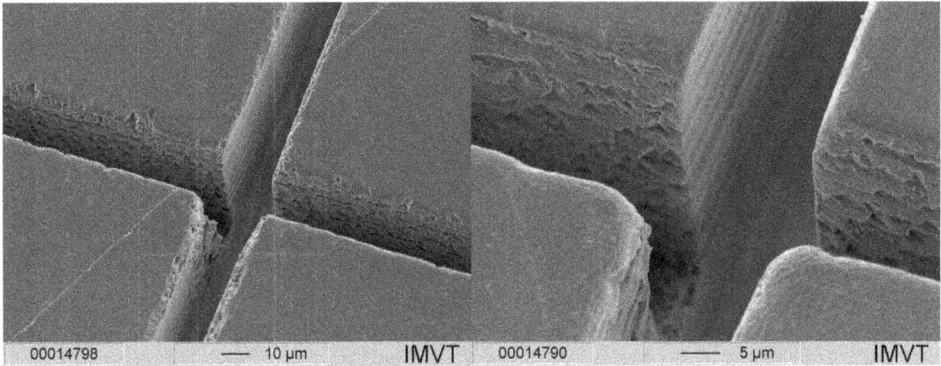

Fig. 6. PMMA trenches about 15 µm width, 150 µm in depth.

Fig. 7. Left: Mold insert for a cell chip made of brass with slotted trenches. Right: Trench at the base 60 µm in width, aperture angle 3° on both walls, 500 µm in depth.

The material removal rate for slotting is slow. Hence, also the cutting temperatures are low and monocrystalline tools can be used also for ferrous materials.

2.3 Thermal aspects, lubrication and cooling

Mechanical machining is connected with heat generation. Except in the case of dry machining, fluids are applied for cooling and lubrication to reduce the friction of the cutting edge with the work piece material and to decrease the thermal load of the cutting edge connected with increased wear and diffusion processes. As fluid, either water-based emulsions or oils are used. When using emulsions, bacterial contamination, aging and ecological aspects can involve issues of health and safety.

The fluid can be flushed or applied as mist. For lubrication by mist, a few milliliters of oil per hour are atomized by pressurized air. Oil mist has been preferred recently due to a number of advantages like reduced costs due to handling of smaller amounts of liquid, less storage and disposal costs, no hygienic problems due to bacterial contamination and less cleaning effort for liquid and the work piece. On the other hand, the right dosing of the oil in the air stream is essential especially in the case of micromachining to prevent the sticking of chips to the tool. The available apparatuses, however, lack in exact dosing systems. Sticking of chips leads to additional and fluctuating tool loads and can be an issue for tool failure. Additionally, the work piece surface quality is worse. For this reason, flush lubrication may be the better choice.

3. Tooling aspects: The role of material substrate, coating technology and tool shape

For micro tools, either hard metal or monocrystalline diamonds are used. Diamond tools are limited to nonferrous and non-carbide forming materials. A perfect cutting edge can be formed either by grinding or ion beam processing [Bor04]. When machining e. g. copper or brass, a very good edge quality without burrs can be achieved.

Hard metal, however, is a composite material. The cutting edge is always jagged (Fig. 8) causing burr formation on ductile materials like most metals.

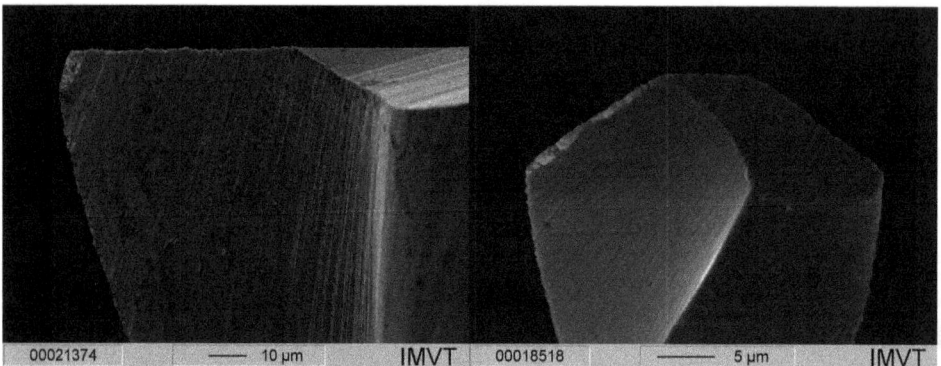

Fig. 8. Left: Imperfect cutting edge of an uncoated hard metal tool with d=0.25 mm (by Dixi). Right: Cutting edge of a 30 μm-drill bit (by Atom).

3.1 Influence of hard metal substrates

In general, hard metals consist of a hard phase and a binder phase. For the hard phase, mainly tungsten carbide is used which is basically responsible for the wear resistance. However, also small amounts of tantalum carbide, niobium carbide, chromium carbide, vanadium carbide and titanium carbide are added. These act as grain growth inhibitors during transient liquid-phase sintering and improve the high-temperature properties [Yao_Wc, Sad99]. Pure carbides cannot be sintered to full density because they would decompose at the necessary high temperatures. Furthermore, they are brittle, and crack propagation resistance is poor. Hence, already small defects in the surface would cause tool failure although a pure tungsten carbide would be desirable under the aspect of wear resistance. Instead, metals exhibiting a limited solubility for carbides at higher temperatures are used as binders. Mostly, cobalt (fcc structure) is used but also nickel and iron are possible. Fig. 9 shows the solubility of Co for WC.

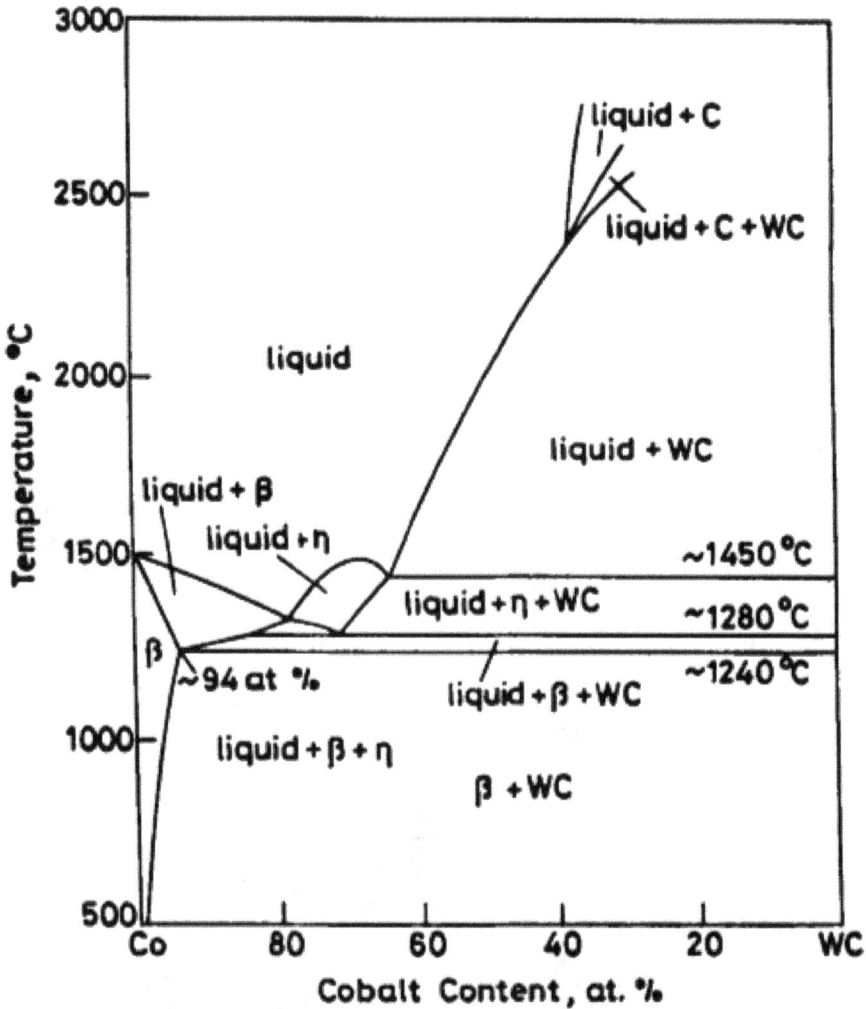

Fig. 9. Pseudo-binary phase diagram of WC-Co from [Upa98].

The mechanical properties of hard metals depend on the binder metal content, which affects mainly hardness and wear resistance, as well as on the average grain size, which is responsible for the flexural strength (Fig. 10).

Especially regarding micro milling, it is very important that the cross section of a tool consists of a sufficient number of hard particles to guarantee isotropic mechanical properties and long tool lifetime. Hence, submicron tungsten carbide powders with an average particle size of 0.2 µm were developed. It is obvious that for practical reasons the critical tool diameter depends on the micro structure of the substrate used. Tools made from submicron hard metal below 30 µm in diameter will not exhibit isotropic properties since a few dozen of hard particles should form the cross section at least.

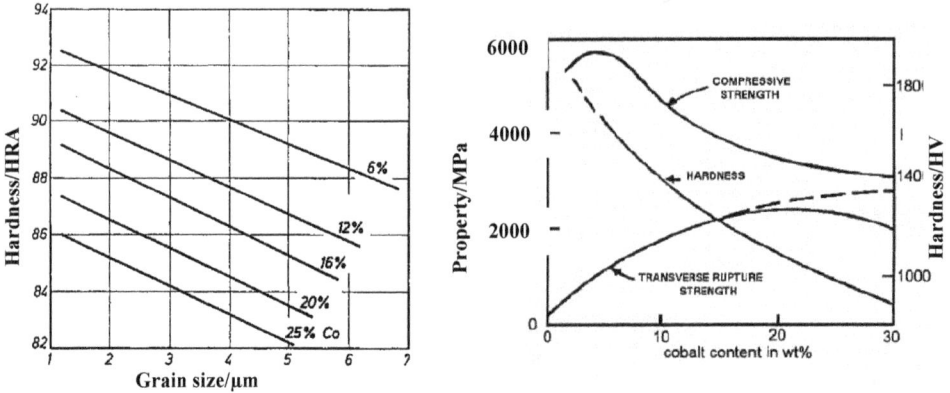

Fig. 10. Left: Hardness versus grain size for hard metals depending on cobalt content [Wei96]. Right: Dependence of mechanical properties on cobalt content [Exn70].

The wear of the tools is mainly controlled by the binder phase content. The binder phase content is adapted to the type of application: For continuous cut low binder content is sufficient. For interrupted cut or fluctuating load, higher binder content is recommended.

3.2 Coatings

In the last five years, progress has been made in achieving a low, uniform thickness of wear-resistant coatings. Previously, only tools larger than 0.3 mm in diameter were coated since the rounding of the cutting edges due to the coating thickness led to increased cutting forces which annihilated the gain of improved wear resistance (Fig. 11) [Klo05].

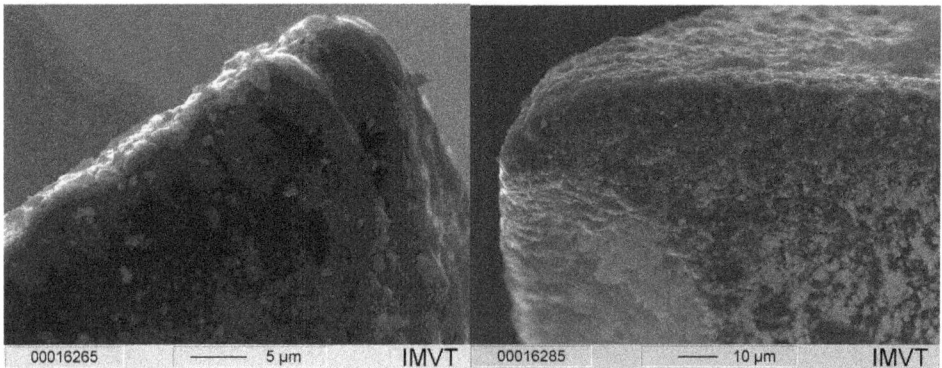

Fig. 11. Rounding of cutting edge by a DLC-coating at an end mill of d=0.4 mm by Karnasch.

Today, tools down to 30 μm in diameter are coated (Fig. 12). The coating is quite uniform and below 1 μm in thickness so rounding of the cutting edge can be neglected.

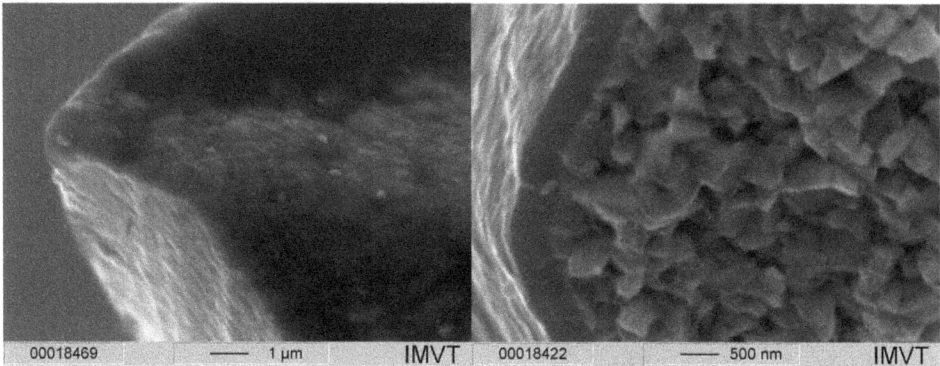

Fig. 12. Coated cutting edge of a 30 μm end mill by Hitachi. Right: Cross section illustration coating thickness and hard metal microstructure.

However, the coating process seems not to be stable all the time. The reproducibility and the results may vary from batch to batch. The formation of droplets certainly must be avoided to prevent coating results having worse machining properties like the ones displayed in Fig. 13 [Klo05].

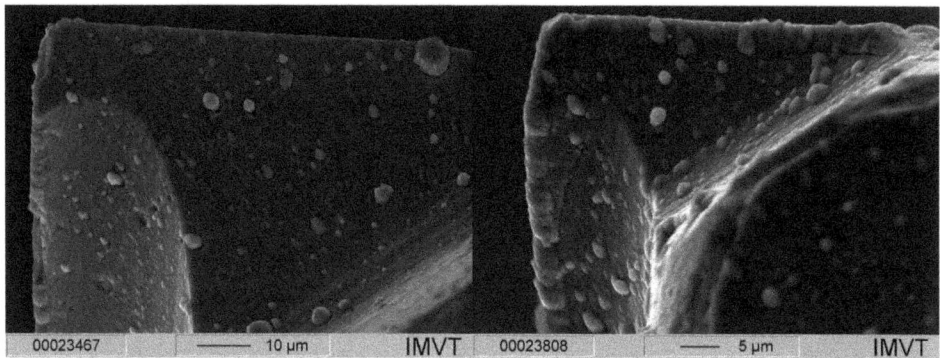

Fig. 13. Droplet formation on coated micro end mills.

Another issue is the adhesion of the coating. By SEM investigation, micro tools with flaking of coating layers were detected not only at the cutting edge but also in smooth substrate areas for different batches (Fig. 14). An appropriate surface processing is a prerequisite to prevent faults and varying quality of micro tools.

Obviously, an inspection of micro tools by SEM is advisable to guarantee machining results of a constant and good quality.

Different coatings influence the wear resistance of the tool, the rounding of the cutting edge, and the friction between work piece and tool. Monolithic, gradient or layered compositions of coatings are known.

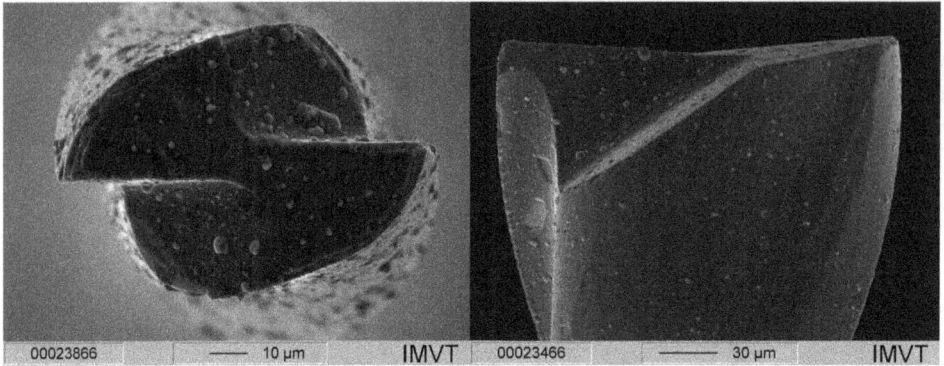

Fig. 14. Faults of adhesion and uniformity of coatings.

It is obvious that the price for micro tools increases strongly with decreasing tool diameter. Apparently the yield of the manufacturing process by grinding decreases significantly for small-diameter tools. Sometimes, undetected cracks cause tool failure. Fig. 15 shows a new 30 µm end mill broken during ultrasonic cleaning for SEM. It seems that cracks and impurities (top of Fig. 15) are present in the cross section, probably originating from the manufacturing process and covered by the coating.

In general, the life time of micro tools is unpredictable and depends strongly on the material machined. Also, the approach of the micro tool to the work piece to get the zero level and the maintenance of a constant engagement across the surface can be an issue due to variation of the flatness of the work piece.

Fig. 15. Left: Cross section of a 30 µm end mill broken when sonicated for SEM analysis. Right: Detail.

3.3 Adapted tool shape for micro milling

During the past years, attention was paid to optimizing the shape of micro end mills to meet the specific demands of the micro cutting process. Especially for small-diameter end mills, bending, tool deflection and the avoidance of chatter marks on the work piece are of interest to improve the stability of the process.

Not only was the fluted length reduced to increase the tool shaft cross section and stiffness. Also, the geometry at the intersection of the constant tool shaft diameter and the conical part where the bending moment is maximal was rounded to prevent crack initiation [Uhl06]. Some companies use a specially shaped fluted tip to eliminate chatter marks on the work piece (Fig. 16).

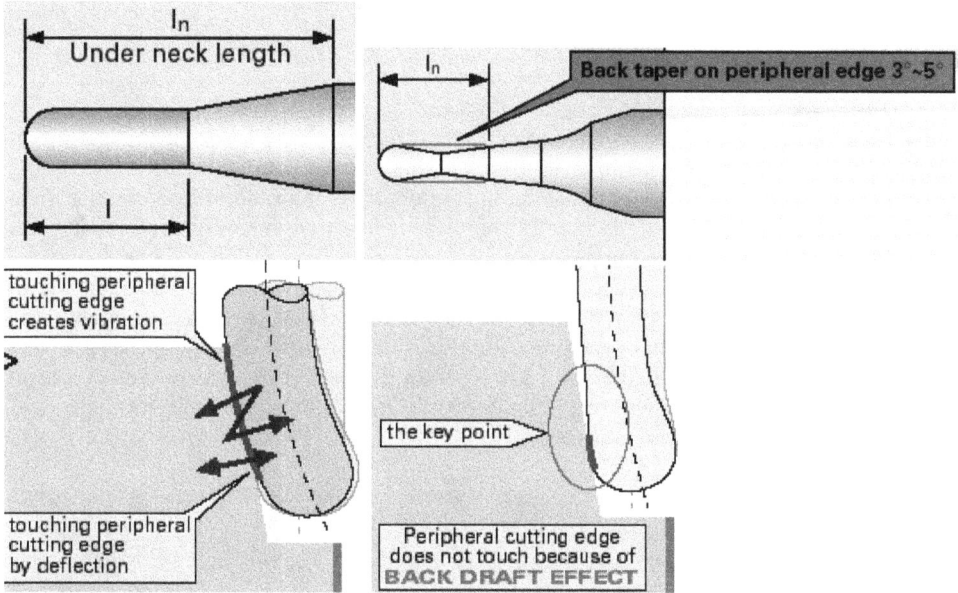

Fig. 16. Comparison of different tool shapes. Left: Conventional design. Right: Design adapted for micro milling [Hte_Ep1].

4. Machining strategies in respect of micro tool needs

4.1 Tolerance issues

Dealing with features of less than 0.01 mm, attention should be paid to tool and machine manufacturing tolerances that are relevant to manufacturing expenses.

In micromachining, tools are often engaged with the full width but not to a certain degree that leads to high load promoting tool deflection. For large formats where a good surface quality of the superficies surfaces is essential, tool change following the depth of the microstructure or caused by tool wear should be avoided since an offset due to tool diameter variation or fluctuating run-out cannot be eliminated.

Quality control of micro features is mostly carried out by optical microscopy. The accuracy of the method should be kept in mind concerning optical resolution depending on magnification and numeric aperture as well as pixel size of the CCD camera used. Regarding the absolute feature size, it can be necessary to shift the microscope table or to stitch multiple pictures for measuring reasons. Specifying tolerances in a range where measuring accuracy or other reasons prevent proving is useless and may increase manufacturing expenses exponentially.

Mostly, quality control is carried out by optical microscopy only at the surface level by edge detection but not at a certain depth. Using tactile devices such as fiber probes [Wer], limitations according to their relevant dimensions must be taken into account.

Generally, tolerances should be one order of magnitude larger than the measuring accuracy and the achievable roughness. Mostly, roughness values for the arithmetic average R_a or highest and lowest peaks within a certain distance like R_t are specified or predefined. With mechanical micro structuring, R_a-values are in the range of 0.2 µm. Typically for micro milling, R_t is 7-10 times higher than R_a, namely in the range of 1-2 µm.

4.2 CAM-software and machine controller issues

Often, the CAM routines are not able to handle multiple structures according to the special needs of micromilling. For example, the tool path is not generated to meet sequential machining of multiple features but machining is often done in a randomized manner. As a consequence, pins or holes are machined irregularly as the tool moves over a certain area. Lift-off the tool and moving to the next spot take additional time and may cause deviations due to thermal drift when the machining time is very long (the structure displayed in Fig. 18 was machined in three frames, 8h each). Moreover, additional tool loads and bending occurs due to unnecessary sinking in at each new spot. It is obvious that dipping in has a strong impact on the wear and the lifetime of micro tools. In the case of the structure displayed in Fig. 17, sequential machining was forced by insertion of additional frames dividing the field into 19 fields with three lines of pins each.

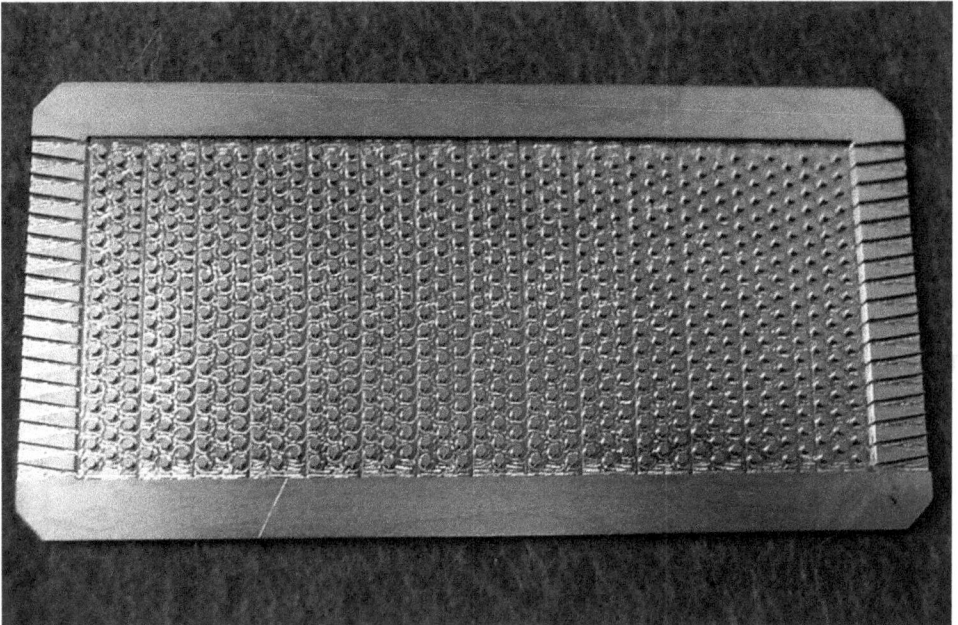

Fig. 17. Multiple pin array of a fixed-bed reactor with 732 pins, diameter 0.8mm, height 0.8mm, distance in between 0.8 mm, machined in titanium grade 2 using a 0.6 mm micro end mill.

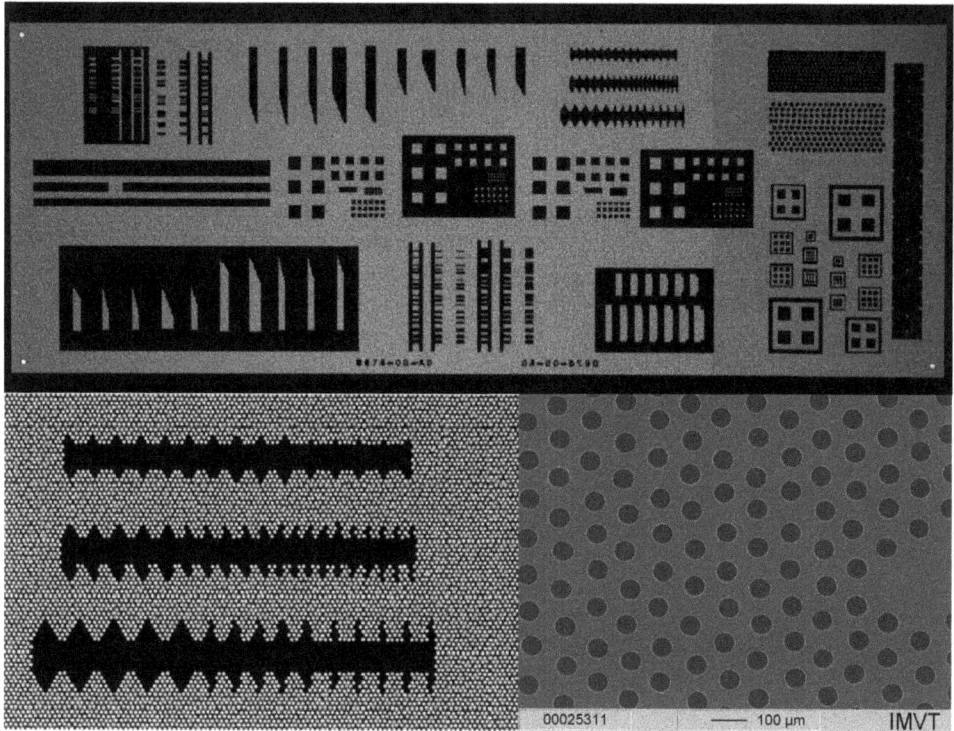

Fig. 18. Top: Sputter mask with approximately 114.500 holes, 50 μm in diameter, made of lead-free brass with a thickness of 100 μm. Bottom: Detail views.

Also, the possibilities of defining machining strategies sometimes are not sufficient for micro milling. Using routines for simple 2D structures, it is not possible to combine a ramp for sinking in the tool and to approach to a contour tangentially to avoid a stop mark from bending of the tool and cutting clear when it stops for turnaround as can be seen in Fig. 19. A smooth tool movement without changes in the feed rate is required. Perpendicular approach of the tool to micro features must be avoided. Unfortunately, it is not easy to meet all these requirements at once. Especially for micro machining of prototypes it is often necessary to make a test piece for preliminary inspection.

The NC unit of the machine must be able to process sufficient numbers of instructions per second. A comparison of different machine control units ranging from 250 to 1000 cycles/s is given in [Wis_Co]. Together with the definition of the accuracy (e. g. cycle 32 for Heidenhain, see [Hei]) requiring the machine to meet the exact NC data path, the drop of the feed rate caused by tiny details can be dramatic. Here, the influence of high axis acceleration becomes evident. Although already written some years ago, [Rie96] gives a good overview of the interaction of CAM data, data processing and NC-settings.

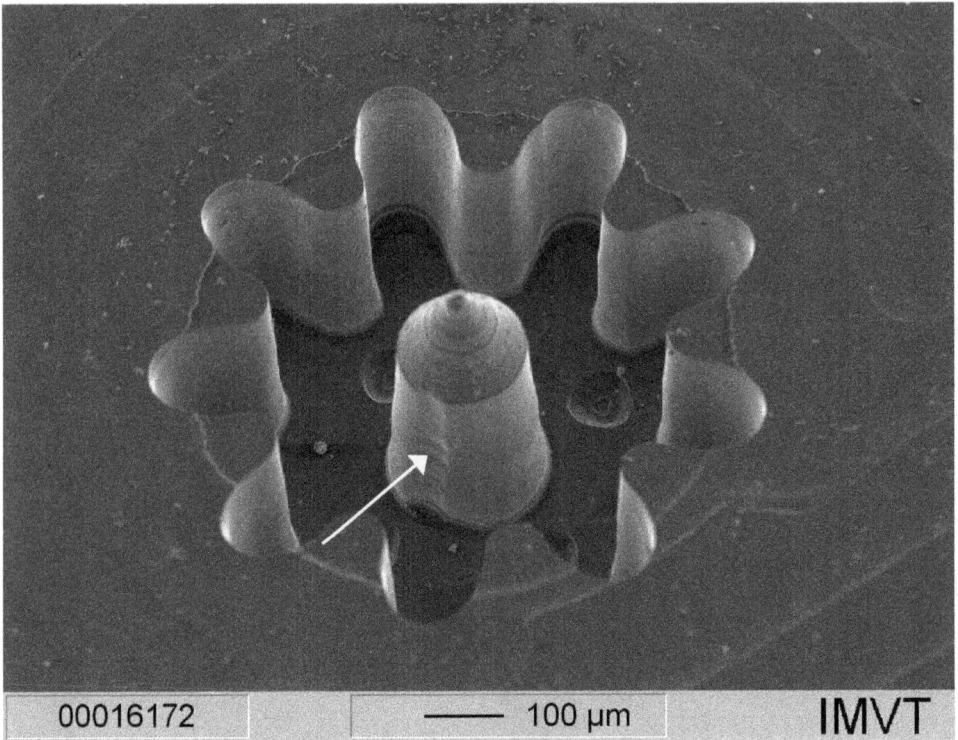

Fig. 19. Micro gearwheel top of teeth diameter: 800 μm, depth: 300 μm, diameter of column: 160 μm, height of the cone: 140 μm, smallest detail: 100 μm. Mark from clear cutting of the micro end mill at the perimeter of the pin caused by machining strategy and low tool stiffness.

4.3 Machine issues
4.3.1 Thermal effects
Especially for large numbers of microstructures, the thermal stability of machines is very important. A constant room temperature within 1 Kelvin and absence of direct solar irradiation are advised. Strict sequential machining of microstructures is a must to prevent irregularities. Often, this has to be forced by additional design work introducing multiple frames to prevent irregular machining.

The construction of the machine and the materials used also have an impact on thermal stability. For the machine bed, KERN uses polymer concrete with a low thermal coefficient of expansion of 10-20*E-06/K [Epu_Cr] and much better vibration damping properties than cast iron [Ker_Ev]. Taking a closer look at the historic development of this class of machines, progress in spindle clamping is evident. Since the machine concept is similar to a c-shape-rack and high-strength aluminium is used for the spindle clamping, the shape and fixing position of the clamping to the machine have a high impact on thermal drift due to the high thermal coefficient of expansion of 23*E-06/K of aluminum. For this reason, we changed the original clamping of an older machine by one made of Invar (α=1.7*E-06/K). Other suppliers use granite and a portal architecture for their machines [Kug_Mg, Ltu] for low thermal shift.

4.3.2 Clamping and measurement of micro end mills

The detection of tool length and tool diameter by laser [Blu_Na] or mechanical dipping onto a force sensor [Blu_Zp] is problematic for very small tool diameters. Laser measurement is normally only possible above 100 µm tool diameter. According to [Blu_Ha], the limit was recently shifted down to 10 µm diameter using special laser diodes. Mechanical dipping ends at 50 µm tool diameter.

For such small tools, a very high true running accuracy is essential to make sure both cutting edges are engaged at the same load. Collet chucks must be closed applying a certain torque. Thermal shrinking is superior to mechanical clamping. True running accuracy for thermal shrinkage [Die_Tg, Schun_Ce] or hydro stretch chucks [Schun_Tr] is about 3 µm, however, collet chucks are in the range of 5 to 10 µm only [Far, Ntt_Er].

Finally, a number of interfaces from tool to the spindle are adding up. For minimization of the run-out it is favourable to use vector-controlled spindles to ensure the same orientation of the chuck inside the spindle.

4.3.3 Spindle speed

Most machines on the market possess spindles with relatively low rotational speeds of 40-60.000 rpm [Ker_Ev, Mak_22]. For micro machining, often very high numbers of revolution are necessary to achieve reasonable material removal rates. However, much more importance should be attached to questions like tool life, true running accuracy [Weu01, Bis06], the stability and the dynamic behaviour of the machine.

The stability and damping behaviour of the machine are important to avoid vibrations and chatter marks on the work piece surface as well as additional stress of the micro tool due to vibrations. Often, polymer concrete with a very good damping behaviour superior to that of grey cast iron is used for the machine base [Epu_Fi].

Especially for micro features, the dynamic behaviour, namely the acceleration of the axes, the velocity to the NC-control unit and the maximum number of instructions per seconds are important to maintain a programmed feed rate. In this context, also the definition of how accurately the machine has to meet the calculated tool path is important. If the tolerance is very low, the servo-loop can cause an extreme breakdown of the feed rate. This leads to squeezing of the cutting edges, increased tool wear or even tool rupture. In the last decade, the acceleration could be improved from about 1.2 m/s^2 to more than 2 g (20m/s^2) [Wis_Ma] also by using hydrostatic drives [Ker_Ac].

Especially high-frequency spindles lack sufficient torque at lower speed as well as an easy-to-operate tool handling system. Mostly, three jaw chucks are used. Measurement of true running accuracy is a must in this case for ensuring a constant engagement of the normally two cutting edges of a micro end mill. Since the feed rate per tooth is far below 1 µm due to machine limitations and since the true running accuracy and cutting edge rounding are not taken into account, it is questionable if very high numbers of revolution in the range of 100.000 rpm and more that are stated e. g. in [Rus08] are appropriate. Instead, a minimal feed per tooth is required to obtain chip formation at all [Duc09].

Often, machining parameters like rotational speed and feed rate cannot be extrapolated. For instance, a speed of 15.000 rpm with a feed rate of 90 mm/min worked fine for micro drilling using a 50 μm drill bit for the sputter mask displayed in Fig. 18 but 40.000 rpm and 240 mm/min did not.

4.4 Design rules

Referring to the tool shapes with only a short fluted length as displayed in Fig. 3 and Fig. 16, new specific problems can occur. Whereas in Fig. 20 no shape distortion of the spinneret can be observed, a similar negative microstructure (Fig. 21) shows a strong distortion at a depth of 1 mm. Obviously, it is caused by insufficient chip removal from the narrow trenches. The chips are not conveyed by flutes up to the surface level and stick to the tool since oil mist instead of flushing was used for lubrication and cooling.

Fig. 20. Positive spinneret made of brass using Hitachi EPDRP-2002-2-09 with 1° slope, height 2.8 mm.

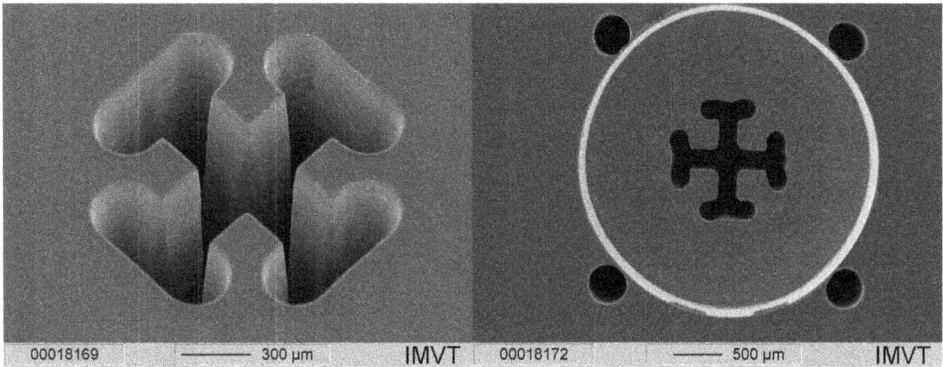

Fig. 21. Left: Surface level of a negative spinneret made of brass with 1° slope, final depth 2.8 mm using Hitachi EPDRP-2002-2-09 and oil mist. Right: Distortion of the same microstructure at a level of -1 mm due to insufficient chip removal.

For serial production, all machining parameters can be optimized for a certain design to gain maximum output from the process but for prototype or small-scale production the effort exceeds the saving of machining time extremely.

5. Material concerns in mechanical micro machining

5.1 Machinable materials

Micro milling or slotting is a very variable process in terms of material classes possessing a high material removal rate. With some limitations on ceramic materials, all kinds of materials like metals, polymers and ceramics can be machined. However, the kind of material machined has a huge impact on machining time, tool wear, surface quality and burr formation.

For micro process devices, often highly corrosion-resistant materials are used. It is not possible to compare the machining behaviour of normal tool steels that are used e. g. for molds for injection molding with aluminum- and copper alloys, with tough materials like stainless steels, nickel base alloys, titanium and tantalum or with brittle materials like ceramics. Mostly, the recommendations given by the suppliers for infeed, lateral engagement, feed rate and number of revolutions depending on tool diameter and tool length are not appropriate for micro tools. Often, there is no defined engagement width but the tool is engage with its full diameter. Trial and error must be applied to find optimal parameters. Mostly it is a good idea to work with low infeed but higher feed rate instead of using the recommended infeed to keep the tool wear low, especially for tough materials.

Ductile materials tend to form burrs at the edges of micro structures. Depending on the resistance of a certain material against chipping and its strength, cold work hardening can be an issue. The machining strategy must be adapted to prevent deformation of very thin and high walls like displayed for stainless steel in Fig. 22. The structure was made of different materials, namely aluminum (Fig. 22), stainless steel (1.4301, Fig. 24) and MACOR (Fig. 25), a machinable ceramic consisting of about 45 % borosilicate glass and 55 % mica acting as micro crack propagators [Mac]. While MACOR and aluminum were easy to machine, stainless steel machining was very challenging. Machining of only a few trenches to the final depth led to cold work hardening. Subsequently, bending of narrow walls and tool deflection occurred (Fig. 23). Finally, the microstructure was machined successfully in stainless steel using three ball-nose tools made by Hitachi with lengths of 1, 2 and 3 mm and a diameter of 0.4 mm. For the first two tools, 36.000 rpm and a feed rate of 1800 mm/min were applied. The infeeds were 0.03 and 0.021 mm, respectively. For the 3 mm long tool the parameters were reduced to a speed of 32.000 rpm, a feed rate of 1600 mm/min and the infeed to 0.011 mm. With the first tool, all channels were machined with the same infeed to 0.6 mm depth followed by machining to a depth of 1.9 mm with the second and to the final depth with the third tool. Flushing with lubricant oil was applied. The wear of the tools was estimated not to be critical for any of the materials.

Fig. 22. Matrix heat exchanger made of aluminum, 14 in 15 comb-shaped interlaced micro channels, 23 mm long each. Channels are 0.4 mm in width; depth at beginning is 2.9 mm, ending at 0.6 mm, wall thickness 0.2 mm.

Fig. 23. Tests of the microstructure displayed in Fig. 22 made of stainless steel 1.4301 without optimization of the machining strategy using a radius end mill. Distortion of the thin walls and tool deflection can clearly be seen.

Fig. 24. Details of the final heat exchanger made of stainless steel 1.4301. No burr formation at the surface level but some lateral burrs.

Fig. 25. Microstructure of the matrix heat exchanger made of MACOR. Very good shape stability at the edges without flaws.

5.2 Burr removal from ductile materials

Micro milling of ductile materials is often accompanied by burr formation, especially at the edges of the microstructures. Burrs can be removed e. g. mechanically using small tools, preferably with sharp edges but consisting of a softer material. For steel e. g. spicular tools made of brass are suitable. For microstructures e. g. made of PMMA or PTFE, wood can be used. The disadvantage of this method is the high manual effort. Mostly, it is used only for single channels e. g. for microfluidic devices. For more complex designs of metallic parts, an electrochemical approach, namely electropolishing, is preferred. It can remove burrs from metals possessing a homogeneous microstructure like austenitic stainless steels, nickel and some copper base alloys. Homogeneity means that no precipitations at grain boundaries or a different second phase are present affecting the electrochemical behaviour and forming an electrochemical element in an electrolyte. For instance, in the case of brass, electropolishing works only for lead-free grades. For tool steels with a carbon content of more than 0.1 %, achievement of a good surface quality through electropolishing is not possible because the microstructure consists of a ferritic or martensitic matrix with embedded carbide particles of different chemical compositions. However, with a one order of magnitude smaller inhomogeneity, e. g. in the presence of small precipitations in the grains as in dispersion-strengthened alloys, electropolishing works very well (Fig. 26).

In the case of copper-based alloys, for example conventional alloyed Ampcoloy 940 and 944 [Amp] and dispersion-strengthened alloys like Glidecop or Discup [Dis_1, Dis_2], comparable mechanical strengths can be achieved. However, the microstructures are very different. Whereas Glidecop and Discup can be electropolished, Ampcoloy cannot.

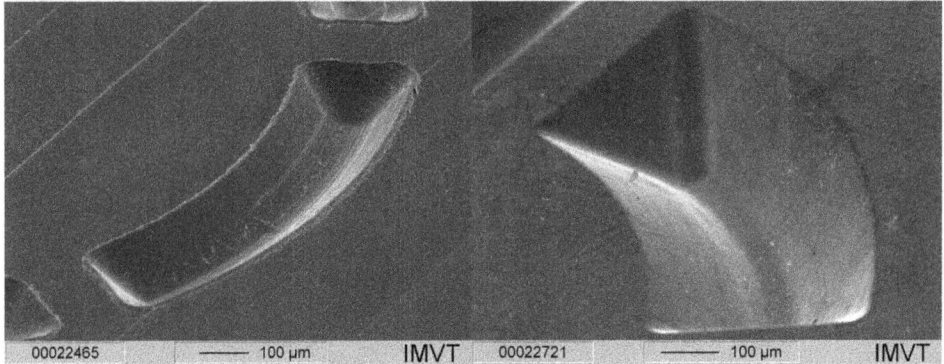

Fig. 26. Micro milled structure made of a dispersion strengthened cooper alloy (Glidcop Al-60, [Gli]). Left: After micromilling. Right: After electropolishing.

Generally, electropolishing removes material according to the field line density. At the burrs and edges, the electric field has the highest density. For monitoring, electropolishing must be stopped and the microstructure evaluated by microscopy. After the burrs are removed, the process must be finished to avoid that edges are rounded. At spots without burrs, edges are eroded from beginning. That means, an uniform burr formation is preferred to only partial burrs. On flat surfaces ghost lines are flattened and roughness is decreased by electropolishing.

5.3 Ceramic materials for micromachining
Beside MACOR, most other ceramic materials like alumina, zirconia and so on can be machined in the CIP (cold isostatic pressed) or presintered state with acceptable tool wear (Fig. 27). At temperatures below normal sinter temperature sintering starts with neck formation between single powder particles. Depending on the residual porosity, the strength of the blanks and tool wear may vary in a wide range. However, the adhesion is much lower than at full density. After machining, the parts are sintered to full density assuming a certain shrinkage. The value of shrinkage must be known or determined by experiments and be taken into account to meet the exact dimensions. By doing so, accuracy within +/- 0.1 % can be achieved.

Another approach consists in using shrink free ceramics [Gre98, Hen99] e. g. based on intermetallic phases like $ZrSi_2$ undergoing an internal oxidation into $ZrSiO_4$ accompanied by an expansion compensating the shrinkage from pore densification. By adjusting the composition of the blend of low-loss binder, inert phase and $ZrSi_2$, the final dimension can be controlled very exactly.

Generally, the material removal rate for ceramics is rather high since a higher infeed and feed rate can be applied. However, machines must be equipped for machining ceramics to protect guideways and scales from damage by abrasive particles.

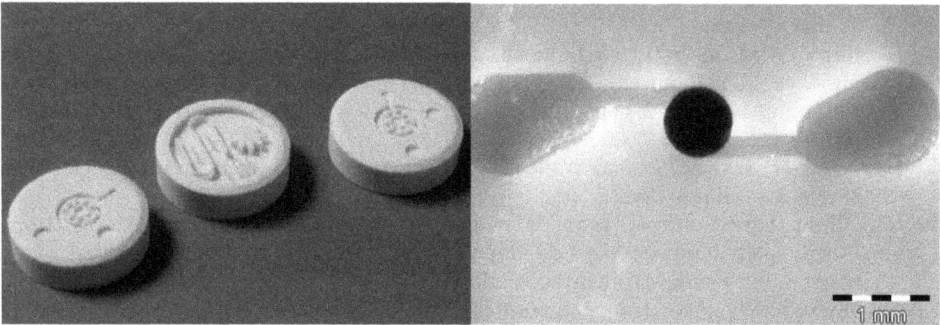

Fig. 27. Microstructures made of shrink free $ZrSi_2O_4$ (left) and zirconia (right)

6. Conclusion

In this chapter, the recent developments in micromachining were outlined. Especially improvements of machine tool, spindles, clamping technology and tool production can be stated within the last five years, having a big impact on productivity.

In general, micromachining is a very flexible and cost efficient technique, not only for large scale series but also for prototyping and applicable for a wide range of materials.

Due to mechanical and material scientific reasons, further miniaturization of tools seems not to be promising in terms of stability and cost efficiency. Instead, attention should be paid to improvement of reliability of the cutting process and the adaption of machining routines to the specific requirements of sensitive micro tools.

Tolerances in micromachining should be always specified according the real practical demand, with respect to measuring accuracy as well as to achievable surface roughness values.

Especially for replication techniques like micro injection molding and hot embossing, burr formation can be an issue. For some ductile metallic materials the removal of burrs at microstructures can be achieved by electropolishing. Basically, the micro structure of the material has an impact on machinability and surface quality of microstructures after machining and electropolishing. Hence, a homogeneous microstructure is superior to heterogeneous materials.

7. Acknowledgement

All examples of microstructures displayed in this chapter were made by D. Scherhaufer, T. Wunsch and F. Messerschmidt. Only their professionalism and persistence enabled successful microstructuring of many different prototype designs made of a wide variety of materials.

We acknowledge support by Deutsche Forschungsgemeinschaft and Open Access Publishing Fund of Karlsruhe Institute of Technology.

8. References

Amp, data sheets of different Amcoloy-alloys, date of access: 16.08.2011, available from: http://www.ampcometal.com/common/datasheets/ampco_alloy_brochure.pdf

Ato_Ad, Atom Saito Seisakusho micro drill bit ADR-0002, date of access: 16.08.2011, available from: http://www.atom21.co.jp/english/detail.php?act=view&page=0&id= adr&u=mm

Bis06, G. Bissacco, H. N. Hansen, L. De Chiffre: "Size Effects on Surface Generation in Micro Milling of Hardened Tool Steel", *CIRP Annals - Manufacturing Technology*, vol. 55, Issue 1, pp. 593-596, 2006

Blu_Ha L. Halder, answer be email from 13/07/2011: tools down to 10 μm in diameter with special laser diode detectable

Blu_Na, Blum LaserControl Nano NT, date of access: 16.08.2011, available from: http://www.blum-novotest.de/de/messkomponenten/produkte/optische-werkzeugmessung/lasercontrol-nt.html

Blu_Zp, Blum Probe Z-Pico, date of access: 16.08.2011, available from: http://www.blum-novotest.de/uploads/media/Z-Pico_EN.pdf

Bor04, S. Borek, K. Schauer, M. Füting, A. Heilmann: „Endbearbeitung von Mikrofräsern mittels Ionenstrahltechniken", *wt Werkstattechnik online*, vol. 94, no. 11/12, pp. 600-604, 2004

Con, Contour, Fine Tooling Ltd., Wedgwood Way, Stevenage, Herts SG1 4QR, UK, date of access: 26.09.2011, available from: http://www.contour-diamonds.com/HTML/ mircromilling.html

Die_Tg, Diebold ThermoGrip, p. 6, date of access: 16.08.2011, available from: http://www.diebold-hsk.de/pdfs/04.pdf

Dis_1, presentation dispersion-strengthened Discup-alloy, date of access: 26.09.2011, available from: http://www.kupfer-institut.de/symposium/media/pdf/Cu-Werkstoffe   durch PM.pdf

Dis_2, information about Discup can be ordered under: http://www.ecka-granules.com/en/ecka-granules/products/catalogue-request/

Duc09, F. Ducobu, E. Filippi, E. Rivière-Lorphèvre: "Chip Formation and Minimum Chip Thickness in Micro-milling", *Proc. of the 12th CIRP Conference on Modeling of Machining Operations*, pp 339-346, 2009

Epu_Cr, overview about different types of polymer concrete possessing different thermal coefficients of expansion, date of access: 15.09.2011, available from: http://www.epucret.de/en/epument-machine-beds-made-of-mineral-casting/materials/

Epu_Fi, Epufill – Composite structures with mineral casting filling, p. 3, date of access: 16.08.2011, available from: http://www.epucret.de/fileadmin/rampf_downloadcenter/EPUCRET-Mineralgusstechnik/Broschueren/EPUFILL/EPU_ Epufill_GB.pdf

Ewa_Ws, Ewag AG, Industriestr. 4, 4554 Etziken, Switzerland, date of access: 26.09.2011, product information availabe from: http://www.ewag.com/de/produkte/ schleifen/ws-11.html

Exn70, H. E. Exner, J. A Gurland: "A review of parameters influencing some mechanical properties of tungsten carbide cobalt alloy", *Powder Metallurgy*, vol. 13, pp. 13-31, 1970

Far, Farion collet chucks, date of access: 26.09.2011, available from:

http://shop.fahrion.de/index.php?id=17&mastersku=HF-A8+f%C3%BCr+CP&
pgsn=060280030&sdzgrouping=FAHRION+450E+FM32&L=0&chash=cdf728&&L=1

Fri08, A. H. Fritz, G. Schulze: *"Fertigungstechnik"*, 8th ed., Springer, p. 264, 2008

Gli, data sheet of dispersion strengthened Cu-alloy Glidecop Al-60, date of access: 26.09.2011, available from:
http://www.spotweldingconsultants.com/GlidCop_AL_60.pdf

Gre98, P. Greil: „Near Net Shape Manufacturing of Polymer Derived Ceramics", *J. of. Europ. Ceram. Soc.*, vol. 18, pp. 1905-1914, 1998

Ham_38, Hartmetall-Werkzeugfabrik Andreas Maier GmbH, 88477 Schwendi-Hörenhausen, series HAM 382 Micro Prima, date of access: 26.09.2011, product information available from:
http://www.ham-tools.com/fileadmin/templates_ham/
Download/HAM_Micro.pdf, oral information and already bought: diameter 30 μm

Hei, technical info, permissible contour deviation, p. 3, date of access: 16.08.2011, available from:
http://www.heidenhain.de/fileadmin/pdb/media/img/636_225-21.pdf

Hen99 V. D. Hennige et al.: "Shrink-free ZrSiO4-ceramics: Characterisation and Applications", *J. of. Europ. Ceram. Soc.*, vol. 19, pp. 2901-2908, 1999

Hte_Em, Hitachi Tool Engineering Europe GmbH, Itterpark 12, 40724 Hilden, series EMM, date of access: 26.02.2011, product information available from:
http://www. hitachitool-eu.com/download/brochures/view/emm.pdf

Hte_Ep1, Brochure 413, EPDBP/EPDRP, Hitachi Tool Engineering Europe GmbH, Itterpark 12, 40724 Hilden, Germany, 2005, date of access: 26.09.2011, available from:
http://www.hitachimetals.com/product/cuttingtools/pdf_solid/epdb_epdbp_lo
w_res.pdf

Hte_Ep2, Coated 100 μm-end mills, AR=10 by Hitachi Tool Engineering Europe GmbH, series EPDS, date of access: 16.08.2011, date of access: 16.08.2011, product information available from:
http://www.hitachitool-eu.com/download/ brochures/view/epds.pdf

Ker_Ac, acceleration of KERN-machines, date of access: 15.09.2011, available from:
http://www.kern-microtechnic.com/page.php?page_id=53&lid=1

Ker_Ev, Usage of polymer concrete and spindles by KERN Micro- und Feinwerktechnik, date of access: 26.09.201, available from:
http://www.kern-microtechnic.com/ upload/media/kern_evo_e.pdf

Klo05, F. Klocke, J. v. Bodenhausen, K. Arntz: „Prozesssicherheit bei der Mikrofräsbearbeitung", *wt Werkstattstechnik online*, issue 95, vol. 11-12, pp. 882-886, 2005

Kön90, W. König: „Fertigungsverfahren –Drehen, Bohren, Fräsen-", vol. 1, 3rd ed., VDI-publishing, ISBN 3-18-400843-6, p.63, 1990

Kön99, W. König: „Fertigungsverfahren –Schleifen, Honen, Läppen-", vol. 2, 2nd ed., VDI-publishing, ISBN 3-18-400180-X, p.11, 1990

Kug_Mg, Kugler, MICROGANTRY nano 3/5X, date of access: 26.09.2011, available from:
http://www.kugler-precision.com/index.php?page=99&modaction=detail
&modid=188

Ltu, LT-Ultra, UP-Machine, MSC-1100, date of access: 26.09.2011, available from:
http://www.lt-ultra.com/index.php/de/up-maschine/msc-1100

Mac, MACOR, machinable glassy ceramic, date of access: 26.09.2011, available from:
http://www.precision-ceramics.co.uk/mcomp.htm

Mak_22, spindle speed of Makino V22, date of access: 15.09.2011, available from: http://www.makino.com/machines/V22/Graphite/

Med, Medidia GmbH, Alte Poststr. 23, 55743 Idar-Oberstein, date of access: 26.09.2011, available from: http://www.medidia-diamond-tools.com/page/?menu=220

Möß, Mößner GmbH, Kelterstr. 82, 75179 Pforzheim, date of access: 16.08.2011, available from: http://www.hamoedia.de/html/m-k-d-wz.html

Nst_Di, Monocrystalline diamond end mills by NS Tool Co. Ltd., Japan, date of access: 26.09.2011, available from: http://www.ns-tool.com/cgi-bin/e/docat/dt.php?j =1&s=73

Nst_Em, Coated 100 μm-end mills, AR=10 NS Tool Co. Ltd., Japan, date of access: 26.09.2011, available from: http://www.ns-tool.com/cgi-bin/e/docat/dt.php?j=1&s=9

Ntt_Er, ERC precision collet chuck, date of access: 15.09.2011, available from: http://www.nttooleurope.com/products/erc/

Rie96, Asko Riehn: "The Challenges of High Speed Cutting", published in "*Modern Machine Shop*", May 1996, date of access: 15.09.2011, available from: http://findarticles.com /p/articles/mi_m3101/is_n12_v68/ai_18389412/

Rus08, Rusnaldy, T. J. Ko, H. S. Kim: „An experimental study on microcutting of silicon using a micromilling machine", *Int. J. Adv. Manuf. Technol.*, vol. 39, pp. 85-91, 2008

Sad99, R. K. Sadangi et al.: "Grain growth inhibition in Liquid Phase Sintered of Nanophase WC/Co Alloys", *Int. J. of powder metallurgy*, No. 35, vol. 1, pp. 27-33, 1999

Schun_Ce, Schunk, CELSIO SSF HSK-A 32, p. 78, date of access: 16.08.2011, available from: http://www.schunk.com/schunk_files/attachments/WZH-Systeme__HSK__DE_EN.pdf

Schun_Tr, Schunk, HSK-E 25 Tribos Polygonal Toolholder, p. 130, date of access: 16.08.2011, available from: http://www.schunk.com/schunk_files/attachments/TRIBOS-RM_Aufnahmeschaft_HSK-E_DE_EN.pdf

Uhl06, E. Uhlmann, D. Oberschmidt, K. Schauer: „Innovative Fräswerkzeuge für die Mikrozerspanung", *wt Werkstattstechnik online*, issue 96, vol. 1-2, pp. 2-5, 2006

Upa98, G. S. Upadhyaya: „*Cemented Tungsten Carbides – Production Properties and Testing*", Noyes Publications, U.S., ISBN 0-8155-1417-4, p. 25, 1998

War00, G. Warnecke, K. Eichgrün, L. Schäfer: „Zuverlässige Serienbauteile aus Hochleistungskeramik", *Ingenieur-Werkstoffe*, Nr. 2, April 2000

Wei96, F. Weiland: „Cerametal S.à.r.l., eine Spitzenadresse nicht nur in Europa. Hartmetall und Hartmetallwerkzeuge auf und mit Welt-Niveau" *Revue Technique Luxembourgeoise*, vol. 2, 1996

Wer, Werth fiber probe, date of access: 26.09.2011, available from: http://www.werth.de/ index.php?id=76&L=1

Weu01, H. Weule, V. Hüntrup, H. Tritschler: „Micro-Cutting of Steel to Meet New Requirements in Miniaturization", *CIRP Annals - Manufacturing Technology*, vol. 50, Issue 1, pp. 61-64, 2001

Wis_Co, Comparison of the velocity of different NC`s, date of access: 26.09.2011, available from: http://www.wissner-gmbh.de/Page/Technologie/ steuerungen.htm

Wis_Ma, acceleration of WISSNER-machines, date of access: 26.09.2011, available from: http://www.wissner-gmbh.de/Page/GAMMA/gamma303.htm

Yao_Wc, Z. Yao, J. J. Stiglich, T. S. Sudarshan: „*Nanograined Tungsten Carbide-Cobalt (WC/Co)*", p. 8, date of access: 26.09.2011, available from: http://www.matmod. com/Publications/armor_1.pdf

8

Micro Eletro Discharge
Milling for Microfabrication

Mohammad Yeakub Ali, Reyad Mehfuz,
Ahsan Ali Khan and Ahmad Faris Ismail
International Islamic University,
Malaysia

1. Introduction

Miniaturization of product is increasingly in demand for applications in numerous fields, such as aerospace, automotive, biomedical, healthcare, electronics, environmental, communications and consumer products. Researchers have been working on the microsystems that promise to enhance health care, quality of life and economic growth. Some examples are micro-channels for micro fuel cell, lab-on-chips, shape memory alloy 'stents', fluidic graphite channels for fuel cell applications, miniature actuators and sensors, medical devices, etc. (Madou, 2002; Hsu, 2002). Thus, miniaturization technologies are perceived as potential key technologies.

One bottleneck of product miniaturization is the lack of simpler and cheaper fabrication techniques. Currently the common techniques are based on silicon processing techniques, where silicon-based materials are processed through wet and dry chemical etching. These techniques are suitable for microelectronics, limited to few silicon-based materials and restricted to simple two dimensional (2D) or pseudo three dimensional (2.5D) planar geometries. Other fabrication processes, such as LiGA (lithography, electroforming and molding), laser, ultrasonic, focused ion beam (FIB), micro electro discharge machining (EDM), mechanical micromilling, etc. are expensive and required high capital investment. Moreover these processes are limited to selected materials and low throughput (Ehmann et al. 2002).

A less expensive and simpler microfabrication technique is sought to produce commercially viable microcomponents. Micro electro discharge (ED) milling, a new branch of EDM, has potential to fabricate functional microcomponents. The influences of micro ED milling process parameters on surface roughness, tool wear ratio and material removal rate are not fully identified yet. Therefore, modeling of ED milling process parameters for surface roughness, tool wear ratio and material removal rate are necessary.

1.1 Micro electro discharge machining

EDM has been successfully used for micromachining with high precision regardless of the hardness of work piece material. It uses the removal phenomenon of electrical discharges in a dielectric fluid. Two conductive electrodes, one being the tool and the other the workpiece, are immersed in a liquid dielectric. A series of voltage pulses are applied between the electrodes, which are separated by a small gap. A localized breakdown of the dielectric occurs and sparks are generated across the inter-electrode gap, usually at regions where the local electric field strength is highest. Plasma channels towards workpiece are formed during the discharge and high speed electrons come into collision with the workpiece. Each spark erodes a small amount of work material by melting and vaporizing from the surface of both the electrodes. The momentary local plasma column temperature can reach as high as 40,000 K (DiBitonto et al., 1989).

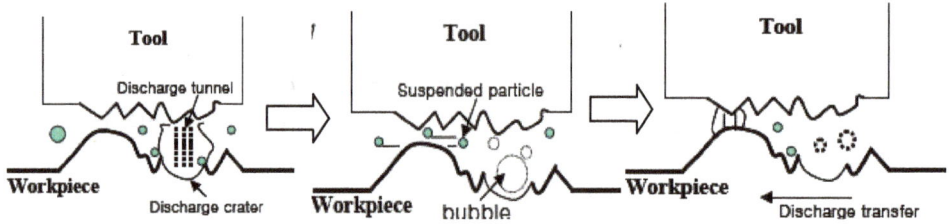

Fig. 1.1. Schematic of EDM principle (Kim et al., 2005)

This high temperature causes the melting and vaporization of the electrode materials; the molten metals are evacuated by a mechanical blast, resulting in small craters on both the tool electrode and work materials. It is understood that the shock waves, electromagnetic and electrostatic forces involved in the process are responsible for ejection of the molten part (debris) into the dielectric medium. The repetitive impulse together with the feed movement (by means of a servo mechanism) of the tool electrode towards the workpiece enables metal removal along the entire surface of the electrodes. Figure 1.1 shows the schematic of EDM principle.

Production of micro-features using EDM pays significant attention to the research community. This is due to its low set-up cost, high accuracy and large design freedom. Compared to lithography based miniaturization, micro EDM has the clear advantages in fabricating complex 3D shapes with higher aspect ratio (Lim et al., 2003). Moreover, all conductive materials regardless of hardness can be machined by EDM. Conventional EDM is especially useful to produce molds and dies. Micro EDM is now basically focused on the fabrication of miniaturized product, like molds and dies, with greater surface quality. In this endeavor new CNC systems and advanced spark generators are found as great assistance (Pham et al., 2004). Current micro EDM technology used for manufacturing micro-features can be categorized into four different types:

a) Die-sinking micro-EDM, where an electrode with micro-features is employed to produce its mirror image in the workpiece.

b) Micro-ED drilling, where micro-electrodes (of diameters down to 5–10 μm) are used to 'drill' micro-holes in the workpiece.

c) Micro-ED milling, where micro-electrodes (of diameters down to 5–10 μm) are employed to produce complex 3D cavities by adopting a movement strategy similar to that in conventional milling.

d) Micro-wire EDM, where a wire of diameter down to 20 μm is used to cut through a conductive workpiece.

Precision micro holes required for gas and liquid orifices in aerospace and medical applications, pinholes for x-ray and nuclear fusion measurements, ink-jet printer nozzles and electron beam gun apertures can be fabricated with high precision accuracy and with surface roughness less than 0.1 μm using the micro EDM process (Dario et al., 1995). Although micro EDM plays an important role in the field of micromachining, it has disadvantages such as high tool-electrode wear ratio and low *MRR*. The wear of electrode must be compensated either by changing the electrode or by preparing longer electrode from the beginning or fabricating the electrode in situ for further machining (Asad et al., 2007).

1.1.2 Micro electro discharge milling

In micro ED milling the material is eroded by non-contact thermo-electrical process, where a series of discrete sparks occur between the workpiece and the rotating tool electrode. The workpiece and tool are immersed in a dielectric fluid. The dielectric fluid is continuously fed to the machining zone to convey the spark and flush away the eroded particles. The work feeding system of micro ED milling is similar to mechanical end milling process. Like mechanical end milling, here the workpiece is fed to the tool electrode while the tool is in rotation. The tool movement is controlled numerically to achieve the desired three-dimensional shape with high accuracy. Figure 1.2 shows the schematic of micro ED milling.

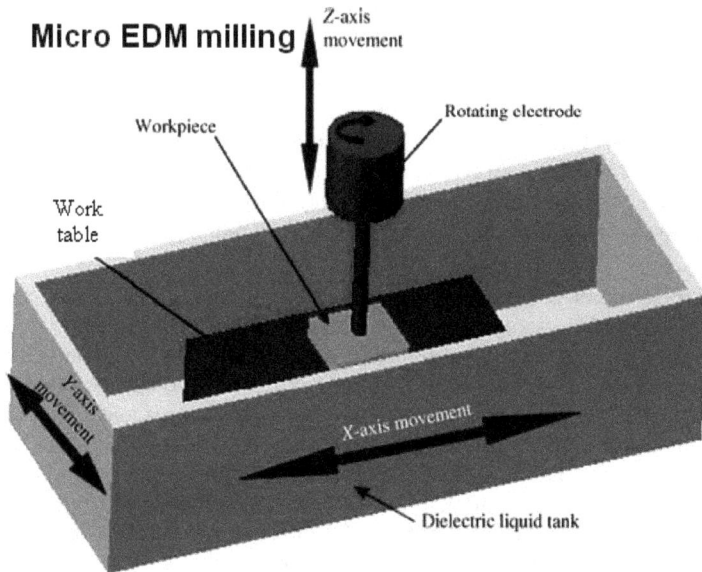

Fig. 1.2. Schematic of micro ED milling (Murali and Yeo, 2004)

Process parameters	
Open voltage	Pulse on time
Capacitance	Pulse off time
Discharge current	Pulse frequency
Resistance	Tool rotation speed
Feed rate	Dielectric flow rate
Polarity	

Table 1.1. Micro ED milling process parameters

1.1.3 Process parameters

Micro ED milling has a number of process parameters, which are influential on the machining performance. Depending upon the circuit type used in the machine, the importance of parameters is also varied. Table 1.1 shows different micro ED milling process parameters. Some researches were conducted to optimize different EDM process parameters for R_a, R_y, TWR and MRR (Puertas and Luis, 2004; Lin et al., 2006). But almost all of them were on conventional macro EDM. Because of the stochastic thermal nature of the EDM process, it is difficult to explain all of those effects fully. The full understanding of micro ED milling process parameters are yet to be developed because the process itself is a new branch of manufacturing. For electric discharge the discharge energy (E) generated in the discharge circuit can be expressed roughly as follows:

$$E = \frac{1}{2}(C + C')V^2 \qquad (1.1)$$

Where,
C = capacitance of capacitors,
C' = total stray capacitance,
V = discharge voltage

According to the equation (1.1), there are two ways to reduce the discharge energy: by reducing the discharge voltage or by reducing the total capacitance (C + C'). However, low voltage results in an unstable discharge. Therefore, the total capacitance should be controlled primarily. The total stray capacitance (C') in the circuit plays a significant role here. The following subsections discuss the influence of different micro ED milling parameters on surface roughness, tool wear and material removal rate.

1.1.4 Surface roughness

Variation of discharge energy affects surface roughness. It is experimentally identified that with the increase in discharge energy either by increasing voltage or capacitance, R_a increased in die sinking EDM, wire EDM (WEDM) and WEDG (Uhlmann et al., 2005; Han et al., 2006; Chiang, 2007). A large discharging energy usually causes violent sparks and results a deeper erosion crater on the surface. Accompanying the cooling process after the spilling of molten metal, residues remain at the periphery of the crater to form a rough surface (Liao et al., 2004). Theoretical model for R_y or R_{max} was developed for transistorized pulse generator type EDM (Chen and Mahdivian, 2000). It showed that with the increase in current and pulse duration, R_y increased. Surface roughness of ceramic due to EDM was also investigated. EDM of boron carbide (B_4C) showed that R_a and R_y increased with pulse duration but decreased with current (Puertas and Luis, 2004). On the other hand, EDM of high speed steel showed that with the increase of current, R_a increased and with the increase of pulse duration it decreased (Lin et al., 2006). It was suggested to use low conductive or higher resistive dielectric to ensure low R_a and R_y (Liao et al., 2004). Use of rotating electrode was helpful for high shape accuracy and higher MRR but it caused higher surface roughness (Her and Weng, 2001).

Formation of crack is a challenge in EDM. The presence of high crack density is not desirable because it leads early product failure. Investigation of EDM parameters showed that higher pulse-on duration increased both the average white layer thickness and the induced stress (Lee and Tai, 2003; Ekmekci et al., 2005). These two conditions tend to promote crack formation. When the pulse current was increased, the increase in material removal rate caused a high deviation of thickness of the white layer. Thus with the increase in pulse current, crack formation decreased. It was also verified that lower thermal conductivity of workmaterial caused higher crack formation. There was no analysis reported on the effect of feed rate on machined surface.

1.1.5 Tool wear

Tool wear is responsible for shape inaccuracy and dimensional instability. In micro ED milling at higher feed rate tool wear was found decreasing (Lim et al., 2003). It was also observed that tool wear was increasing with feed rate but after reaching the peak it started decreasing (Kim et al., 2005). High discharge energy usually caused high TWR (Uhlmann et al., 2005). In EDM of high speed steel and boron carbide (B_4C), tool wear was found to be proportional to current and inversely proportional to pulse duration (Puertas and Luis, 2004; Lin et al., 2006). It was also found that tool wear showed an optimum peak value while pulse duration had increased (Ozgedik and Cogun, 2006). However, use of shorter pulse on-time was suggested to achieve lower TWR (Son et al., 2007).

The negative polarity of the tool gives a lower tool wear than that of positive polarity in the range of low to medium discharge current values. At high current settings, the polarity has no significant effect on the tool wear. A slight decrease in TWR was observed with increasing current in negative polarity. In the case of positive polarity, TWR decreased significantly with increasing current (Lee and Li, 2001).

The type of dielectric fluid is also responsible for tool wear rate. Kerosene is frequently use as the dielectric medium. In EDM, carbide layer formed on the workpiece surface with the use of kerosene, which has a higher melting temperature than the oxide layer formed with the use of distilled water. The carbide layer formed by kerosene needed higher pulse energy for melting and evaporation which caused high tool wear (Chen et al., 1999). The use of low resistive deionized water as a dielectric fluid reduced the tool wear as compared to kerosene (Chung et al., 2007).

1.1.6 Material removal rate

Experimental investigation in conventional EDM found that MRR increased with the discharge current. With the increase in pulse duration MRR showed a peak and then downfall (Chen and Mahdivian, 2000; Lin et al., 2006;). Increase in discharge energy by discharge voltage or capacitance, resulted in higher MRR (Kim et al., 2005). It was also observed that material removal depth was inversely proportional to feed rate (Lim et al., 2003). MRR can be increased in both positive and negative polarity by rotating the tool-electrode (Her and Weng, 2001). Higher MRR was obtained for low ultrasonic vibration frequency combined with tool rotation (Ghoreishi and Atkinson, 2002).

Dielectric flushing system is also responsible for the variation in MRR. The low material removal rate in static condition is mainly due to improper flushing of the molten and evaporated workpiece material from the machining gap (Ozgedik and Cogun, 2006). With the increase in flushing flow rate, MRR increased. Use of kerosene and tap water as the dielectric fluid showed higher MRR than distill water (Chen et al., 1999).

1.1.7 Tool electrode

EDM tool electrodes should be conductive. A list of different tool electrode materials is shown in Table 1.2. In micro range, the replacement of tool during machining is unacceptable because it leads to positional inaccuracy. For this, tool with low wear ratio are needed in micro EDM. Because of its low wear ratio, tungsten or tungsten carbide rods or tubes of diameters within the range 100-400 μm are preferred as tool electrodes for micro ED drilling and milling (Pham et al., 2004). The handling of tungsten and tungsten- carbide rods is difficult as they can be easily damaged. Therefore, sub-systems are incorporated into micro EDM machines for on-the-machine manufacture and holding of the required micro-electrodes. The most common sub-systems are ceramic guides and dressing units such as WEDG.

Electrode material	Volume wear ratio (tool/workpiece)	Machinability	Cost	Best application
Brass	1 : 1	Easy	Low	Holes, slits
Copper	1 : 2	Easy	Moderate	Holes
Copper-Tungsten	1 : 8	Moderate	High	Large areas
Zinc	1 : 2	Easy	Low	Die cavities on steel
Steel	1 : 4	Easy	Low	Through holes into non-Fe materials
Tungsten	1 : 10	Poor	High	Small slots, holes
Tungsten carbide	1 : 10	Poor	High	Small slots, holes

Table 1.2. Features of different tool electrode materials

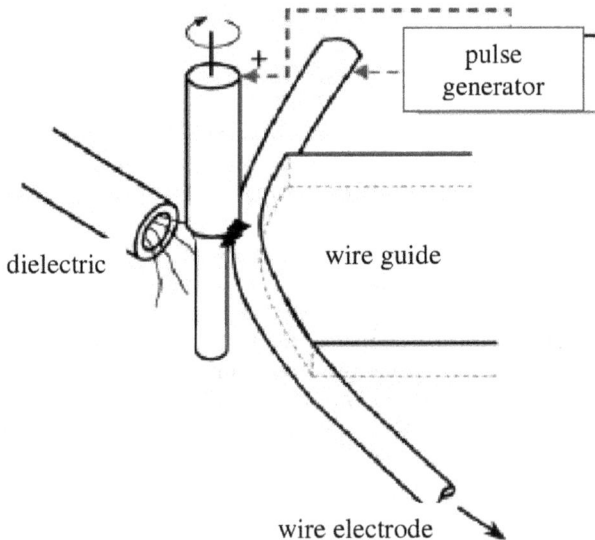

Fig. 1.3. Principle of WEDG with one wire guide (Fleischer et al., 2004)

The tool electrode can be dressed or prepared either by etching, grinding or lithography process. However, when the fabricated tool is chucked on the machine, an alignment error between tool center and spindle center and run-out error of tool which has an influence on the variation of the discharge gap occurs and cause serious problems in micro machining and assembly of micro parts. For this, on-machine tool dressing is always recommended. WEDG is an on-machine tool dressing method. The principle of WEDG was developed in the mid 1980s. The WEDG process is using a traveling wire between one or two guides. The wire is the tool electrode of 50 and 200 μm diameter. In Figure 1.3 the principle of the WEDG-unit with one wire guide is shown, where rotating workpiece is fed downwards and the discharge area is limited to the front edge of the wire. The straightness of the ground part depends only on the variation of the gap distance at the machining point. Minimum 5.5 μm tool was fabricated by WEDG from experiment (Kim et al., 2005). But in WEDG it is difficult to control the shape, dimension and roughness of the ground electrode (Lim et al., 2003).

1.1.8 Workpiece materials
In general, all conductive materials can be machined by the process that follows EDM principle. Different work materials for EDM are as follows:
a)   Metals: copper, iron, alloys etc.
b)   Semiconductor materials: silicon (Madou, 2002)
c)   Conductive ceramics: Alumina, Zirconia, Silicon Nitride, Boron Carbide (Puertas and Luis. 2004; Schoth et al., 2005)
d)   Conductive polymers (King et al., 1999) Metal matrix composites (Xingchao, Z et al., 2007)

2. Experimental design

The experiment was designed based on 3-level full factorial statistical model, where the three micro ED milling process parameters were feed rate, capacitance and voltage. The experiment was designed to analyze the individual influences, interaction effects and quadratic effects of these parameters on R_a, R_y, TWR and MRR. Table 2.1 shows the experimental conditions. Total number of 32 experiments had been conducted to complete the analysis. Six experiments at the same combination were conducted to reduce the curvature effect. The response parameters were statistically analyzed by applying factorial ANOVA at probability value of 0.05. This experimental study has been briefly published before (Mehfuz and Ali, 2009)
The definition of response parameters are as follows:
R_a (μm)= *The arithmetic average deviation from the mean line*
R_y (μm)= *Maximum peak-to-valley roughness height*

$$TWR = \frac{Volume\ of\ tool\ wear}{Volume\ of\ workpiece\ wear}$$

$$MRR\ (mg\ /\ min) = \frac{Weight\ of\ material\ removed}{Machining\ time}$$

Controlled Parameters		Experimental conditions		
		Level		
	Factors	1	2	3
Feed rate (μm/s)	A	2.00	4.00	6.00
Capacitance (nF)	B	0.10	1.00	10.00
Voltage (volts)	C	80	100	120

Fixed Parameters	Experimental conditions
Tool electrode	Tungsten
workpiece	Be-Cu alloy
Tool electrode dia (mm)	0.50
Spindle speed (rpm)	2000
Polarity	Workpiece +
Dielectric fluid	EDM-3 (synthetic oil)
Machining length (mm)	13.00
Machining depth (mm)	0.20

Table 2.1. Experimental Conditions of micro ED milling

2.1 Tool and workpiece materials
The workpiece material was beryllium-copper (Be-Cu) alloy (Be = 0.4 %, Ni = 1.8 % Cu = 97.8 %), which is industrially known as Protherm. Some features of Be-Cu alloy are as follows:
a)   High thermal conductivity (245.00 W/m^{o}C at 20°C).
b)   Resistance to high temperature.
c)   Excellent corrosion resistance, which enables it to become a suitable mold material.
d)   Good machinability and polishability.
The tool electrode material was tungsten (W). It has the following features:
a)   Low tool wear ratio,
b)   Suitable for small holes or slot,
c)   High cost,
d)   Machinable by WEDG.

2.2 Experimental procedures and measurements
In this subsection the equipments used to conduct the experiments and the experimental procedures are described. A commercial multi-purpose miniature machine tool (DT-110 Mikrotools, Singapore), shown in Figure 3.1, was used to conduct the micro ED milling experiments. The machine contained RC circuit to generate small but high frequency

discharge energy. EDM-3 synthetic oil was used as the dielectric fluid. The following subsections discuss the different steps of the experiments.

2.2.1 Sample preparation

Before the experiments, the workpieces were prepared by grinding and polishing. Workpieces were ground by using (NAGASEI SGW 52, Japan) surface grinding machine. Next, the workpiece was grounded manually by using 240, 320, 400 and 1000 grade of sand papers respectively. Then the workpieces were polished by 3 μm diamond suspensions and lubricant. The polishing steps were needed to obtain mirror surface finish. After the final polishing, the workpiece were immersed in acetone and cleaned for 15 minutes by ultrasonic cleaning machine, (BRANSON 2510). Tungsten tool electrodes of 500 μm diameter were also cleaned by the same ultrasonic cleaning machine to clean. Ultrasonic cleaning is useful for the removal of micro sized dirt particle. Then both the workpieces and tool-electrodes were weighed on an electric balance (B204-S Mettler Toledo, Switzerland) with a resolution of 0.1 mg. The measured weights before and after machining were compared. The following shows the specifications of a sample workpiece and tool-electrode after the final polishing. The prepared samples were then used for micro ED milling as discussed in following subsection.

Specification of the workpiece:

Thickness = 5 mm
Length = 50 mm
Width = 15 mm

Specification of the tool-electrode:

Diameter = 500 μm
Length = 50 mm

2.2.2 Experiments by micro electro discharge milling

At the beginning of the micro ED milling set-up, the tool electrode was clamped in the spindle, while the workpiece was clamped on the worktable. After clamping, the spindle rotation was fixed at a speed of 2000 rpm, the worktable was fed and dielectric fluids were flown to the sparking zone. When discharge occurred between the tool-electrode and the workpiece, a small portion of workpiece melted and evaporated due to high local temperature, creating a crater on the work surface. The workpiece cooled rapidly by the influence of dielectric fluid. Figure 2.1a schematically depicts the experimental setup. The experiment was done to fabricate a microchannel for the dimensions of 13 mm length, 0.5 mm width and 0.2 mm depth. Figure 2.1b shows the schematic of the microchannels after micro ED milling. Computer numerical coding was used to conduct the experiment. The machining time was obtained from the auto stopwatch of the integrated computer. After the experiment, both the tool-electrode and workpiece were cleaned again in ultrasonic cleaning bath and dried.

After this, the tool-electrode and workpiece were weighed to find out the weight loss due to machining. The differences of weight before and after machining were used to calculate the *TWR* and *MRR*. After weighing, the samples were put into a dry box to avoid the moisture contact. These samples were used for SEM investigation and measurement of R_a and R_y as discussed in the following subsections.

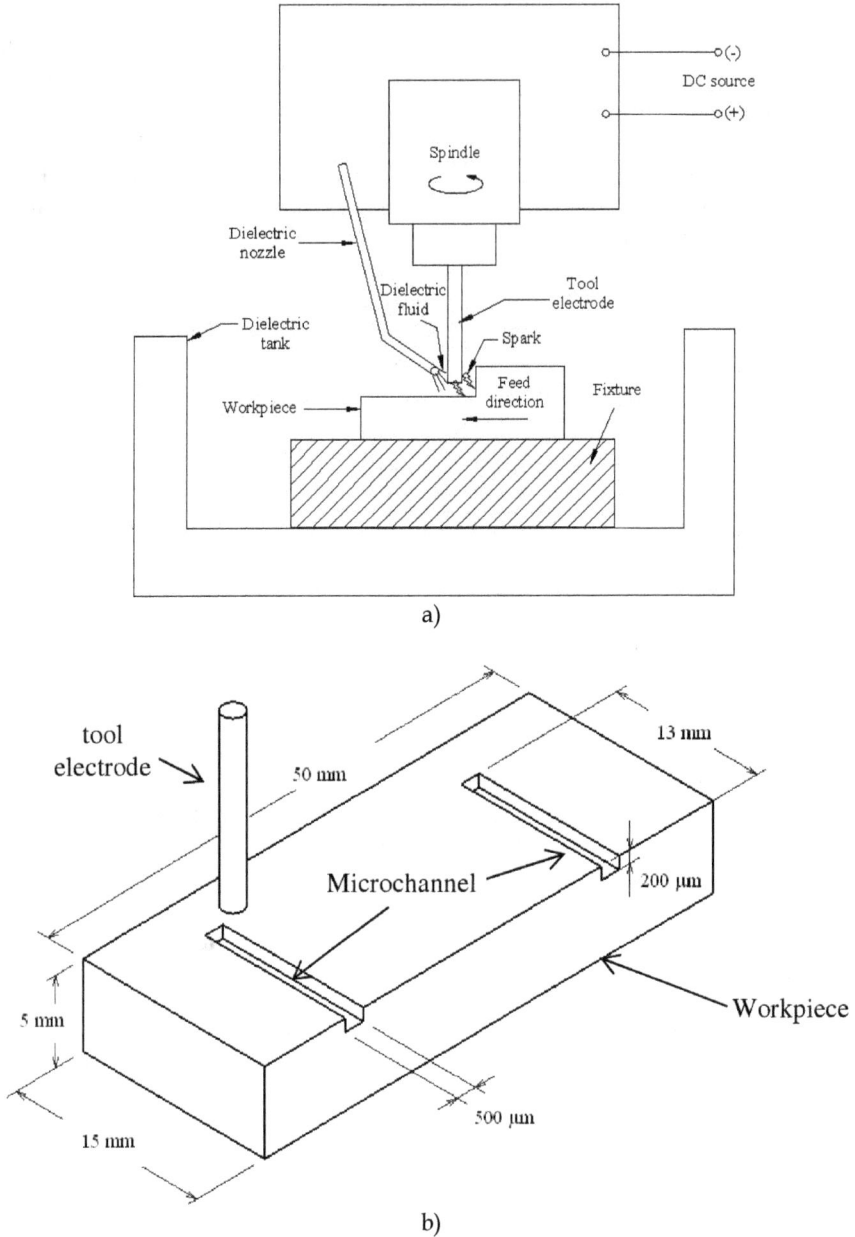

a)

b)

Fig. 2.1. Schematic of experiment: (a) micro ED milling process, (b) specification of the microchannel

(a)   (b)

Fig. 2.2. SEM micrographs of a microchannel, a) top view, b) isometric view

2.2.2.1 SEM Inspection

The work sample was cleaned by ultrasonic cleaning and sputter coated with carbon by Coater (SC 7640). Though the work material was conductive, it was coated to get high quality images. The machined surfaces were then inspected by scanning electron microscope (SEM) (JEOL JSM-5600). Figure 2.2 shows SEM micrographs of a microchannel produced by micro ED milling.

2.2.2.2 Measurement of surface roughness

Both the R_a and R_y were measured by using a precision surface profiler (Mitutoyo, Surftest SV-500, Japan). This profiler provided surface roughness along a line with a resolution of 10 nm. As the length of the microchannel was 13 mm long, the roughness was measured in three segments. Each segment was 3.5 mm long. Average of these three measurements was taken as the surface roughness value.

3. Results and discussions

The experimental results of R_a, R_y, TWR and MRR, are analyzed using variance approach (ANOVA) to check the adequacy of the developed statistical model. The analysis ultimately showed the main and interaction effects of the process variables on the responses. Main effect was the direct effect of an independent variable while interaction effect was the joint effect of two independent variables on the response.

3.1 Calculation of analysis of variance

ANOVA is a calculation procedure to allocate the amount of variation in a process and determine if it is significant or is caused by random noise. ANOVA tables are constructed for R_a, R_y, TWR and MRR respectively. The statistical significance of each effect by means of the comparison of the mean squares (MS) with an estimation of the experimental error was performed. In the table, SS represents sum of squares while DF represents the number of degrees of freedom. The column corresponding to MS is obtained simply by dividing SS by its corresponding DF. In contrast, column of F value is calculated as the quotient of each of the MS of the effects divided by the value of the MS corresponding to the residual. Column of

Exp No.	Feed rate (µm/s)	Capacitance (nF)	Voltage (volt)	R_a (µm)	R_y (µm)	TWR	MRR (mg/min)
1	2.00	0.10	80.00	0.04	0.31	0.121	0.02
2	4.00	0.10	80.00	0.04	0.35	0.044	0.07
3	6.00	0.10	80.00	0.04	0.36	0.133	0.09
4	2.00	1.00	80.00	0.10	0.78	0.154	0.04
5	4.00	1.00	80.00	0.12	0.91	0.066	0.09
6	6.00	1.00	80.00	0.10	0.89	0.165	0.09
7	2.00	10.00	80.00	0.44	3.23	0.182	0.06
8	4.00	10.00	80.00	0.44	3.21	0.089	0.10
9	6.00	10.00	80.00	0.48	3.53	0.220	0.11
10	2.00	0.10	100.00	0.05	0.49	0.165	0.03
11	4.00	0.10	100.00	0.05	0.40	0.049	0.09
12	6.00	0.10	100.00	0.06	0.54	0.157	0.09
13	2.00	1.00	100.00	0.19	1.28	0.194	0.06
14	4.00	1.00	100.00	0.17	1.20	0.065	0.16
15	6.00	1.00	100.00	0.18	1.38	0.186	0.14
16	2.00	10.00	100.00	0.53	3.99	0.239	0.10
17	4.00	10.00	100.00	0.54	4.08	0.098	0.35
18	6.00	10.00	100.00	0.53	4.05	0.216	0.13
19	2.00	0.10	120.00	0.05	0.44	0.198	0.06
20	4.00	0.10	120.00	0.07	0.52	0.065	0.15
21	6.00	0.10	120.00	0.08	1.23	0.166	0.10
22	2.00	1.00	120.00	0.19	1.81	0.227	0.07
23	4.00	1.00	120.00	0.17	1.22	0.086	0.36
24	6.00	1.00	120.00	0.18	1.37	0.198	0.14
25	2.00	10.00	120.00	0.56	3.11	0.261	0.16
26	4.00	10.00	120.00	0.62	3.79	0.128	0.41
27	6.00	10.00	120.00	0.54	3.77	0.247	0.15
28	4.00	1.00	100.00	0.22	1.30	0.058	0.10
29	4.00	1.00	100.00	0.22	1.53	0.072	0.49
30	4.00	1.00	100.00	0.15	1.06	0.067	0.24
31	4.00	1.00	100.00	0.19	1.47	0.062	0.13
32	4.00	1.00	100.00	0.16	1.28	0.075	0.20

Table 3.1. Design matrix of the experiment and measured responses

Probability values gives the probability values associated with values that take the variable of a function of distribution F.

$$SS_{total} = SS_{error} + SS_{treatments}$$

$$SS_{total} = SS_{error} + SS_{treatments}$$

$$SS_{total} = SS_{error} + SS_{treatments}$$

$$Model\ F-value = \frac{MS\ model}{MS\ residual}$$

3.1.1 Surface roughness

Table 3.2 shows the analysis of variance for R_a. Using this analysis a second order quadratic model was developed, which is shown below in Equation (3.1). The model was developed for 95% level of confidence. The model F-value of 213.06 implies the model is significant. There is almost no influence, 0.01%, of noise on the model developed. By checking F value and P-value it is clearly seen that Factor B (capacitance), Factor C (voltage) and Factor B2 are most influential on R_a. The P-value of each of these factors indicates the confidence level is more than 99.00%, which shows their very strong influence. The P - value of interaction effects of BC shows the confidence level is above 95.0% and thus shows very good influence on R_a. The P-value of Factor A (feed rate) and interaction Factor AC has insignificant influence over R_a as it provides very high P values. The lack of fit F-value of 0.53 implies that the lack of fit is not significant compare to the pure error. The high P-value of lack of fit, 85.35%, indicates the model is fit, while the very low P-value of the model, 0.01%, indicates that the model is significant. The specific power transformation was chosen within the confidence level, which was suggested by the Design Expert software toolbox using Box-Cox plotting. In this case, natural log power transformation was suggested. Thus the developed statistical quadratic equation for R_a is:

$$Ln(R_a) = -7.956 - 0.044f + 1.423C + 0.087V - 0.111C^2 - 0.0004\,V^2 + 0.0006fV - 0.0007CV \quad (3.1)$$

Source	SS	DF	MS	F Value	Prob > F	For 95% level of confidence
Model	24.88	7	3.55	213.06	< 0.0001	significant
A	0.02	1	0.02	1.18	0.2885	
B	23.82	1	23.82	1427.90	< 0.0001	
C	0.59	1	0.59	35.44	< 0.0001	
B2	4.69	1	4.69	281.06	< 0.0001	
C2	0.17	1	0.17	10.47	0.0035	
AC	0.01	1	0.01	0.43	0.5194	
BC	0.07	1	0.07	4.42	0.0461	
Residual	0.40	24	0.02			
Lack of Fit	0.27	19	0.01	0.53	0.8535	not significant
Pure Error	0.13	5	0.03			
Cor Total	25.28	31				

Table 3.2. Analysis of variance for main and interaction effects on R_a

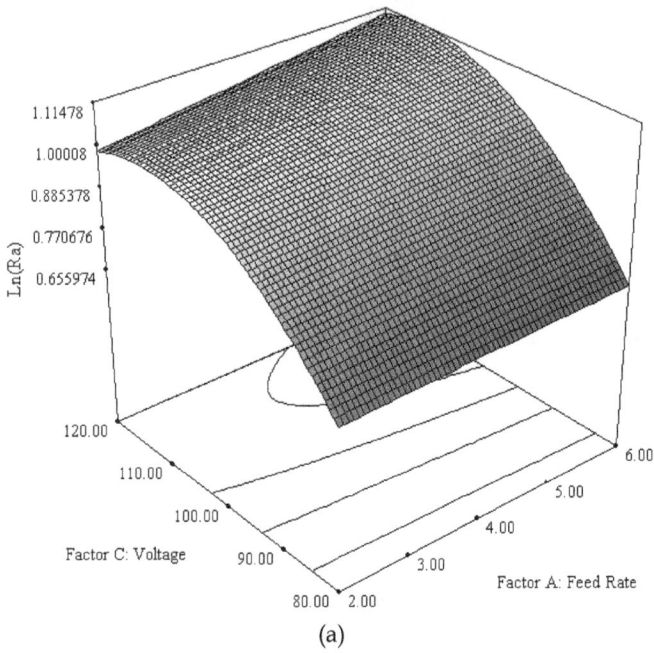

(a)

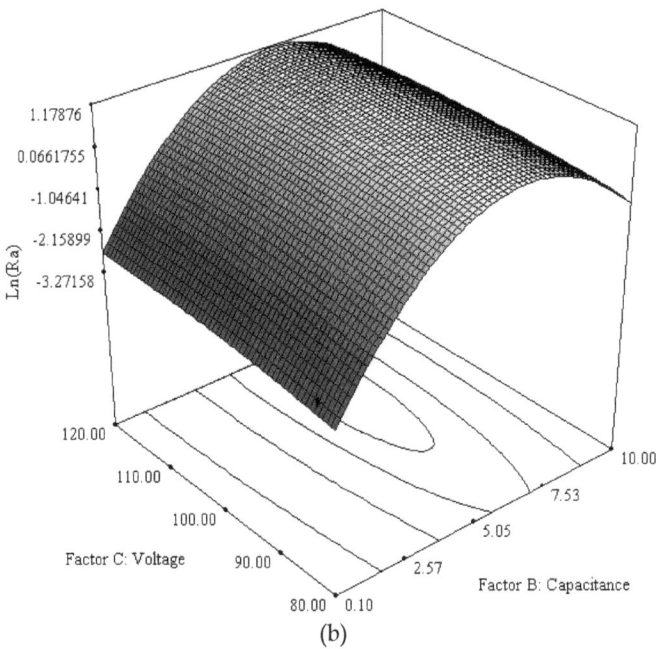

(b)

Fig. 3.1. Estimated response surface of R_a (µm): (a) $Ln(R_a)$ vs. f and V when C = 1.0 nF, (b) $Ln(R_a)$ vs. C and V, when f = 4.0 µm/s

Figure 3.1 shows the effect of feed rate-voltage and capacitance-voltage on R_a respectively. Capacitance and voltage are strongly influential on R_a. R_a increases along with the increase in voltage, which is same as the conventional EDM. With the increase of capacitance from 0.1 to 5 nF the R_a value increases but for further increase of capacitance the R_a values reduce. As the capacitance increased, large energy dissipated which erodes more materials with stronger spark. This strong spark erodes materials with high amount of debris from both the tool electrode and workpiece creating uneven crater. As these debris are trapped in between the plasma channel, it causes unwanted spark. Thus, a high amount of discharge energy is employed to spark with debris, while work material is effectively removed by a small portion of discharge energy. Thus lower R_a is obtained.

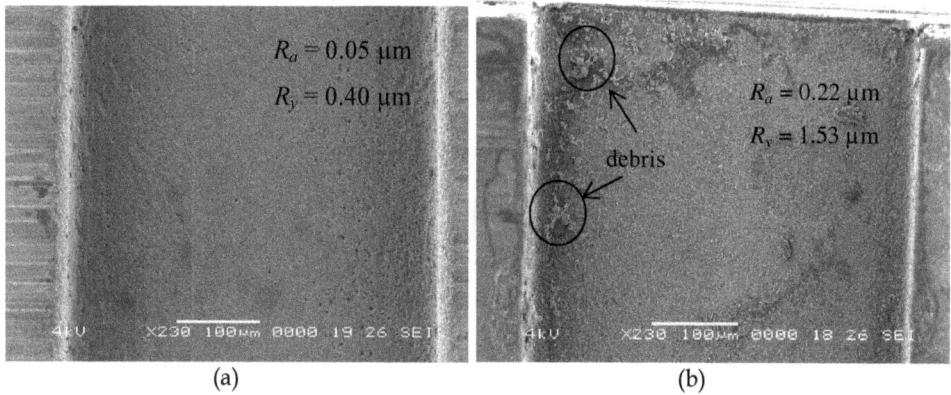

(a)   (b)

Fig. 3.2. SEM micrograph of the surface texture of micro ED milled surface when feed rate = 4 μm/s and voltage = 100 volts, with varying capacitance of (a) 0.1 nF and (b) 1.0 nF.

Figure 3.2 shows SEM surface texture of two machined surfaces with varying capacitance. In the Figure 4.2a the surface texture was found very smooth (R_a = 0.05 μm) with 0.1 nF capacitance. Figure 4.2b shows the surface texture (R_a = 0.22 μm) with the capacitance of 1 nF. Here, the surface was found rougher with increased amount of unwashed debris. Similarly the developed statistical quadratic equation for maximum peak-to-valley height roughness, R_y is:

$$Ln(R_y) = -4.981 - 0.077f + 1.264C + 0.07V - 0.091C^2 - 0.0003\,V^2 + 0.0012fV - 0.0014CV \quad (3.2)$$

3.1.2 Tool Wear Ratio

A second order quadratic model was developed which is the mathematical expression of TWR:

$$TWR = 0.285 - 0.203f + 0.031C + 0.002V + 0.029f^2 - 0.003C^2 +$$
$$0.0002fC - 0.0003fV \quad (3.3)$$

Figure 3.4 shows the effect of feed rate-capacitance and feed rate-voltage on TWR. Capacitance and voltage are found as the strong influencer of TWR, like R_a, along with the

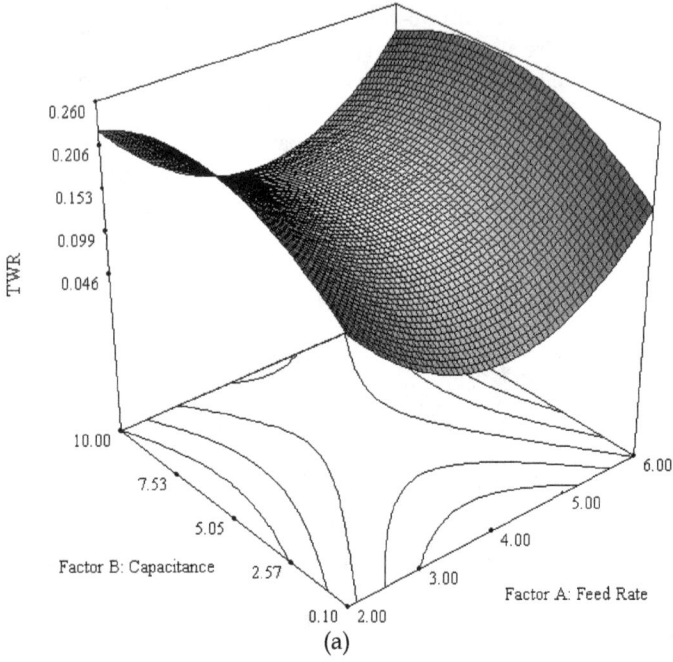

(a)

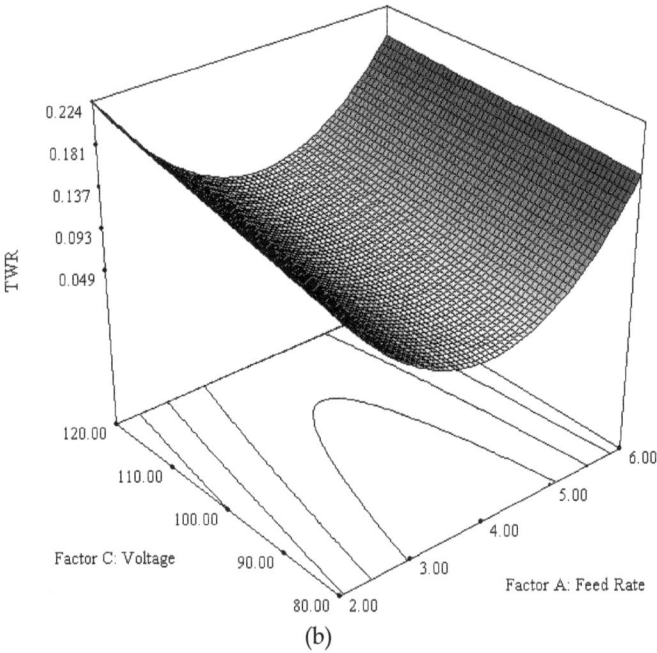

(b)

Fig. 3.4. Estimated response surface of *TWR*: (a) *TWR* vs. *f* and *C* when *V* = 100 volts, (b) *TWR* vs. *f* and *V*, when C = 1.0 nF

interaction effect of feed rate-voltage. *TWR* decreases with the increase in feed rate. For further increase in feed rate *TWR* starts rising. In the first phase, for a particular spark energy discharge if the feed rate is so small it will have high electron emission rate from the tool electrode, resulting high *TWR*. As the feed rate increase the spark energy is more involved in material erosion, which reduces *TWR* and reaches to minimum. In the second phase, for further increase of feed rate from the optimum, the unflashed eroded materials cause unwanted spark with the tool, which results more tool wear. Thus high *TWR* is obtained. With the increase in capacitance large energy dissipated which produces stronger spark resulting high work material erosion. Higher spark energy produces higher amount of debris. These debris sticking on the workpiece, trapped in and causes unwanted spark. The unwanted sparks erodes materials from the tool electrode, which results high tool wear. Thus, higher capacitance results higher *TWR*. As significant amounts of spark energy are employed in sparking with debris, effectively a lower amount of work material is eroded. *TWR* is a ratio of tool wear and workpiece wear, so decreases in material removal mean increase in *TWR*.

3.1.3 Material Removal Rate

The developed *MRR* statistical quadratic model is:

$$\frac{1}{\sqrt{MRR}} = 16.85 - 3.68f - 1.24C - 0.070V + 0.28f^2 + 0.09C^2 + 0.04fC + 0.01fV \qquad (3.4)$$

Figure 3.5 shows the effect of feed rate-capacitance and feed rate-voltage on $\frac{1}{\sqrt{MRR}}$.

Therefore, the decrease of $\frac{1}{\sqrt{MRR}}$ with the increase of any of the factors means the increase in *MRR*. *MRR* increases with the increase in feed rate. For further increase in feed rate *MRR* starts falling slightly. As the feed rate increase, the spark energy is more involved in material erosion, which increases *MRR* till reaches to the optimum. For further increase of feed rate from the optimum, the unflashed eroded materials cause unwanted spark with the tool, which changes the tool shape making it uneven and rough. The spark discharge from this uneven tool results less *MRR*. With the increase in capacitance large energy dissipated which erodes more work materials with stronger spark. With the material erosion, the unflashed debris sticking on the workpiece is trapped in between the machining zone, which causes unwanted spark with the tool-electrode. Thus a great portion of discharge energy occupies with unwanted sparking, while the remaining erodes the work material. Hence, effectively a lower amount of work material is eroded.

3.2 Multiple response optimizations

This subsection discusses the optimization of four output responses, R_a, R_y, *TWR* and *MRR*. The developed non-linear models Equations (3.1) to (3.4) are used for multiple response optimization. Desirability function approach was used for optimization. The method finds operating conditions, e.g. feed rate, capacitance and voltage that provide the "most desirable" values of the responses, e.g. R_a, R_y, *TWR* and *MRR*. The following subsection describes the desirability function approach.

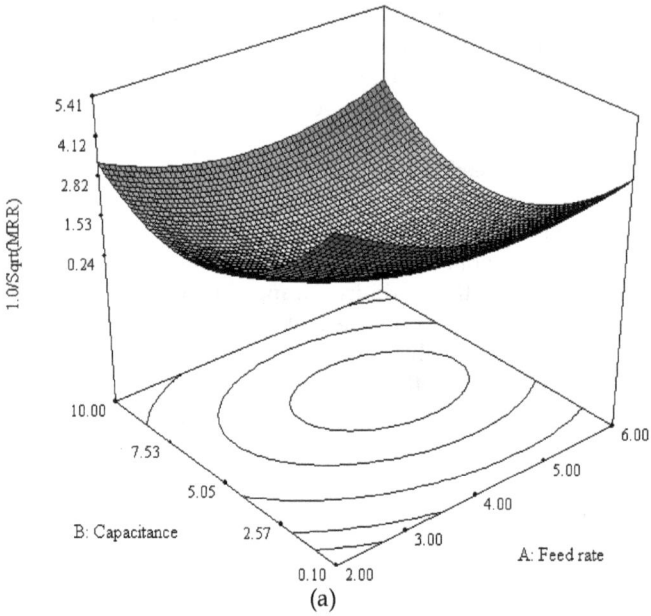

(a)

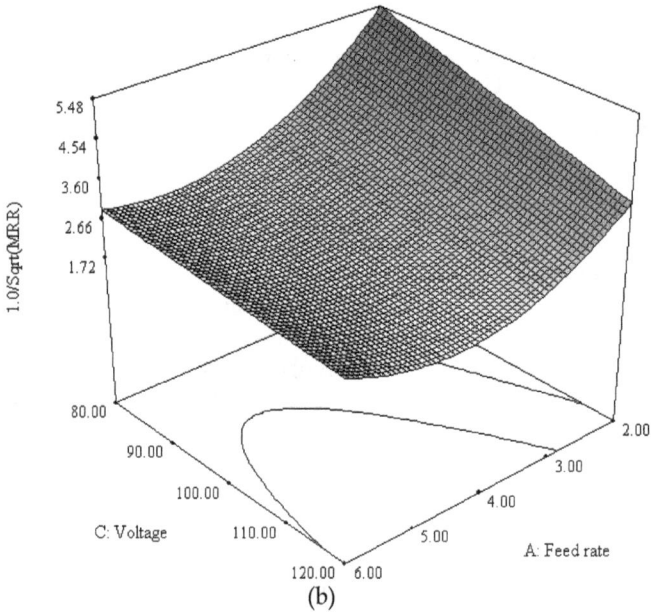

(b)

Fig. 3.5. Estimated response surface of MRR (mg/min): (a) $\dfrac{1}{\sqrt{MRR}}$ vs. f and C when $V = 100$

volts, (b) $\dfrac{1}{\sqrt{MRR}}$ vs. f and V, when $C = 1.0$ nF

3.2.1 Desirability function approach

In the analysis the objective function, $D(Y_i)$, called the desirability function, reflects the desirable ranges for each response $Y_i(x)$ (where, i = R_a, R_y, TWR, MRR). For each response, a desirability function $d_i(Y_i)$ assigns numbers between 0 and 1 to the possible values of Y_i.

$d_i(Y_i)$ = 0 representing a completely undesirable value of Y_i and

$d_i(Y_i)$ = 1 representing a completely desirable or ideal response value.

The individual desirabilities are then combined using the geometric mean, which gives the overall desirability D:

$$D = \sqrt[n]{(d_1 \times d_2 \times ..... \times d_n)} = (d_1 \times d_2 \times ..... \times d_n)^{\frac{1}{n}} \tag{3.5}$$

where n is the number of responses in the measure. From the equation (4.5) it can be noticed that if any response Y_i is completely undesirable $(d_i(Y_i) = 0)$, then the overall desirability is zero. In this case, the geometric mean of overall desirability is as follows:

$$D = (d_{R_a} \times d_{R_y} \times d_{TWR} \times d_{MRR})^{\frac{1}{3}} \tag{3.6}$$

Depending on whether a particular response Y_i is to be maximized, minimized, or assigned a target value, different desirability functions $d_i(Y_i)$ can be used. In this case, R_a, R_y and TWR are needed to be minimized while MRR are needed to maximized. Following are the two desirability functions:

$$d_i(Y_i) = \begin{cases} 0 & if \quad Y_i(x) \le L_i \\ \left(\dfrac{Y_i(x) - L_i}{T_i - L_i}\right)^s & if \quad L_i \le Y_i(x) \le T_i \\ 1.0 & if \quad Y_i(x) > T_i \end{cases} \tag{3.7}$$

$$d_i(Y_i) = \begin{cases} 1.0 & if \quad Y_i(x) < T_i \\ \left(\dfrac{Y_i(x) - U_i}{T_i - U_i}\right)^s & if \quad T_i \le Y_i(x) \le U_i \\ 0 & if \quad Y_i(x) > U_i \end{cases} \tag{3.8}$$

where,
L_i = Lower limit values
U_i = Upper limit values
T_i = Target values
s = weight (define the shape of desirability functions)

Feed rate (µm/s)	Capacitance (nF)	Voltage (volts)	R_a (µm)	R_y (µm)	TWR	MRR (mg/min)	Desirability
4.79	0.10	80.00	0.04	0.34	0.044	0.08	88.06 %

Table 3.3. Values of process parameters for the optimization of R_a, R_y, TWR and MRR

Equation (3.7) is used when the goal is to maximize, while to minimize Equation (3.8) is needed. The value of s = 1 is chosen so that the desirability function increases linearly towards T_i. Table 3.3 shows the process parameters obtained after multiple response optimization. For the shown values of process parameters, it is 88.06% likely to get the R_a 0.04 μm, R_y 0.34 μm, TWR 0.044 and MRR 0.08 mg/min. Any other combination of the process parameters will either statistically less reliable or give poor results of at least one of the responses. The analysis was done by using computer software, Design Expert.

3.3 Verification of optimized values
Experiments were conducted to verify the result obtained from the multiple response optimization. The actual values obtained from the experiments are compared with the predicted values in Table 3.4. From the table it can be noticed that the predicted values of R_a shows no error with the actual, while TWR shows the maximum error. In 88.06% desirability, the percentages of error were found lesser for TWR and MRR. The bar charts of Figure 3.6 shows the comparison of predicted and actual values.

Desirability	Responses	Predicted	Actual	% Error
88.06%	R_a (μm)	0.04	0.04	0.00
	R_y (μm)	0.34	0.36	5.56
	TWR	0.044	0.053	16.98
	MRR (mg/min)	0.08	0.09	11.11

Table 3.4. Verification of multiple response optimization

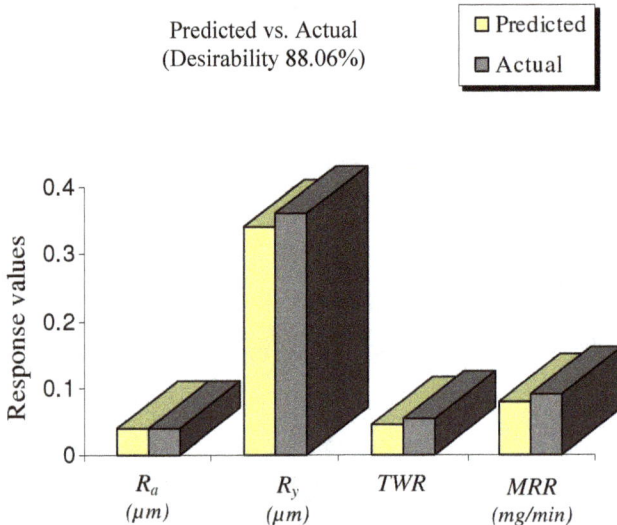

Fig. 3.6. Comparison of predicted vs. actual responses: (a) at desirability of 88.06%

4. Application of micro ed milling: Micro swiss-roll combustor mold

Micro swiss-roll combustor is a heat re-circulating combustor. It uses hydrocarbon fuels to generate high density energy. The advantage of a micro swiss-roll combustor is that it provides high density energy by reducing heat loss [Ahn and Ronney, 2005; Kim et al., 2007]. The generated heat inside the micro swiss-roll combustor is entrapped and re-circulated. Thus, high density energy is obtained. One of the challenges in micro combustor design is to reduce the heat loss. To reduce the heat loss by reducing surface-to-volume ratio, wall thickness should be as small as possible [Ahn et al. 2004]. The application of micro swiss-roll combustor includes portable electronics, such as cell phone, laptop, space vehicles, military uses, telecommunication, etc.

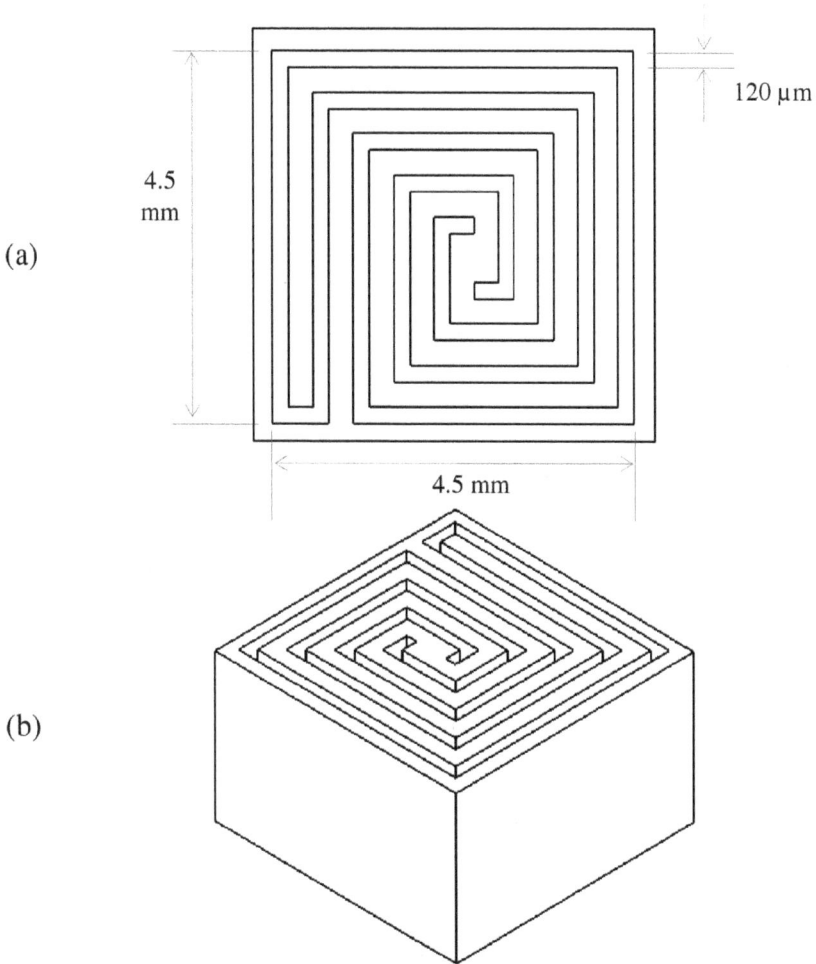

Fig. 4.1. Proposed design of micro swiss-roll combustor mold cavity (a) top view and (b) isometric view

The proposed design of the micro swiss-roll combustor mold cavity is shown in Figure 5.1. Beryllium-copper alloy (Protherm) was selected as the mold material, because of its high thermal conductivity, high heat and corrosion resistance. The microchannel of the mold cavity was fabricated by using a tungsten tool-electrode of 100 μm diameter. The minimum gap between two microchannels was 380 μm. The preliminary drawing and the numerical code (NC) of the design was generated by using CATIA V.5 R14 computer aided drafting software.

4.1 Fabrication of tool electrode by WEDG

Commercially available 300 μm diameter cylindrical tungsten rod was first dressed to 100 μm diameter by WEDG. Later this rod was used as a tool-electrode in micro ED milling to fabricate microchannels. Figure 4.2a illustrates the mechanism of WEDG. Figure 4.2b is the picture taken during the experiment and Figure 4.2c illustrates the SEM image of fabricated tool electrode. Computer numerical coding was used to control the size and shape of the required tool-electrode. The process parameters used are shown in Table 4.1. The parameter values were selected after preliminary studies.

Parameters	Values
Wire speed (mm/s)	20
Wire tension (%)	20
Capacitance (nF)	1
Voltage (volts)	100
Threshold (volts)	30
Polarity	Wire –ve
Spindle speed (rpm)	3000
Machining length (mm)	3
Di-electric medium	EDM-3 synthetic oil

The dimensions of the proposed micro-swiss roll combustor mold are:
(length × width × depth) = (4.5 mm × 4.5 mm × 1.0 mm)

Table 4.1. Experimental condition of WEDG for dressing of tool-electrode

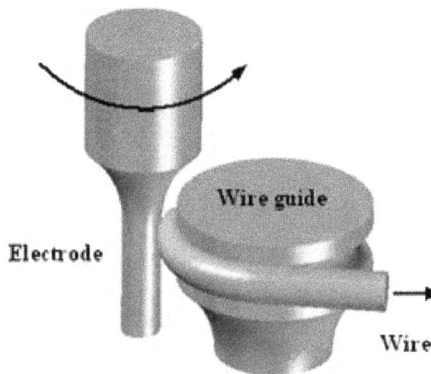

Fig. 4.2. Schematic of tool-electrode dressing by WEDG

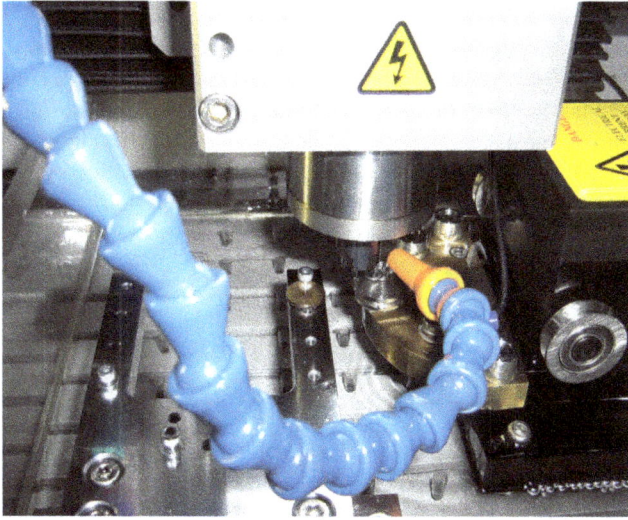

a)

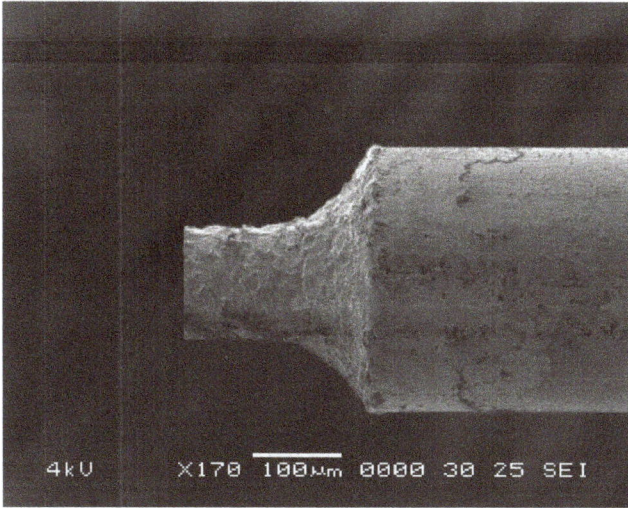

b)

Fig. 4.3. Tool-electrode dressing by WEDG: (a) picture during WEDG and (b) SEM image of tool-electrode after dressing

4.2 Fabrication of micro mold cavity

The micro swiss-roll combustor mold cavity was fabricated by micro ED milling. Be-Cu alloy plate of 6 mm thickness was used as the workmaterial. The tool-electrode of 100 μm diameter was used, which produced microchannels of 120 μm width and 1 mm depth. Channel width comprises of the tool diameter and spark gap. Layer by layer approach was chosen to get better dimensional accuracy. The thickness of each layer was 200 μm. Figure

4.4 explains the layer by layer approach. The gap between two microchannels was 380 µm. After machining each 500 µm, the tool-electrode was dressed by WEDG to reduce the shape inaccuracy due to tool wear. The whole machining was done using computer numerical control. Figure 4.5a is the picture during experiments, Figure 4.5b shows the final product and Figure 4.5c shows the SEM micrographs of the window A in Figure 4.4b. The process parameters obtained from the multiple responses optimization were used in the microfabrication. The experimental condition is shown in Table 4.2.

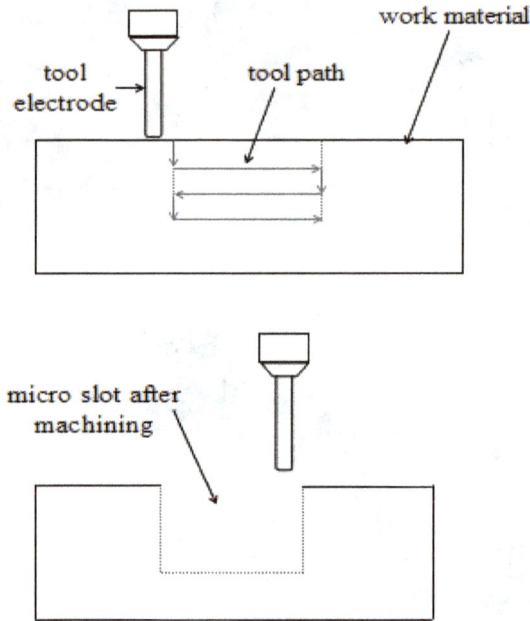

Fig. 4.4. Layer by layer machining: (a) before machining, b) after machining

Parameters	Values
Feed rate (µm/s)	4.79
Capacitance (nF)	0.1
Voltage (volts)	80
Threshold (volts)	30
Tool electrode dia (µm)	0.10
Spindle speed (rpm)	2000
Di-electric medium	EDM-3 synthetic oil
Depth per pass (µm)	200
Machining length per tool dressing (µm)	500

Table 4.2. Micro ED milling parameters for micro swiss-roll combustor mold

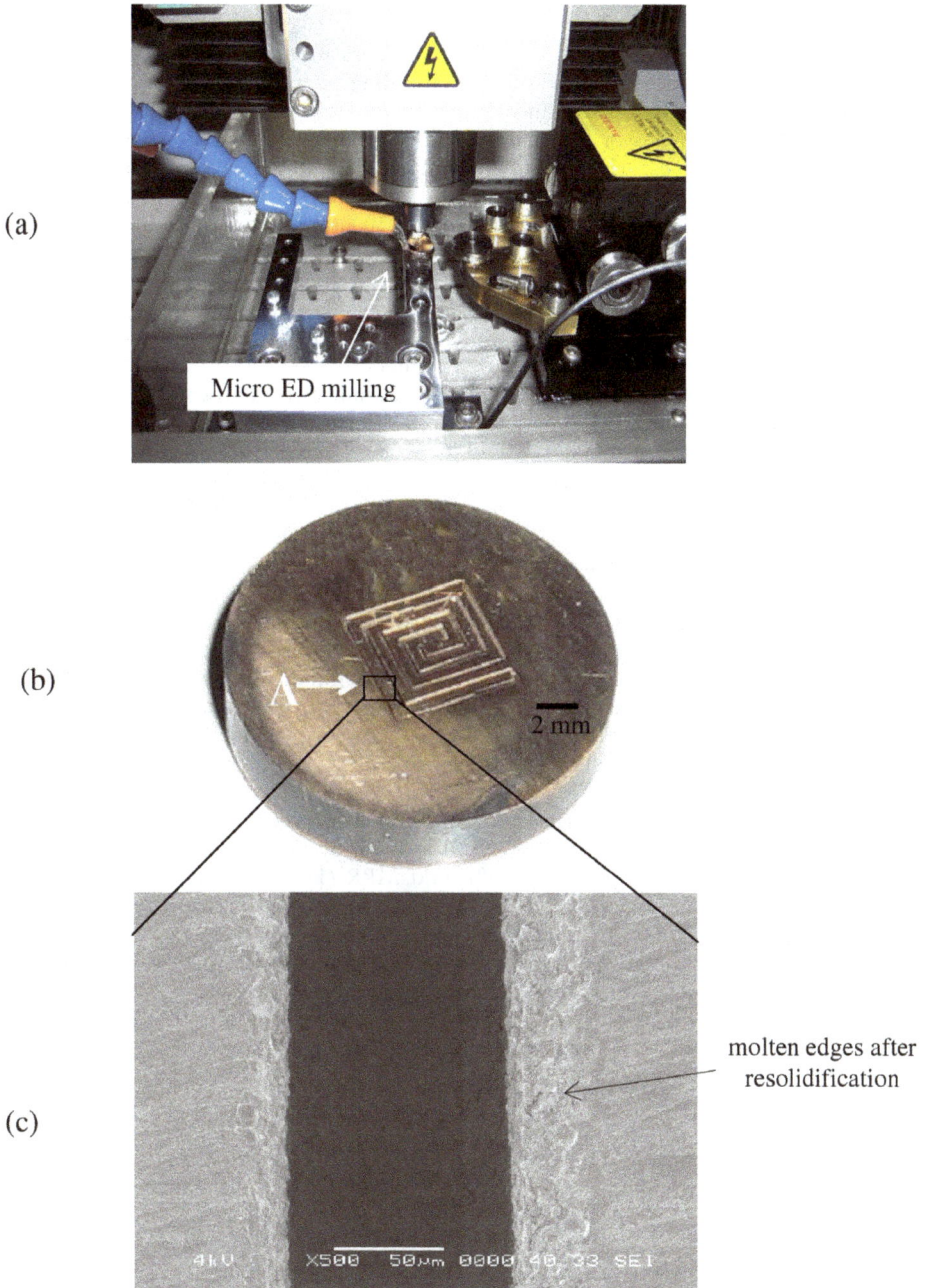

Fig. 4.5. Fabrication of micro swiss-roll combustor mold cavity by micro ED milling: (a) picture during micro ED milling, (b) fabricated micro swiss-roll combustor mold cavity, (c) SEM micrographs of window A in Figure 4.5 b.

5. Conclusion

Micro ED milling is shown as a potential fabrication technique for functional microcomponents. Influences of three micro ED milling parameters, feed rate, capacitance and voltage, were analyzed. Mathematical models were developed for output responses R_a, R_y, TWR and MRR. Analysis of multiple response optimization was done to get the best achievable response values. The micro ED milling process parameters obtained by the multiple response optimization were used in the fabrication of micro mold cavity. WEDG was used to dress the tool-electrode to a diameter of 100 µm. The final product was a micro swiss-roll combustor mold cavity. In brief, this research showed the followings:

1.   Capacitance and voltage have strong individual influence on both the R_a and R_y, while the interaction effect of capacitance and voltage also affects the roughness greatly. Ususally higher discharge energy results higher surface roughness. The unflushed debris sticking on the workpiece causes higher R_a and R_y. At very high discharge energy the entrapped debris inside the plasma channel creates unwanted spark with the tool-electrode. Thus only a small portion of discharge energy involves in material erosion process, which results low R_a and R_y.
2.   Capacitance and voltage plays significant role on TWR along with the interaction effect of feed rate and voltage. At high discharge energy large amount of debris are produced, which causes high TWR by generating unwanted sparks with the tool-electrode.
3.   Feed rate, capacitance and voltage have strong individual and interaction effects on MRR. Usually, MRR is higher at high discharge energy. But the presence of high amount debris in the plasma channel often creates unwanted spark with the tool electrode. Thus only a portion of energy involves in workmaterial removal, which reduces MRR.
4.   Multiple response optimization shows 88.06% desirability for minimum achievable values of R_a, R_y, TWR and maximum achievable MRR, which are 0.04 µm, 0.34 µm, 0.044, 0.08 mg/min respectively when the feed rate, capacitance and voltage are 4.79 µm/s, 0.10 nF and 80.00 volts respectively. The achieved R_a and R_y values are in the acceptable range for many MEMS applications.
5.   The result of multiple response optimization was verified by experiment. The percentages of errors for R_a (0.0%), R_y (5.56%) at 88.06% desirability were found within the acceptable range. For TWR (16.98%) and MRR (11.11%), it was found relatively unsteady. Low resolution (0.1 mg) of electric balance could be a reason behind this.
6.   A micro swiss-roll combustor mold cavity was fabricated by using the WEDG dressed tool. Optimized and verified micro ED milling process parameters were used for fabrication. The final product has the channel dimension of 0.1 mm.
7.   Combination of micro ED milling and molding can be a suitable route for the mass replication of miniaturized functional components at a lower cost.

6. Acknowledgement

This research was jointly funded by grant FRGS 0207-44 from Ministry of Higher Education, Malaysia and EDW B11-085-0563 from International Islamic University Malaysia.

7. References

Ahn, J., Ronney, P. D., (2005). *Plastic mesocombustor*. 4th Joint U.S. Sections Meeting, Combustion Institute, Philadelphia, PA.
        http://carambola.usc.edu/Research/MicroFIRE/PlasticMesoCombustors.pdf

Ahn, J. Eastwood, C., Sitzki, L., Ronney, P. D., (2004). *Gasphase and catalytic ombustion in heat-recirculating burners*. Proceedings of the Combustion Institute. (30).

Asad, A. B. M. A., Masaki, T, Rahman, M., Lim, H.S., Wong, Y.S., (2007). *Tool-based micro-machining*. Journal of Materials Processing Technology. (192–193) 204–211.

Bao, W. Y., Tansel, I. N., (2000). *Modeling micro-end-milling operations. Part 1: analytical cutting force model*. International Journal of Machine Tool and Manufacture. (40) 2155-2173.

Bao, W. Y., Tansel, I. N., (2000). *Modeling micro end milling operations. Part III: Influence of tool wear*. International Journal of Machine Tool and Manufacture. (40) 2193-2211.

Benavides, G. L., Bieg, L. F., Saavedra, M. P., and Bryce, E. A., (2002). *High aspect ratio meso-scale parts enabled by wire micro-EDM*. Microsystem Technologies (2002) (8) 395–401.

Cao D. M., Jiang, J., Yang, R., and Meng, W. J., (2006). *Fabrication of high-aspect-ratio microscale mold inserts by parallel μEDM*. Microsystem Technologies (2006) (12) 839–845.

Chen, Y., Mahdivian, S.M., (2000). *Analysis of electro-discharge machining process and its comparison with experiments*. Journal of Materials Processing Technology (104) 150-157.

Chen SL, Yan BH, Huang FY (1999) *Influence of kerosene and distilled water as dielectrics on the electric discharge machining characteristics of Ti-6Al-4V*. Journal of Materials Processing Technology (87)107–111.

Chiang, K. (2007). *Modeling and analysis of the effects of machining parameters on the performance characteristics in the EDM process of Al2O3+TiC mixed ceramic*. International Journal of Advanced Manufacturing and Technology. DOI 10.1007/s00170-007-1002-3

Chung, K. D., Kim, B. H. and Chu, C. N., (2007). *Micro electrical discharge milling using deionized water as a dielectric fluid*. Journal of Micromechanics and Microengineering (17) 867–874.

Dario, P., Carrozza, M. C., (1995). *Non-traditional technologies for micro fabrication*. Journal of Micromechanics and Microengineering. (5) 64-71.

DiBitonto, D. D., Eubank, P. T., Patel, M. R., Barrufet, M. A., (1989). *Theoretical models of the electrical discharge machining process 1. a simple cathode erosion model*. J Appl Phys. (66) 4095–4103.

Dryer, F. L., Yetter, R. A., (2006, November 10). *Research: chemical energy conversion and power generation at the microelectromechanical systems (MEMS) scale*. http://www.princeton.edu/~cml/html/research.html.

Ehmann, K. F., DeVor, R. E., and Kapoor, S. G., (2002). *Micro/meso-scale mechanical manufacturing- opportunities and challenges*. Proceedings of JSME/ASME International Conference on Materials and Processing. (1) 6-13.

Ekmekci, B., Elkoca, O., and Erden, A., (2005). *A comparative study on the surface integrity of plastic mold steel due to electric discharge machining*. Metallurgical and Materials Transactions B. (36B) 117-124.

Fleischer, J., Masuzawa, T., Schmidt, J., Knoll, M., (2004). *New applications for micro-EDM*. Journal of Materials Processing Technology. (149) 246–249.

Ghoreishi M, Atkinson J (2002) *A comparative experimental study of machining characteristics in vibratory, rotary and vibro-rotary electrodischarge machining*. Journal of Materials Processing Technology. (120) 374–384.

Han, F., Jiang, J., Yu, D., (2006) *Influence of machining parameters on surface roughness in finish cut of WEDM"*, International Journal of Advanced Manufacturing Technology. DOI 10.1007/s00170-006-0629-9.

Her, M-G. and Weng, F-T., (2001) *Micro-hole machining of copper using the electro-discharge machining process with a tungsten carbide electrode compared with a copper electrode*. International Journal of Advanced Manufacturing and Technology (17) 715–719.

Hsu, T-R., (2002). *MEMS & microsystems- design and manufacture*. McGraw Hill.

Kim, N. I., Yokomori, T., Fujimori, T., Maruta, K., (2007). *Development and scale effects of small swiss-roll combustors*. Proceedings of Combustion Institute, doi :10.1016/ j.proci.2006.08.043.

Kim, Y. T., Park, S. J., Lee, S. J., (2005). *Micro/meso-scale shapes machining by micro EDM process*. International Journal of Precision Engineering and Manufacturing (6:2).

King, J. A., Tucker, B. D., Vogt, D., (1999). *Electrically and thermally conductive nylon 6,6*. Polymer Composites, (20) No. 5.

Lee, H. T., Tai, T. Y., (2003). *Relationship between EDM parameters and surface crack formation*. Journal of Materials Processing Technology (142) 676–683.

Lee, S. H., Li, X. P., (2001). *Study of the effect of machining parameters on the machining characteristics in electrical discharge machining of tungsten carbide*. Journal of Materials Processing Technology (115) 344–358.

Liao, Y. S., Huang, J. T., Chen, Y. H., (2004). *A study to achieve a fine surface finish in wire-EDM*. Journal of Materials Processing Technology. (149) 165–171.

Lim, H. S., Wong, Y. S., Rahman, M., Lee, M. K. E., (2003). *A study on the machining of high-aspect ratio micro-structures using micro-EDM*. Journal of Materials Processing Technology. (140) 318-325.

Lin, Y. C., Cheng, C. H., Su, B. L., Hwang, L. R., (2006). *Machining characteristics and optimization of machining parameters of SKH 57 high-speed steel using electrical-discharge machining based on taguchi method*. Materials and Manufacturing Processes. (21:8) 922 – 929.

Lin, Y., Matson, D. W., Kurath, D. E., Wen, J., Xiang, F., Bennett, W. D., Martin, P. *devices on polymer substrates for bioanalytical application* M., Smith, R. D., (1999). *Microfluidics*. Pacific Northwest National Laboratory.

Madou, M. J., (2002). *Fundamentals of micro fabrication*. The Science of Miniaturization. (2nd ed.) Florida, USA: CRC Press LLC.

Murali, M., and Yeo, H. S., (2004). *Rapid biocompatible micro device fabrication* by Micro Electro Discharge Machining. Biomedical Microdevices. (6:1) 41-45.

Mehfuz, R., Ali, M. Y. (2009). *Investigation of Machining Parameters for the Multiple Response Optimization of Micro Electro Discharge Milling*. The International Journal of Advanced Manufacturing Technology, (43), 264-275.

Orloff, J. (1997), *Handbook of charged particles optic*. CRC Press, New York, USA, 319-360.

Ozgedik, A. and Cogun, C., (2006). *An experimental investigation of tool wear in electric discharge machining*. International Journal of Advanced Manufacturing Technology (27) 488–500.

Pham, D. T., Dimov, S. S., Bigot, S., Ivanov, A., Popov, K., (2004). *Micro EDM- recent developments and research issues*. Journal of Materials Processing Technology. (149) 50-57.

Puertas, I., Luis, C. J., (2004). *A study of optimization of machining parameters for electrical discharge machining of boron carbide*. Materials and Manufacturing Processes, (19:6) 1041 – 1070.

Rajurkar, K. P., Pandit, S. M., (1986). *Formation and ejection of EDM debris*. ASME Trans Journal of Engineering Industry. (108) 22–26.

Ronney, P. D., (2003). *Analysis of non-adiabatic heatrecirculating combustors*. Combustion and Flame. (135) 421-439

Schoth, A., Fo"rster, R., Menz, W., (2005). *Micro wire EDM for high aspect ratio 3D microstructuring of ceramics and metals*. Microsystem Technologies. (11) 250–253.

Snakenborg, D., Klank, H., Kutter, J. P., (2004). *Microstructure fabrication with a CO_2 laser system*. Journal of Micromechanics and Microengineering (14) 182-189.

Son, S. M., Lim, H. S., Kumar, A. S. and Rahman, M., (2007). *Influences of pulsed power condition on the machining properties in micro EDM*. Journal of Materials Processing Technology. (190) 73–76.

Uhlmann, E., Piltz, S., Jerzembeck, S., (2005). *Micro-machining of cylindrical parts by electrical discharge grinding*. Journal of Materials Processing Technology (160) 15–23.

Xingchao, Z., Yong, Z., Hao, T., Yong, L., Xiaohua, C., Guoliang, C., (2007). *Micro-electro-discharge machining of bulk metallic glasses*. Proceedings of HDP'07.

9

Release Optimization of Suspended Membranes in MEMS

Salvador Mendoza-Acevedo[1], Mario Alfredo Reyes-Barranca[1],
Edgar Norman Vázquez-Acosta[1], José Antonio Moreno-Cadenas[1]
and José Luis González-Vidal[2]
[1]CINVESTAV-IPN, Electrical Engineering Department,
[2]UAEH, Computing Academic Area,
México

1. Introduction

Releasing part(s) of micro-fabricated devices using etching techniques is one of the fundamental post-processing steps in micro-machining and it is important to have a comprehensive concept on how it can be done, since the final result will significantly influence the electrical and mechanical performance of devices. As micro--electromechanical systems (MEMS) are three dimensional structures, they are obtained by eliminating materials commonly used in this technology, such as crystalline silicon, polycrystalline silicon, silicon dioxide and silicon nitride.

Micro-machining can be undertaken by etching the bulk of the substrate or sacrificial layers deposited at the surface of the wafer. Bulk etching removes great quantities of material and is usually applied to obtain thin membranes for pressure sensors, etching from the back surface of the substrate. On the other hand, surface micro-machining is based in the elimination of sacrificial layers deposited at the surface of a substrate underneath the layers that should be active. The present study will deal only with bulk etching.

Therefore, as several materials are used to obtain finally the typical sensors and actuators in MEMS devices, etching must be selective depending on the material that should be removed. There are available methods for material etching based either on gaseous or liquid etchants. The former employs complex and expensive equipment, but with good yield, and the latter is a low cost process, but care should be taken since in some cases it uses toxic or even corrosive solutions and facilities are needed to exhaust or protect from the vapours produced during etching.

However, both kinds of etching processes are widely used in MEMS and CMOS technology and this fact can be conveniently used to match both technologies, so complete MEMS devices can be integrated in the same substrate, reducing fabrication costs. In particular, wet etching can be either isotropic if material is removed uniformly in all crystallographic directions or anisotropic if etching is selective to a given crystallographic plane.

With regard to anisotropic etching, two solutions are commonly used like potassium hydroxide (KOH) and trimethyl ammonium hydroxide (TMAH). Both solutions are regularly used in MEMS technology to obtain structures, such as thin membranes and cantilevers. The present study gives emphasis to the properties of TMAH and how it acts in silicon upon different geometries designed to obtain a thin membrane used in semiconductor gas sensors (SGS) with the objective to reduce etching time so damage to materials different to silicon can be minimized.

As the convenience to keep compatibility between MEMS technology and CMOS integrated circuits technologies has been demonstrated, it is important to apply etching techniques without damaging the layers used as masks. Within the criteria that should be considered is the etching time since optimum processes must be applied so well defined tridimensional structures can be perfectly obtained at any time.

Based on crystallographic concepts and the way TMAH acts during silicon etching, the main purpose of the present study is to demonstrate the effect in etching time using a given geometry for suspended membranes. It will be shown that improvement can be achieved using a specific geometry outline, compared with other options. Knowledge of the influence of size and orientation of the geometric elements on the anisotropic etching made on wafers with (100) orientation can help to optimize the designing process, as will be shown. Simulations were made with the help of specialized software and experimentally confirmed.

2. MEMS etching process compatible with CMOS technology

Micro--electromechanical systems (MEMS) are devices designed for specific sensing of actuating functions based on tridimensional structures and mechanisms that can be fabricated with a set of known micro--machining steps being compatible with mature CMOS technologies used in micro--electronics for the fabrication of silicon integrated circuits.

Actually, those systems find wide application in diverse disciplines since their main functions as sensors and actuators at micro- and nano scale allow, for instance, size reduction of measurement instrumentation. Besides, the development followed by these devices has significantly influenced the creation of elements, such as optics for telecommunications, RF devices, analytic instrumentation, biomedics, optic systems for image processing, micro--fluidics, mechanical supports, etc.

It is clear that these tridimensional micro-structures (sensors and actuators) need an electronic circuit to operate properly as they interact with the environment to complete a desired function based on input or output signals. Configurations like analogue-to-digital converters, digital memories, artificial neural networks, temperature controllers, etc., are some of the circuits helping in tasks like signal reading or conditioning in domains like digital, analogue or mixed electronic circuits. Therefore, MEMS are considered systems that consist on several blocks with specific features integrated around to deliver or perform a particular function.

One way this integration can be done is to interface the sensors or actuators with separated micro--modules. Each of them (tridimensional structures and modules) are fabricated in a separate chip then connected and packaged together. The advantage of this alternative is the independence in the fabrication technology of each block, applying particular convenient steps to meet specific purposes, i.e. etching on one side and micro-electronics on the other.

Then, both are packaged and interconnected to configure the complete system. However, a great disadvantage that this alternative presents is that stray capacitance is added affecting the performance of the device. Also, packaging of the modules can be highly complex adding the chance of device failure and yield reduction as a consequence (Korvink & Paul, 2006).

Therefore, a good choice is to monolithically integrate the system in the same substrate, where sensors or actuators are placed next to the electronics in the same technological process where MEMS and circuitry are fabricated, eliminating extra interconnections, reducing the area of the system and increasing the yield. A clever design can give high compatibility among the different blocks used in the system and the packaging process.

Nevertheless, this apparent simplicity is true only for MEMS fabricated using compatible CMOS technologies. However, care should be taken to eliminate or reduce damage to layers and materials used as protection masks. So, the main goal is to process the chip without risk or damage to the masking layers when micro-machining the typical three dimensional structures needed as sensors or actuators.

Due to this limitation, it is important to optimize the geometric design of the structures in order to assure the physical integrity of the different layers used in the fabrication of an integrated circuit, as well as to keep the compatibility of MEMS fabrication with CMOS technologies (Baltes, 2005). One of the main advantages of this alternative is the reduction in production costs since a high number of devices can be fabricated in batch run.

It should be remembered that micro-machining is, in general, a set of techniques and tools used to obtain tridimensional elements and structures with high precision and good repeatability by adding or removing layers in a controlled way.

A basic technique in MEMS is volumetric wet etching, when an etching is made to a substrate thick enough with a solution prepared with certain elements that react with it, eliminating part of the substrate. The amount of material etched away depends on the kind and conditions of the solution used, like temperature, concentration, etching time, stirring and the crystallographic orientation of the substrate.

Materials used as protective masks also play an important role in the micro-machining process so the desired structure can be readily obtained. Hence it is important to keep in mind the type of substrate and layers that will be used in the fabrication of the integrated circuit that will contain MEMS, since this will indicate which solution must be used for micro-machining (Hsu, 2002).

Usually, volumetric wet etching is used with silicon substrates for the fabrication of structures like micro-cavities, thin membranes, through holes, beams and cantilevers, taking advantage of the structural layers included in CMOS integrated circuits' technology.

This kind of wet etching can be classified as isotropic and anisotropic. The first has a uniform etch rate to the substrate in all crystallographic directions of silicon, and the second is selective on the crystallographic direction, that is, the etch rate is higher on those directions whose atom density is not too high.

3. Suspended membranes and applications

The case presented here deals with membranes fabricated with anisotropic wet etching of a silicon substrate. These kinds of structures are also thin layers that can operate as sensors and mechanical support for the circuitry. Within the most common applications for these membranes, there are piezoresistive pressure sensors, micro-hotplates and pyroelectric sensors, among others (Barrettino et al., 2004a, 2004b; Capone et al., 2003; Chen et al., 2008; Gaitan et al., 1993; Tabata, 1995)

In general, these structures are fabricated etching the substrate from the back side of a silicon wafer, where no electronic devices are present. This method allows perfect protection of devices placed at the front of the wafer when the compatibility of the layers used is limited, using a mechanical mask or a protective film.

Likewise, this kind of process simplifies the design of the structures that are to be etched, since this step does not affect the geometric configuration of the circuitry at the other side of the wafer. However, a main disadvantage is the large area used for the definition of the membrane due to the characteristics of the anisotropic etching (Madou, 2001). Typical shapes obtained with anisotropic etching using trimethyl ammonium hydroxide and water (TMAHW) are shown in Fig. 1.

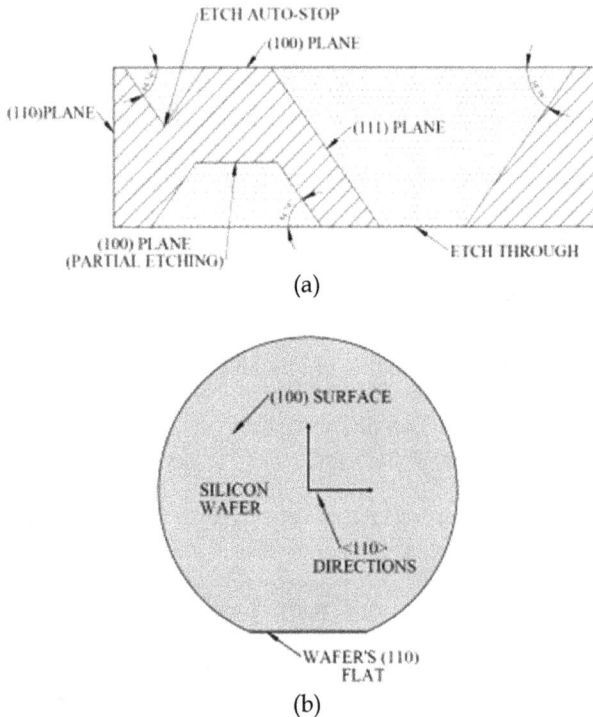

(a)

(b)

Fig. 1. Anisotropic etching of a silicon substrate.

Nevertheless, it is obvious that this technique takes a long time to complete the etching because the wafer is usually too thick and the etch rate of TMAHW is around 1 micron per minute. Also, this back side etching requires simultaneous alignment at the top and back of the wafer (double alignment) in order to perfectly define the needed structure. So this step introduces an extra restriction to achieve the thin membrane.

On the other hand, this membrane can also be obtained with an etching process made at the front surface of the wafer. This is called front bulk etching, a process also frequently used in MEMS. Although it is made at the front surface of the wafer, it is still considered bulk etching since no surface sacrificial layers are removed to obtain the thin membrane, whereas bulk silicon beneath the defined membrane area is etched away. As with the back etching, with the front silicon etching process it is also needed to protect those areas that will not be part of the tridimensional structure.

Since the front part of the wafer is where the CMOS electronic devices are placed, metal layers are present as well, and they should be protected against the etching solution. Therefore, here the solution should be modified with additives so this solution can selectively etch only silicon when a etching CMOS post-process is applied to an integrated circuit chip. Some commercial products are available and used for protection purposes as an alternative. With this front etching, the design topology of the desired structures must be modified to expose only the silicon areas that must be etched away, taking care not to unprotect the remaining surface where the electronic devices are present.

Comparing back and front etching, it is obvious that the latter takes less time to complete the membrane since etching is carried out just a few microns down from the surface of the silicon wafer, not through almost all the bulk of the substrate. Besides, the area needed to obtain the complete structure is less than that needed with back etching, despite the characteristics of the anisotropic etching, since a shallower inverted truncated pyramid is obtained.

The membrane obtained with front etching is a suspended structure mechanically supported by two or more thin arms, with a central area as the active part of the membrane. Here it can contain circuitry or some other kind of devices having specific functions for the system's operation. The definition of the membrane's area is given with "etching windows" through which selected silicon areas are left exposed.

This can be achieved using appropriate layer layout to generate a CIF of GDS file used for the CMOS integrated circuit fabrication at the silicon foundry. Hence, the solution will only etch the exposed bulk silicon and the rest of the surface will remain protected with an overglass layer commonly used to isolate the integrated circuit from environment contamination before packaging. So compatibility is maintained to a high degree between the steps needed for MEMS fabrication and CMOS technology (Tabata, 1998; Tea et al., 1997).

As mentioned before, the etching solution used in this work for the delineation of the membranes was TMAHW, which is typical for anisotropic etching. This solution has different etch rates depending on the crystallographic orientation of the silicon substrate. Generally, rates for planes $\{100\}$, $\{110\}$ and $\{111\}$ are the most used in this kind of task, although other orientations could also be useful for etching purposes as high etch rates can be achieved.

Etch rates using TMAHW depend also on the temperature and concentration of the reactive. In particular, the study presented here was made with a concentration of 10% of TMAH and 90% of deionized water at 80°C, from which an etch rate of approximately 0.72 µm/min was obtained for a (100) plane.

TMAH was used since it is highly selective for silicon etching allowing the use of SiO_2 as the protective mask against etching. This is an important issue because SiO_2 is one of the layers used through the fabrication of CMOS integrated circuits and there is no need to add extra layers that are not used in this technology.

Taking advantage of the photolithographic steps with an appropriate knowledge of the fabrication steps, the etching areas can be easily defined. Therefore, compatibility between etching and the layers used in the fabrication of CMOS integrated circuits is maintained.

It should be mentioned that the designed geometry in the mask will determine the final shape of the anisotropic etching. One of the most important features in this process is the way the etching proceeds in time with regard to the corners of the geometry i.e. concave or convex. It was found that with TMAH, when concave corners are aligned with a $\{110\}$ plane, etching will stop at the moment when faces with $\{111\}$ planes coincide, i.e. in the vertex formed by a $\{100\}$ plane.

On the other hand, convex corners will generate {111} planes as well, but in the vertex of the adjacent planes, etching continues below the corner, etching away other planes and releasing the structure so defined. Fig. 2 shows a mask for a cantilever where the corresponding corners are indicated (Kovacs et al., 1998).

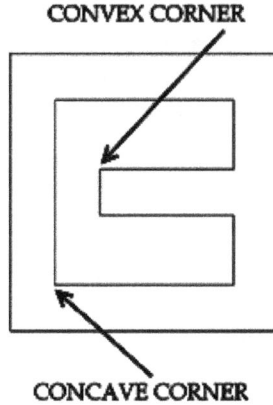

Fig. 2. Geometry of a cantilever illustrating convex and concave corners.

Furthermore, suppose there is a window with an irregular opening, such as the one shown in Fig. 3. The characteristic etching self-alignment with respect to the crystallographic planes due to the anisotropy will be evident.

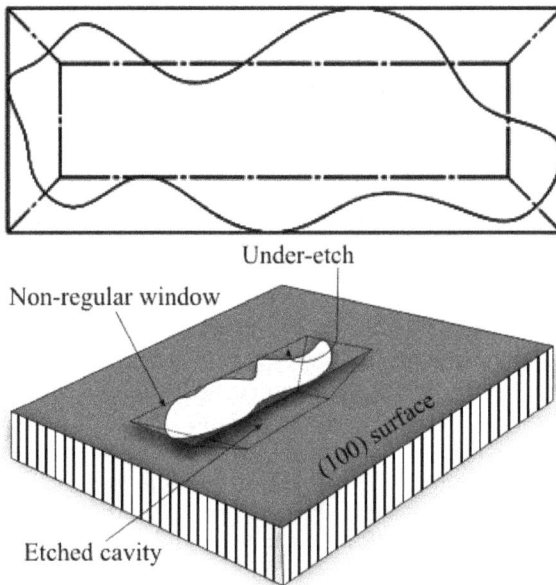

Fig. 3. Etching resulting from an irregular open window.

Dealing with anisotropic etching, there is a feature that is important to consider. When the motifs are aligned with {100} planes, {100} walls will be obtained that are etched as the wafer surface.

4. Geometry and optimization of the suspended membranes

A micro-hotplate was designed to be used in a monolithic CMOS gas sensor which was later fabricated by MOSIS. Then, an anisotropic etching process was performed on the chip using TMAHW, following several formulations that increase the selectivity of the TMAH to avoid damage to the exposed aluminium on the chip caused by the etching solution (Fujitsuka et al., 2004; Sullivan et al, 2000; Yan et al, 2001).
The next figures show the fabricated chip after a TMAHW etching process.

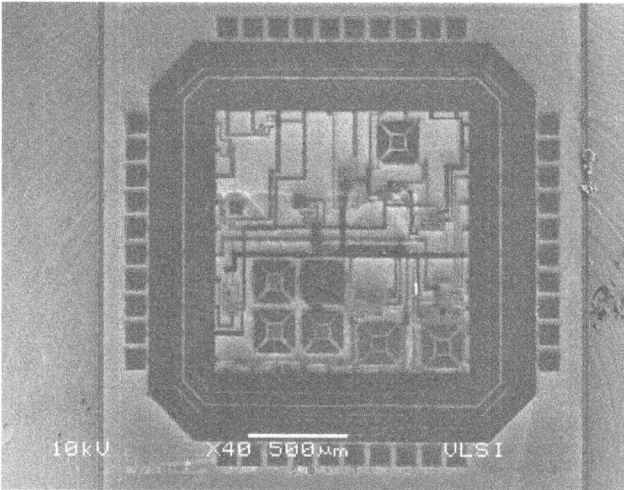

Fig. 4. Fabricated chip after etching.

Fig. 5. Partially etched micro- hotplates.

It was found that the aluminium was sometimes still getting damaged by the solution in an unpredictable way and with a limited repeatability. The damage increased as the etching time was increased, so if the etching time can be reduced by a significant amount, the same applies to the damage of exposed aluminium.

Figure 6 shows photographs from before (left) and after etching, where the exposed aluminium is indicated. The damage can be seen.

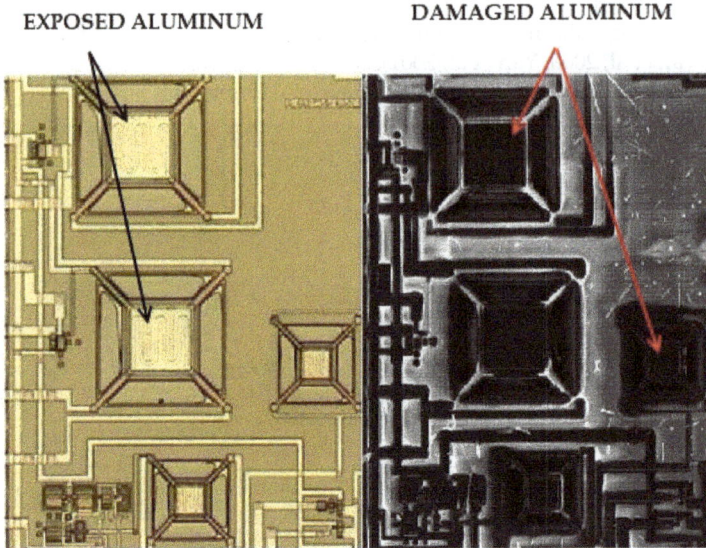

Fig. 6. Comparison between before (left) and after etching.

This motivation is the main objective of this study, which comprises etching and mechanics simulations and the etching of the resulting designs. It should be noted that the designs presented are of micro-hotplates with general applications, as mentioned before.

The most common geometry used for micro-hotplates and suspended membranes are shown in Fig. 7. It can be seen in this figure that the central part of the structure is aligned to {110} planes of the substrate, while the supporting arms have an angle of 45° and 135° with respect to the horizontal reference, therefore aligned to <100> directions (Pierret, 1989). This slope allows other planes to be exposed to the etching solution, hence accelerating the etching process helping to the supporting arms' release. However, this process decelerates when the central part of the membrane is reached, as {111} planes are now exposed at this moment. As already indicated, these planes have the lowest etch rate and in this location, the etching proceeds as with convex corners.

From this moment on, etching takes a longer time until the structure is released. If these effects of the etching solution over the main planes exposed by this geometry are analyzed, alternatives can be found for geometries such that planes with a high etching rate can be readily exposed. For instance, if exposing {111} planes can be avoided or reduced; the consequence will be immediately reflected in a reduction in the etching time.

With this motivation in mind, a study of alternatives for the geometry of the micro-hotplate follows, directed to the reduction of the etching time and the corresponding effects. These two objectives were simulated previous to the experimental process with specialized software for anisotropic etching.

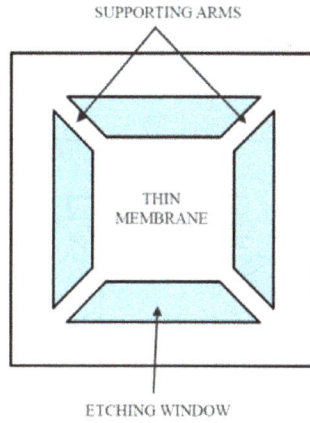

Fig. 7. Common suspended membrane geometry.

4.1 Etching simulations

Features considered in this study for geometry optimization are: a) width of the membrane supporting arms; b) dimensions of the thin membrane; c) orientation of the thin membrane with respect to crystalline planes. Simulations with these considerations were first made with the AnisE software from Intellisuite. The base geometry (A) for the suspended membrane is shown in Fig. 8, having simple dimension ratios among the different elements of the membrane, such as supporting arms, etching windows and membrane area. During simulations, the bulk material considered was silicon and the masking material was exclusively silicon dioxide.

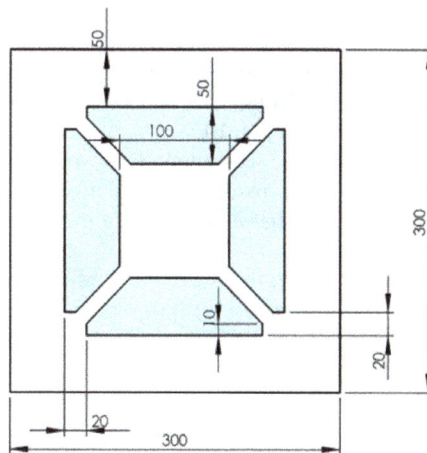

Fig. 8. Dimensions of the base membrane in μm. Geometry A.

First, if the width of supporting arms is increased, it was found that an overlap of the resulting etched areas must exist underneath the arms, proceeding from the exposed silicon windows. This allows for the membrane to be released, otherwise, only four rectangular and separated cavities will be obtained. The required etch overlap is shown in Fig. 9. Due to under etching – always present during the process – this overlap can be a minimum, enough for the supporting arms to be released.

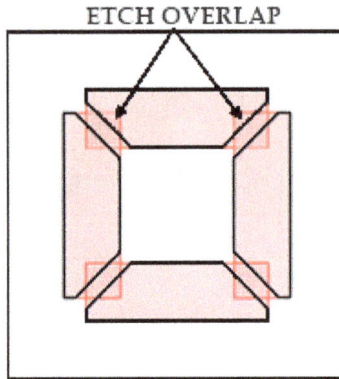

ETCH OVERLAP

Fig. 9. Geometry A. Etching areas (solid lines) and etching overlaps (shadowed).

A 102 min etching time for a complete membrane release was obtained after simulating with the geometry shown in Fig. 8 (Geometry A), with an etch pit depth of about 80μm. It should be noted from this figure that the etch overlaps extend only across the supporting arms, such that when they are released the substrate under the thin membrane presents {111} plane faces to the etching solution, with the same dimensions as the membrane.

Therefore, after release of the supporting arms, the etch rate slows down taking a long time for releasing the thin membrane from the substrate. Then it can be concluded that planes generated at the corners below the supporting arms mainly contribute to the expected etching.

Considering this fact, another geometry (Geometry B) was tested including important overlaps, but that can also avoid features oriented parallel or perpendicular to <110> orientations that can generate {111} planes. It is expected a time reduction in the etching process with this modification, shown in Fig. 10.

As can be seen, the original geometry was rotated 45° with respect to the {110} plane reference, keeping the same area. The result obtained from the simulation of this new geometry was an 18% time reduction, that is, the membrane was completely released in 82 min.

One particularity of the geometry shown in Fig. 10 is the reduction of exposed {111} planes, since with this alternative, edges being parallel or perpendicular to {110} planes are avoided. This reduces both the bulk silicon to be etched away and the etching time.

Next, a new geometry (shown in Fig. 11a) was explored and will be identified as Geometry C. The difference with respect to geometries A and B, respectively, is that although the membrane is also rotated 45°, the supporting arms are aligned along the edges of the membrane. After simulation, a 27% etch time reduction compared to the results from Geometry A was obtained, since the thin membrane was released after 75 min.

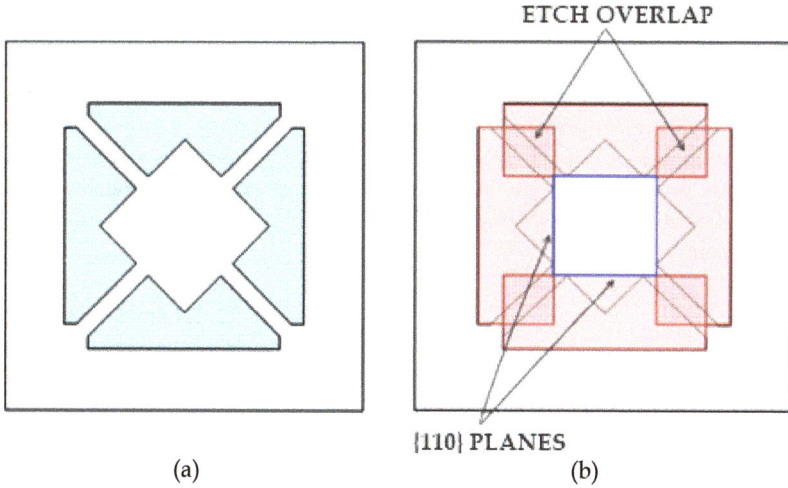

(a)   (b)

Fig. 10. Geometry B. a) Membrane rotated 45° with respect to (110) plane reference; b) Etch overlap.

The reason for the efficiency increase for silicon etching is because with Geometry C there are less {111} planes generated at the perimeter of the thin membrane, allowing the underneath silicon to be etched from the beginning of the process, not after the supporting arms are first released.

According to the simulation, the etched pit is approximately 56μm deep. The difference between the etched depths obtained with geometries A and B can be attributed to the exposure of larger {110} planes, among others, which have a greater etch rate. This is illustrated with the overlaps shown in Fig. 11b.

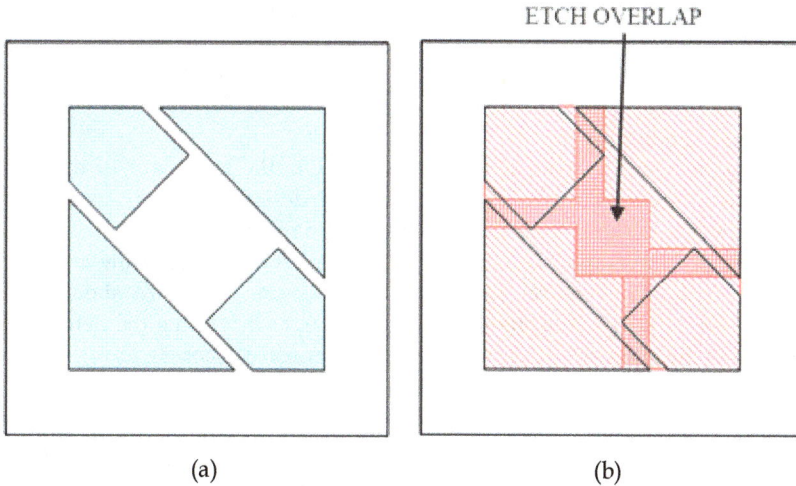

(a)   (b)

Fig. 11. a) Geometry C; b) Etch overlap.

An alternative for this last geometry is presented in Fig. 12a, where additional supporting arms were added. This will be identified as Geometry D. The purpose for these extra supporting arms is to give mechanical support to the thin membrane so any damage can be prevented if an undesired vibration is suddenly present on the chip. After simulation, this modification showed no improvement in etching time, since the membrane was released also in 75 min with a depth of about 56μm for the etched pit. So, compared with Geometry C, it can be considered that the only advantage is the improvement in mechanical support. From Fig. 12b, the difference between the etch overlap areas of Geometry C and Geometry D can be clearly seen.

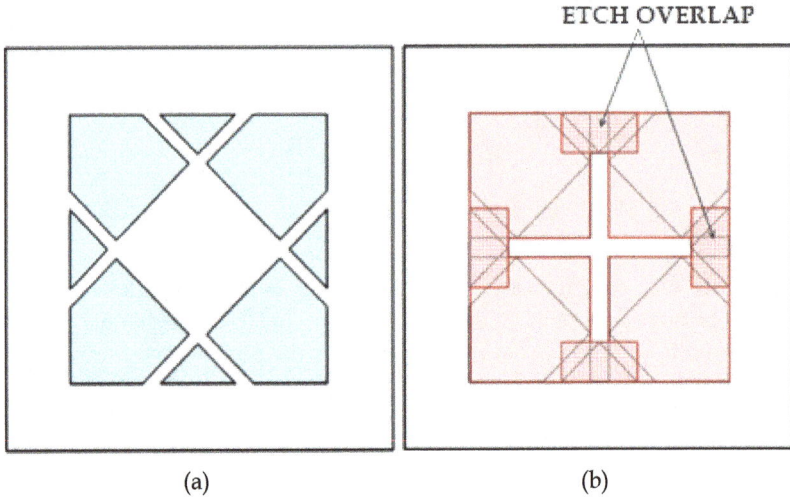

Fig. 12. a) Geometry D; b) Etching areas and etching overlaps.

Although there are no overlaps at the centre, a little substrate area is left (indicated as a thin cross outside the overlaps) that can be rapidly etched away due to its small cross section and the multiple planes present at the vertices of the membrane and the supporting arms.

4.2 Mechanical simulations

Based in a finite element analysis made with COMSOL, the behaviour of the suspended membranes was simulated with each of the geometries described before. Also, in this study it is important to know the weight that the membrane must support. As with restrictions indicated during the mechanical simulation, the extremes of the supporting arms and outer sides of the membrane were set as fixed; the remaining structure should have free movement. The main purpose of the present study was to determine the deformation and stress that exist in the alternative geometries, for comparison purposes.

Geometry A has been widely used and reported in literature and as so, it will be used as the reference geometry to be compared with other geometries. Variables, such as deformation and Von Mises stress, were obtained after simulation in order to evaluate all the membranes, so it can be determined if the proposed modifications introduce some mechanical failure. During simulation, a force equal to the corresponding weight of the membrane was applied considering also the material from which each membrane is made (SiO$_2$) and its thickness (~390nm).

For the case of Geometry A, the maximum deformation obtained was 6.357×10^{-15} µm, with a maximum Von Mises stress of 1.229×10^{-3} MPa, that is significantly below the elastic limit for SiO$_2$ (55 MPa). These results are illustrated in Fig. 13.

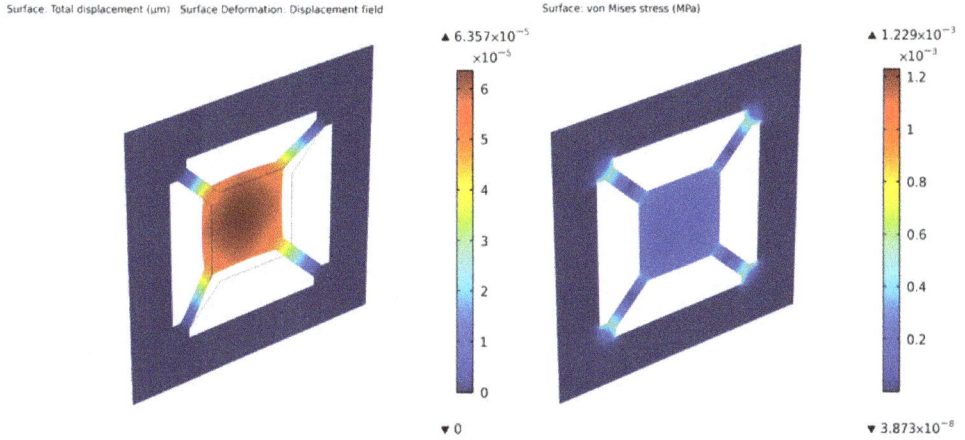

Fig. 13. FEM simulation for Geometry A.

On the other side, the maximum deformation and maximum Von Mises stress obtained in the case for Geometry B were 7.38×10^{-7} µm and 1.523×10^{-5} MPa, respectively. This strain is also below the elastic limit for SiO$_2$. Results are shown in Fig. 14.

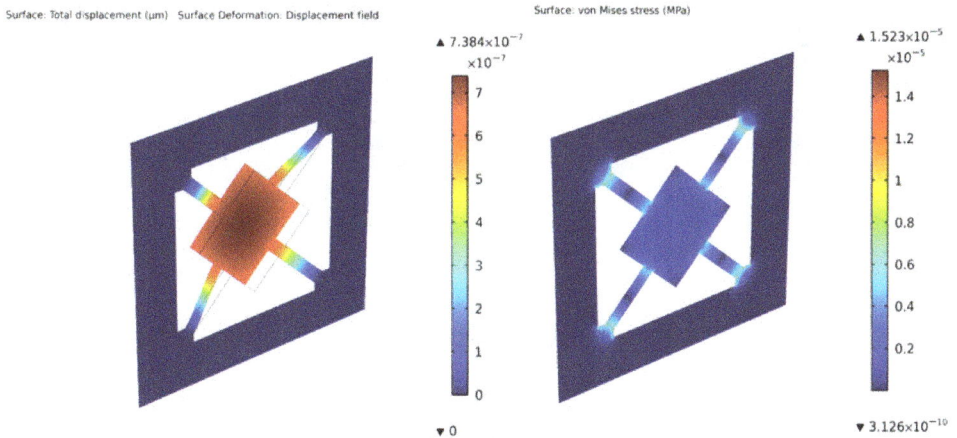

Fig. 14. Deformation and stress for Geometry B.

Next, Geometry C showed a deformation of 2.952×10^{-5} µm with a maximum Von Mises stress of 161×10^{-3} MPa, showing also that it is a good design from the mechanical point of view. These results are shown in Fig. 15.

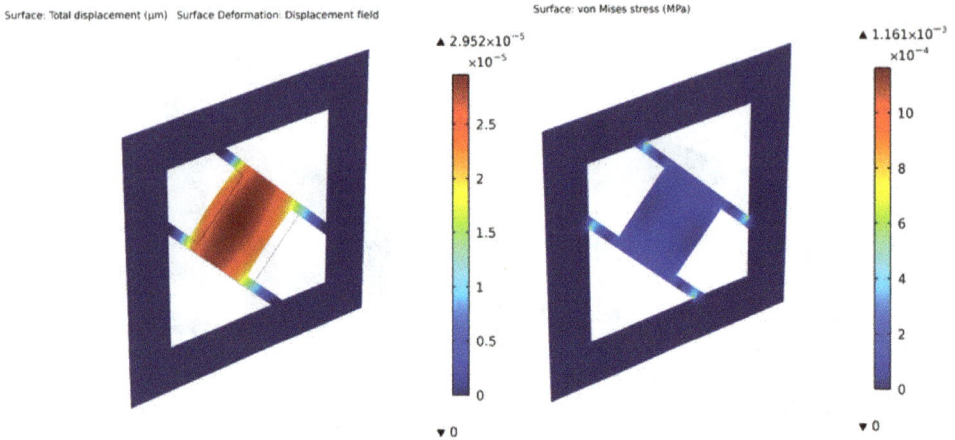

Fig. 15. Simulation results for Geometry C.

Now, Geometry D, having two extra supporting arms, shows a maximum deformation of 2.403×10^{-4} µm with a maximum Von Mises stress of 0.01×10^{-3} MPa located next to the arms' anchors. This is illustrated in Fig. 16.

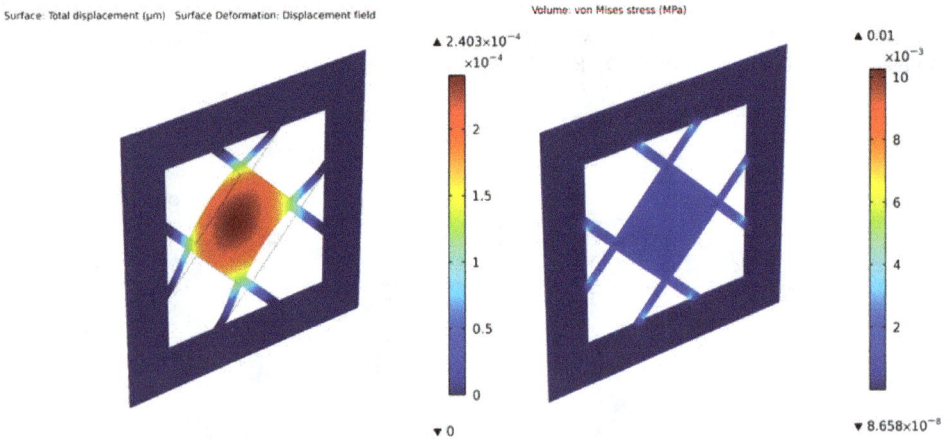

Fig. 16. Mechanical study results of Geometry D.

As is demonstrated, Geometry D shows the highest deformation compared with Geometries A, B and C, but on the other hand, it resulted in the lowest Von Mises strain.

From these results it can be concluded that this geometry is better for the purposes of the present study and also, as will be demonstrated later, with this geometry the supporting arms are released in a considerably shorter etching time.

4.3 Experimental results

Silicon substrates were prepared with a thick silicon dioxide layer (~390nm). Test geometries as those proposed above (A, C and D) were then defined with photolithography. Following, an etching with a 100 ml solution with 10% TMAHW at 80°C added with 1.36 gr of ammonium peroxidisulfate (APS), was done over 25, 50, 75 and 102 min. APS enhanced the sample finishing. This is a common formulation for etching solutions based on TMAHW.

After these times, the samples were checked with a microscope to verify the correct etching.

Fig. 17 shows the advance of the etching process for Geometry A where the characteristic figure predicted during simulation is present at the centre of the membrane caused by the anisotropic attack (far left).

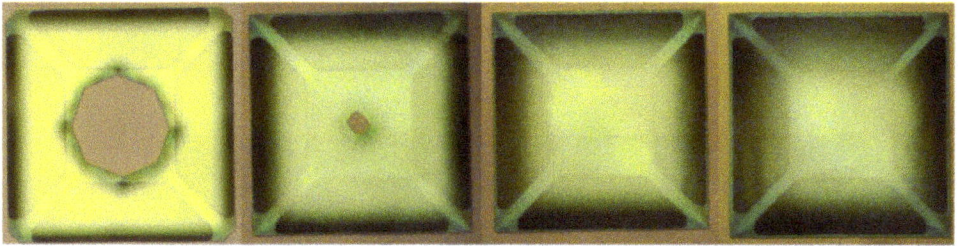

Fig. 17. Geometry A etching photographs.

For Geometry C, Fig. 18 shows the progress of the etching for 25, 50 and 75 min, where the distinctive planes are formed.

Fig. 18. Microphotographs of Geometry C.

In the same way, Geometry D was processed in TMAH and photographs were taken at the prescribed times. Fig. 19. shows how rapidly the flat bottom formed.

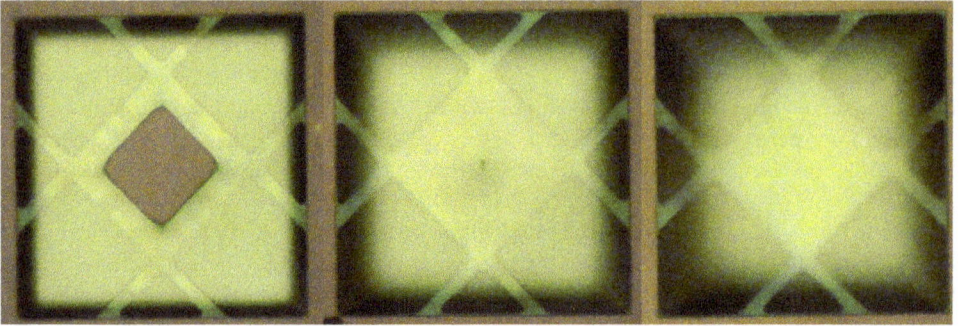

Fig. 19. Geometry D during etching at different times.

Next, results from the experimental etching processes applied are shown and discussed, supported with simulation (left) and SEM images (right).
Geometry A.

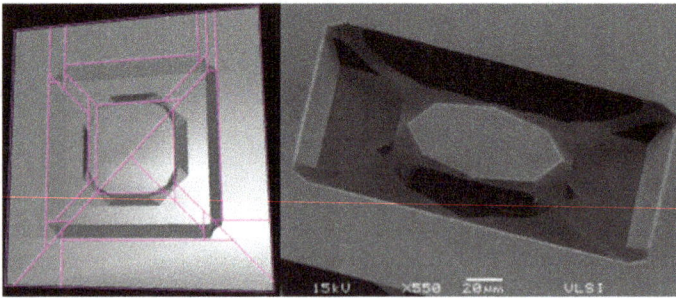

25 minutes: Here it can be seen that after this time, the supporting arms are completely released, but the central bulk of the membrane is just starting to be etched at the corners.

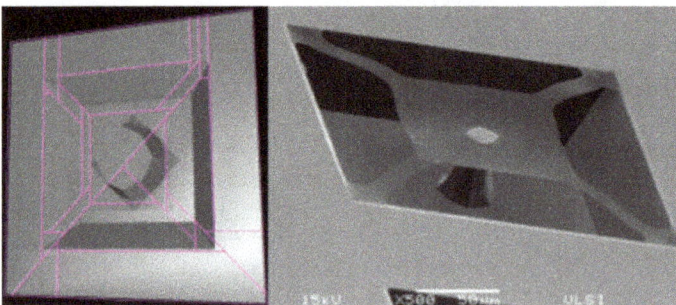

50 minutes: A while later, {111} planes generated due to parallel or perpendicular lines to {110} planes are completely reduced, but there is still contact between the remaining silicon with the membrane.

75 minutes: After this time of etching, a square based pyramid shape is formed at the centre of the membrane, having planes from which the etching can continue thoroughly. At this time, there is a little pyramid still left.

102 minutes: Finally after this time the membrane has been completely released with a bottom cavity surface showing a smooth (100) plane.
Geometry C.

25 minutes: With this geometry, initially the supporting arms are first released exposing {110} planes, that have, as commented before, a high etching rate.

50 minutes: Here it can be seen that a column with {110} facets is formed at the centre of the membrane, so etching can continue easily.

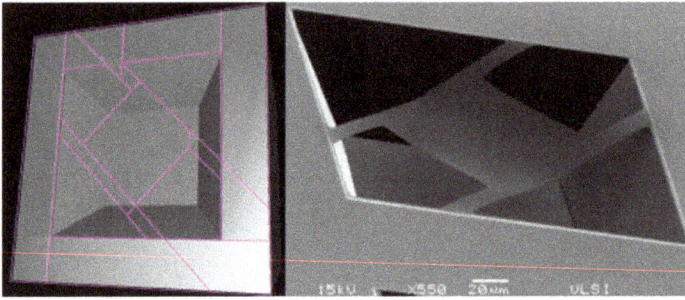

75 minutes: Finally, the membrane was completely released and the cavity has a smooth surface.
Geometry D.

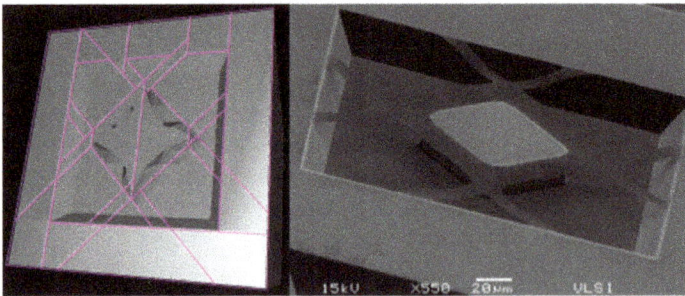

25 minutes: After this initial etching time, the supporting arms were completely released, but a complex structure is still present having convex corners that can slow down the etching process.

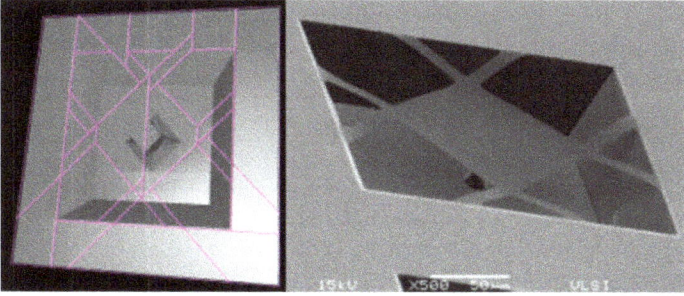

50 minutes: After 25 extra minutes, the substrate of this geometry looks like that obtained after the same time with Geometry C, having also {110} planes with a high etching rate.

75 minutes: Finally, the etching process completely released the membrane also with a smooth cavity bottom.

Comparing the simulation figures and the SEM images from the experimental samples above, it is clear that they are nearly the same, both having a smooth bottom of the cavity, nevertheless, there is a difference in time prediction for the membrane release between theory and experiment.

It may seem clear that the time difference can be attributed to features not considered in AnisE [9] regarding the etching apparatus set, as temperature variations, pH level of the

SUSPENDED MEMBRANE ETCHING			
Structure	**Etching time (minutes)**		
	Simulated	**Experimental**	**Difference**
Geometry A	102	~90	-11%, smooth cavity bottom
Geometry C	75	~72	-0.4%, smooth cavity bottom
Geometry D	75	~72	-0.4%, smooth cavity bottom

Table 1. Suspended membrane etching comparison.

solution and saturation of the solution during the etching process. Despite this difference, the experimental anisotropic etching follows the same behaviour predicted by simulations. Table 1 summarizes these results.

5. Proposed layout for a micro-hotplate

From these results, a new layout for a micro- hotplate has been designed using Geometry D as described above. The micro-hotplate contains a micro-heater made with polysilicon, which will heat the structure when a voltage is applied to the heater terminals. There is a temperature sensor also made with polysilicon which will be connected to a control circuit to maintain the temperature of the micro-hotplate to a given value. Fig. 20 shows this design that must guarantee that the silicon substrate is exposed to the etching solution just where it is desired to accomplish the thin membrane (Marshall et al., 1992).

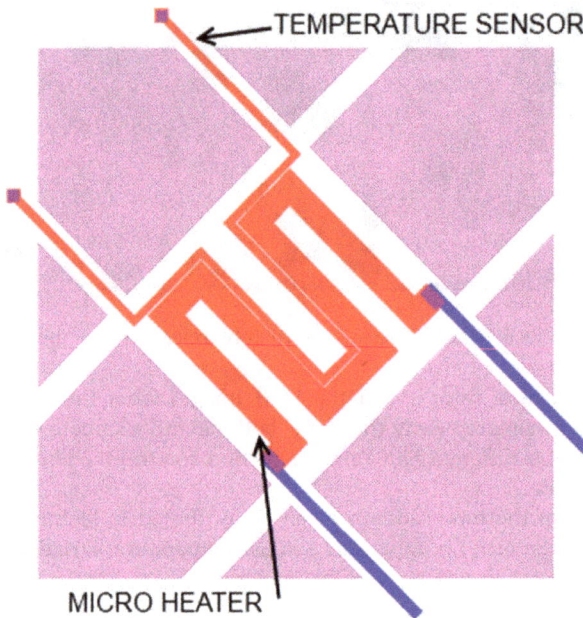

Fig. 20. Layout of micro-hotplate showing temperature sensor and micro-heater.

As mentioned before, this design is intended to be used in a CMOS semiconductor gas sensor that requires a heated thin film to perform the detection. Fabricating a micro-cavity below the heated zone using a MEMS etching process reduces the power needed to achieve the desired temperature and provides thermal isolation to the substrate and signal electronics (Suehle et al., 1993). This can be made following standard CMOS post-process etching steps, keeping compatibility between CMOS technology and MEMS micro-machining.

Using the guidelines explained before, it is expected that this design will help reduce damage to the exposed aluminium pads during the fabrication process due to the reduction of etching time.

6. Conclusions

It was theoretically and experimentally demonstrated that the geometry of a micro-hotplate (as the one used in CMOS compatible micro-heaters for semiconductor gas sensors) can be conveniently modified to reduce the etching time with TMAHW, in the order of 20%. Decreasing the etching time is notably useful if these structures are fabricated with materials that can be damaged by the etching solution used. Therefore, selective solutions with repeatability results should be used to protect the integrity of those layers that have an electrical or structural function and need to be protected during the post-process. The purpose is to keep these devices in the etching solution for the least possible time, protecting them from chemical damage. The geometry analysis proposed, based on crystallographic concepts, is a useful strategy in MEMS design taking advantage of anisotropic etching properties. For the cases here presented and using TMAHW, it is concluded that {111} planes and when possible the <110> directions that generate these planes, should be avoided not only in micro-hotplate membranes, but also in structures with different purposes, such as cantilevers. So, there is an optimization of the etching process, especially if it is included as a post-process for integrated circuits fabricated with CMOS technology compatible with MEMS technology.

7. Acknowledgements

The authors are grateful with Edmundo Rodríguez for the preparation of the etching test masks for photolithography, Benito Nepomuceno for the oxidized silicon substrates preparation and Gaspar Casados for the SEM images of the structures.

8. References

Baltes, H.; Brand, O.; Fedder, G. K.; Hierold, C.; Korvink, J. G. & Tabata, O. (Eds.) (2005). *Advanced Micro & Nanosystems*, CMOS-MEMS, Wiley-VCH Verlag GmbH &Co, ISBN 978-3-527-31080-7, Weinheim, Germany.

Barrettino, D.; Graf, M.; Kirstein, K.; Hierlemann, A. & Baltes, H. (2004). A monolithic fully-differential cmos gas sensor microsystem for microhotplate temperatures up to 450°C, *International Symposium on Circuits and Systems 2004*. Vol.4, pp. 3-6, ISBN 0-7803-8251-X, Vancouver, Canada. 23-26 May 2004.

Barrettino, D.; Song, W. H.; Graf, M.; Hierlemann, A. & Baltes, H. (2004). A Micro-Hotplate-Based Monolithic CMOS Thermal Analysis System. *29th European Solid-State Circuits Conference. ESSCIRC 2004*, pp. 329-332, ISBN 0780379950, Estoril, Portugal.

Capone, S.; Forleo, A.; Francoso, L.; Rella, R.; Siciliano, P.; Spadavecchia, J.; Presicce, D. S. & Taurino, A. M. (2003). Solid State Gas Sensors: State of the Art and Future Activities. *Journal of Optoelectronics and Advanced Materials 2003*. Vol.5, No.5, pp. 1335-1348, ISSN 09317497.

Chen, S.; Zhu, M.; Ma, B. & Yuan, W. (2008). Design and Optimization of a Micro Piezoresistive Pressure Sensor, *Proceedings of the 3rd IEEE International Conference on Nano/Micro Engineered and Molecular Systems*. ISBN: 978-1-4244-1907-4 Sanya, China. Jan. 2008

Fujitsuka, N.; Hamaguchi, K.; Funabashi, H.; Kawasaki, E. & Fukada, T. (2004). Aluminum Protected Silicon Anisotropic Etching Technique using TMAH with an Oxidizing

Agent and Dissolved Si. *RD Review of Toyota CRDL*, Vol.39, No.2, pp. 34-40, ISSN 0385-1508.

Gaitan, M.; Parameswaran, M.; Johonson, R.B. & Chung, R. (1993). Commercial CMOS Foundry Thermal Display for Dynamic Thermal Scene Simulation, *Proceedings of SPIE*, Vol.1969, doi:10.1117/12.154731 Orlando, FL, USA, April 1993.

Hsu, T. R., (2002), *MEMS & Microsystems 1st Ed.*, Mc Graw Hill, ISBN 0-07-239391-2, New York, NY, USA, 2002.

Korvink J. G. & Paul, O. (Eds.) (2006). *MEMS a practical guide to design, analysis and application*, William Andrew, Inc. & Springer-Verlag GmbH & Co. KG, ISBN 0-8155-1497-2 & 3-540-21117-9, NY, USA. & Heidelberg, Germany.

Kovacs, G. T. A.; Maluf, N. I. & Petersen, K. E. (1998). Bulk Micromachining of Silicon, *Proceedings of the IEEE*, Vol.86, No.8, pp. 1536-1551, ISBN: 00189219, Conference Location, Aug 1998.

Madou, M. J. (2001). *Fundamentals of Microfabrication*. CRC-Press, ISBN: 0-8493-9451-1, USA

Marshall, J.C.; Parameswaran, M.; Zaghloul, M.E. & Gaitan, M. (1992). High-Level CAD Melds Micromachined Devices with Foundries. *Circuits & Devices Magazine*, IEEE, Vol.8, No.6, Nov 1992, pp. 10-17, ISSN 8755-3996.

Tea, N. H.; Milanovic, V.; Zincke, C. A.; Gaitan, M.; Zaghloul, M. E. & Geis, J. (1997). Hybrid Postprocessing Etching for CMOS-Compatible MEMS. *Journal of Microelectromechanical Systems*, Vol.6, No.4, Dec. 1997, pp. 363-372, ISSN 1057-7157.

Pierret, R. F. (1989). *Advanced Semiconductor Fundamentals 2nd Ed.*, Addison-Wesley Longman Publishing Co., Inc., ISBN 0201053381, Boston, MA, USA.

Suehle, J. S.; Cavicchi, R. E.; Gaitan, M. & Semancik, S. (1993). Tin Oxide Gas Sensor Fabricated Using CMOS Micro-Hotplates and IN-Situ Processing. *IEEE Electron Device Letters*, Vol.14, No.3, pp. 118-120, ISSN: 07413106.

Sullivan, P.; Offord, B. W. & Aklufi, M. E. (2000). Tetra-Methyl Ammonium Hydroxide (TMAH) Preferential Etching For Infrared Pixel Arrays. *Technical Document 3097*, SPAWAR Systems Center San Diego, San Diego, USA.

Tabata, O. (1995). pH-controlled TMAH Etchants for Silicon Micromachining, *Proceedings of the 8th International Conference on Solid State Sensors and Actuators, and Eurosensors IX*. ISBN: 91-630-3473-5 Stockholm, Sweden, June 1995

Tabata, O. (1998). Anisotropy and selectivity control of TMAH, *Proceedings of the 11th International Workshop on Micro Electro Mechanical Systems, MEMS 98*, ISBN: 0-7803-4412-X Heidelberg , Germany, Jan. 1998.

Yan, G.; Chan, P.C. H.; Hsing, I. M.; Sharma, R. K. & Sin, J. K. O. (2001). An improved TMAH Si-etching solution without attacking exposed aluminum. *Sensors and Actuators A: Physical*. Vol. 89, No. 1-2. Mar. 2001

10

Micro Abrasive-Waterjet Technology

H.-T. Liu and E. Schubert

OMAX Corporation,
USA

1. Introduction

Waterjet technology has come a long way since its commercialization in the late 1970's.[1] It has evolved from merely a rough cutting tool to a precision machine tool, competing on equal footing with established tools such as lasers, mechanical milling and routing tools, EDM, ultrasonics, photochemical etching, and various CNC tools.

The versatile waterjet technology inherently has advantages unmatched by most machine tools. Below is a short list of these advantages.

- Cuts virtually any material, thin and thick
- Fast setup and programming – modern systems are now easy to learn
- Simple fixturing is required – negligible reactionary forces on parts
- Almost no heat generation during cutting – no heat-affected zone (HAZ) on parts
- No residual mechanical stresses
- Safe operations
- No start hole required – one single tool for multi-mode machining
- Narrow kerf – removal of only a small amount of material
- Cost-effective and fast turnaround for both small and large lots
- Environmentally friendly – no hazardous waste byproducts

Figure 1 shows a typical abrasive-waterjet system along with a closeup of a nozzle and representative AWJ-cut parts made of various materials. The advantages and disadvantages of AWJs in comparison with lasers, EDM, plasma, flame cutting, and milling can be found in waterjets.org.[2] Waterjet technology has been adopted rapidly in the manufacturing industry to take advantage of its technological and manufacturing merits.[3] Waterjets can be used to machine noncritical components ready for assembly. For critical components requiring high-precision machining, waterjets have been used extensively as a net-shape tool, particularly for difficult-to-machine materials such as hardened steels, alloys, ceramics, and silicon carbide ceramic matrix composites. The versatility of waterjet technology has led to a broad

---

[1] Waterjet technology generally refers to the use of any of three jets: a water-only jet (WJ), an abrasive-waterjet (AWJ), or an abrasive slurry or suspension jet (ASJ) (Momber & Kovacevic, 1998, Fig. 1.1). For conciseness, the term waterjets is used to refer to all three types of jets unless specified otherwise.

[2] http://waterjets.org/index.php?option=com_content&task=category&sectionid=4&id=46&Itemid=53#advantages_of_waterjet_machining (8 August 2011).

[3] Waterjet machine tools emerged as the fastest growing segment of the overall machine tool industry in the last decade, and this trend is expected to continue (Frost and Sullivan – "The World Waterjet Cutting Tools Markets" Date Published: 30 Aug 2005 (www.frost.com).

range of capabilities from macro- to micromachining in most materials, which is unmatchable by most machine tools (Liu & McNiel, 2010).

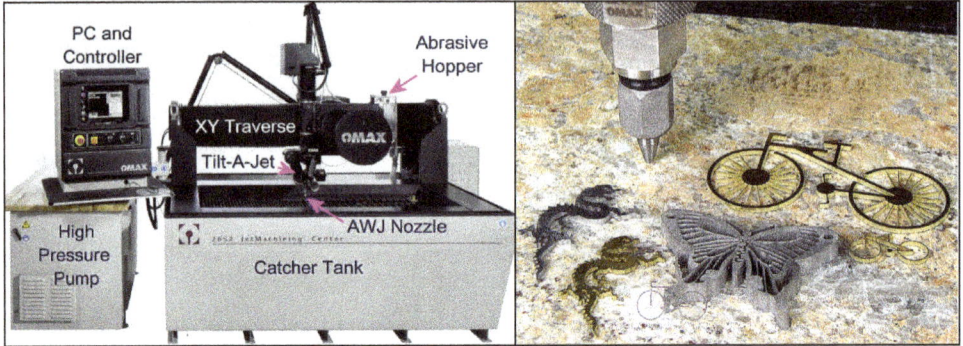

Fig. 1. Typical AWJ system and parts cut with AWJ nozzle (Liu & McNiel, 2010)

The basic erosion mode of waterjets is consistent with micromachining processes.[4] Waterjet machining is accomplished by erosion due to individual high-speed water droplets in a water-only jet or abrasives in an AWJ or ASJ impinging onto a workpiece. The size of machined features, such as the diameter of a hole or the kerf width of a slot, is proportional to the diameter of the jet stream in which the water droplets or abrasives are confined (Liu, 2010). Therefore, the key for developing micro abrasive-waterjet (µAWJ) technology is to reduce the diameter of the jet stream for meso-micro machining. Although the feasibility of applying waterjet technology for micromachining has been demonstrated in the mid-1990's, only limited progress has been made since then. The smallest features machined with commercial AWJs are generally larger than 200 µm. Several issues, such as the change in AWJ flow characteristics, poor flowability and clumping of fine abrasives, and nozzle clogging due to the wetting of abrasives caused by backsplash, have presented challenges for further downsizing AWJ nozzles. For example, micro WJs and low-pressure ASJs have been limited to machining relatively soft materials and to the singulation of SD chips for cellular phones, respectively (Jiang et al., 2005). AWJs have been predominantly used for machining applications.

Realizing the inherent advantages of AWJs and their potential for micromachining, considerable effort has been devoted to studying and seeking solutions to the above issues since the mid-2000's. Investigations have concentrated on understanding the physics of the supersonic/subsonic three-phase microfluidics of the abrasive slurry moving through small-diameter mixing tubes and on issues related to the flow characteristics of fine particles. [5] These efforts have led to the development of novel approaches to resolving issues pertinent to AWJ micromachining. R&D efforts have developed miniature AWJ nozzles and ancillary

---

[4] Strictly speaking, micromachining is defined as machined features in the micron size range. Machined features around 100 µm are often referred to as meso-scale. In practice, however, features with sizes smaller than 150 µm are often loosely referred to as micro-scale. With that understanding, we chose the practical and rather loose definition of meso- and micro-scale for features between 100 µm and 250 µm and less than and equal to 100 µm, respectively.

[5] Supported by OMAX's R&D funds and NSF SBIR Grants #0944229 (Phase I) and #1058278 (Phase II).

devices to machine features 100 μm and smaller. These AWJs have been used to machine miniature parts made from a wide range of materials. As a result, the technical feasibility of further downsizing AWJ nozzles has been demonstrated. This article reviews the current status of μAWJ technology developments.

2. Waterjet technology

As discussed above, waterjet technology generally refers to the use one of any one of three types of jets: a water-only jet (WJ), an abrasive-waterjet (AWJ), and an abrasive slurry or suspension jet (ASJ). A waterjet is generated by pumping high-pressure water through an orifice to achieve supersonic speeds based on Bernoulli's principle. The water is pressurized to 400 MPa and higher on the basis of Pascal's law of hydraulics. Although the basic features of modern waterjet technology were patented in 1939 by Leslie Tirrell (U. S. Patent No. 2,200,587), the lack of a water pump capable of producing the pressures needed for effective cutting limited the applications to surface cleaning and blasting.

Commercially viable 400-MPa pumping systems that took advantage of modern materials and design techniques were not introduced until the late 1970's. These high-pressure WJs, however, were limited to cutting only soft materials. In most cases, the waterjet-cut edges were often frayed with marginal edge quality at best. Abrasive-waterjets entraining abrasives into the WJ stream were developed by several independent R&D teams. [6,7] Abrasive-waterjets were subsequently commercialized in the mid-1980's. For the rest of the 1980's, in the absence of precision control software and hardware, AWJs were mainly used for the rough cutting of materials, particularly those that were difficult to cut using established tools. Since then, much of the research and development has focused on characterizing the emerging technology and realizing its technological and manufacturing merits, as described in Section 1. It was soon realized that, with further advances toward automation, precision, and ease of use, AWJ cutting could have a large market potential in many industrial applications.

Advances in personal computers and peripherals, pumping technology, and materials sciences have greatly benefited the advancement of waterjet technology. In the early 1990's, PC-based CAD/CAM software and controllers were developed and optimized to drive AWJ systems, leading to significant cost reductions for automation while improving the user-friendliness of AWJ machining. The patented "compute first - move later" control algorithm revolutionized AWJ machining (Olsen, 1995). Based on a cutting model (e.g., Zeng & Kim, 1995) and a set of parameters (including nozzle size and pressure; abrasive type, size, and flow rate; material type; path geometry; part thickness; and edge quality), the algorithm calculates the traverse speed of the nozzle along the entire cutting path derived from the CAD drawing of a part. Such a "smart" system greatly reduces the amount of operator input needed to optimize the cutting of a given geometry out of various materials. The path-line-based CAD program is then converted into an executable CAM module (Olsen, 1996). The entire part is machined simply by clicking a "Start" button on the display window without further intervention by the operator.

---

[6]As a team member of one of such R&D group, the senior author developed a tube catcher to dissipate and terminate the residual erosive power of spent abrasives.

[7] Since Tirrell's 1939 patent, several AWJ patents have been issued (refer to U. S. Patent No. 4,555,872).

One of the major improvements in recent years is the development of direct-drive pumps to replace the inefficient and bulky intensifier, resulting in increased efficiency from about 70% to about 90% along with reduced noise and pressure ripples. The simplicity of direct-drive pumps not only reduces operating and maintenance costs but also simplifies maintenance procedures. The availability of advanced materials together with design optimization has continuously extended the operating lives of pump components. Nowadays, the time between rebuilds for an advanced direct-drive pump has been extended to 1000 hours.[8]

Waterjets are generally amenable to 3D machining, although they have some limitations. Spent abrasives still possess considerable residual erosive power and could cause damage to the rest of the workpiece being cut, particularly if the waste stream comes in contact with other parts of the workpiece. For 3D parts with complex geometries, materials downstream and along the path of the cutting AWJ must be protected by a "catcher" to capture the spent abrasives, or by sacrificial pieces to absorb or dissipate the residual erosion power of the spent abrasives. For complex 3D parts, however, there is usually not enough room to place the catcher or sacrificial pieces where they are needed. Most commercial AWJ systems machine 2D and 2D+ parts, although there are commercial 5+-axis AWJ systems available for machining simple 3D parts. Alternatively, novel accessories for 2D AWJ systems have been developed that take into account the jet's geometry and imperfections in the raw materials to improve cutting accuracy, facilitating the fabrication of nearly 3D parts and multimode machining. Brief descriptions of representative accessories and their uses are given in subsequent sections.

2.1 Fundamentals

The PC-based CAD/CAM modules that incorporate a patented controller have been specifically designed to emphasize user friendliness. As a result, the underlying complex physical processes of AWJ flow phenomena are practically transparent to the operators. In this subsection, the fundamentals of AWJ flow phenomena are briefly described.

First, an AWJ is a supersonic, three-phase flow that involves fluid-fluid (air and water), fluid-solid (air, water, and abrasives), and solid-solid (abrasives and materials) interactions. During piercing and cutting, the complex flow is confined in a small space with rapidly changing boundary conditions. In addition, because water is slightly compressible, with a compressibility of approximately 15% at 400 MPa, numerical simulations of such complex flow phenomena have met with little success. CFD models have primarily been used for proof of concept (Liu et al., 1998). Fortunately, a majority of the research work was conducted in laboratory experiments at full scale.

The hydraulic power, P, required to generate a high-speed waterjet through an orifice is proportional to the product of the pressure, p, and the flow rate, Q, or

$$P = p \times Q / c \qquad (1)$$

where c is a constant that is equal to 60 when P is in kW, p is in MPa, and Q is in l/min, for example. The flow rate can then be estimated based on the Bernoulli principle that the water flow rate through an orifice with a cross-sectional area A is

$$Q = c_d A \sqrt{2p / \rho} \qquad (2)$$

---

[8] http://www.omax.com/enduromax.php (8 August 2011).

where $A = \pi d^2/4$, d is the orifice diameter, c_d is the discharge coefficient with a typical value of 0.65, p is the pressure, and ρ is the water density.

A normal diagram relating P, Q, d, and p as derived from Eqs. (1) and (2) with $c_d = 0.65$ is shown in Fig. 2 for a variety of orifice diameters. Knowing any two of the four variables enables determination of the other two. For example, if a cutting pressure of 4000 bar is required using a 0.13 mm orifice, it will draw a flow rate of 0.43 l/min and the stream power will be 1.9 kW (dash-dotted lines). A motor larger than 1.9 kW must be used due to pump inefficiencies.

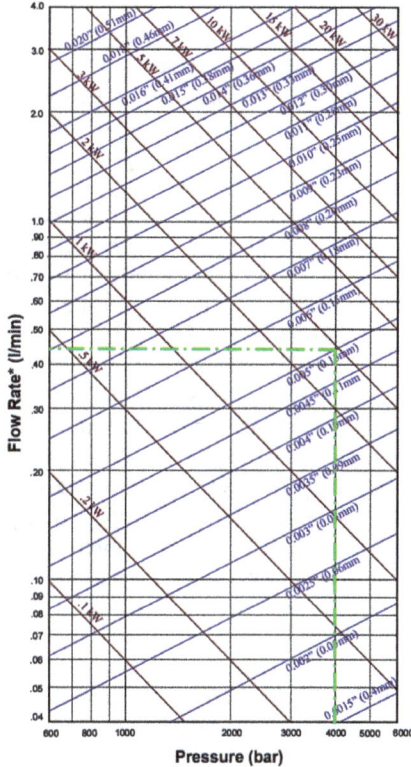

Fig. 2. Normal diagram of power, flow rate, and pressure

2.2 Key components

Abrasive-waterjet systems include both hardware and software components. They are integrated to maximize the cutting speed, user friendliness, and cost effectiveness.

2.2.1 Hardware

A typical AWJ system includes an AWJ nozzle, an abrasive feeding hopper, an X-Y traverse, a high-pressure pump, a motor, a PC, a catcher, and a support tank. Figure 1 illustrates an example of an AWJ system with several key components identified. Depending on the application, the catcher tank that also serves as the support for the X-Y traverse, which usually has a cutting area ranging from about 0.7 m x 0.7 m up to 14 m x 3 m or larger. The X-Y traverse, on which the AWJ nozzle, abrasive hopper, and other accessories may be mounted, has a position accuracy typically from 0.1 mm to 0.03 mm or better.

A high-speed waterjet is formed by using a high-pressure pump, either a hydraulic intensifier or a direct-drive pump, as illustrated in Fig. 3. Early high-pressure cutting systems used hydraulic intensifiers exclusively. At the time, the intensifier was the only pump capable of reliably creating pressures high enough for waterjet machining. A large motor drives a hydraulic pump (typically oil based) that in turn operates the intensifier. Inside the intensifier, hydraulic fluid pumped to about 21 MPa acts on a piston through a series of interconnecting hoses and piping and a bank of complex control valves. The piston pushes a plunger, with an area ratio of 20:1, to pressurize the water to 420 MPa. The intensifier typically uses a double-acting cylinder. The back-and-forth action of the intensifier piston produces a pulsating flow of water at a very high pressure. To help make the water flow more uniformly (thus resulting in a smoother cut), the intensifier pump is typically equipped with an "attenuator" cylinder, which acts as a high-pressure surge vessel. The direct-drive pump is based on the use of a mechanical crankshaft to move any number of individual pistons or plungers back and forth in a cylinder. Check valves in each cylinder allow water to enter the cylinder as the plunger retracts and then exit the cylinder into the outlet manifold as the plunger advances into the cylinder. Direct-drive pumps are inherently more efficient than intensifiers because they do not require a power-robbing hydraulic system. In addition, direct-drive pumps with three or more cylinders can be designed to provide a very uniform pressure output without the use of an attenuator system. Improvements in seal design and materials combined with the wide availability and reduced cost of ceramic valve components now make it possible to operate a crankshaft pump in the 280 to 414 MPa range with excellent reliability. This represents a major breakthrough in the use of such pumps for AWJ cutting. Nowadays, an increasing number of AWJ systems are being sold with the more efficient, quieter, and more easily maintained crankshaft-type pumps.

Abrasive-waterjet systems operating at 600 MPa using intensifier pumps were introduced in the mid-2000's based on the notion that increased pressure means faster cutting. However, such a notion ignores several factors and issues. Specifically, any increase in pressure, for a given pump power, must be matched by a decrease in the volume flow rate, which leads to a decrease in the entrainment and acceleration of abrasives (Fig. 2). In an AWJ cutting system, water is used to accelerate the abrasive particles that perform the cutting operation. It has been shown that the kinetic power of the particles and thus the cutting power of the system is proportional to the hydraulic power of the waterjet. An increase in pressure at the same abrasive load ratio therefore does not yield any gain in cutting performance. Furthermore, high pressure is the enemy of all system plumbing due to material fatigue. As the pressure increases from 400 to 600 MPa, material fatigue significantly reduces the operating lives of components such as high-pressure tubing, seals, and nozzles, leading to considerably higher operating and maintenance costs (Trieb, 2010).[9] Finally, an intensifier pump is 28% less efficient than a direct-drive pump. When the above factors are taken into

---

[9] For example, the maximum von Misses stresses in traditional 3:1 (outside diameter to inside diameter) ratio components will be about 800 MPa to 1200 MPa, respectively. Based on data published in a NASA Technical Note (Smith et al., 1967), for hardened 304 stainless steel, the mean fatigue life will reduce from 35,000 cycles to 5,500 cycles, or a 6.4-fold reduction. As a result, high-pressure components are expected to reduce its life from several years to several months.

consideration, the hydraulic power, rather than the pressure, is the main factor for cutting performance. Real-world experience has consistently demonstrated that the direct-drive 400-MPa pump outperforms the 600-MPa intensifier pump in material cutting tests and in actual operations under the same electrical power (Henning et al., 2011a).

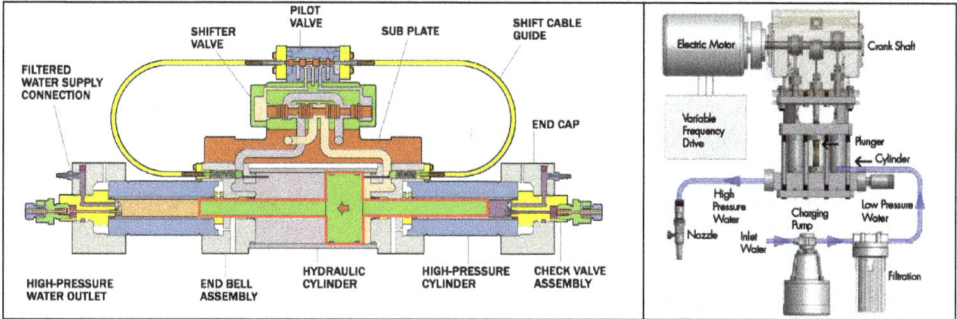

Fig. 3. Two types of high-pressure pumping mechanisms: an intensifier pump (left) and a complete direct-drive pump system (right) (Liu et al., 2010b)

Unlike a rigid cutting tool where material removal is carried out at the contact surface of a fixed-dimension tool and the workpiece, the AWJ is a flexible stream that diverges with the distance travelled. Consequently, AWJ machining has anomalies that must be compensated for with dedicated hardware components together with software control. For example, AWJ-cut edges are tapered depending on the speed of cutting. On the other hand, the spent abrasives still possess considerable erosive power to remove material along their paths. As a result, a catcher or sacrificial pieces must be used to capture spent abrasives or to prevent them from causing collateral damage to the rest of the workpiece. Therefore, AWJs would not be applicable to machine certain complex 3D parts when the placement of the catcher or sacrificial piece to protect the workpiece exposed to spent abrasives becomes impractical or impossible unless controlled depth milling or etching is used to machine blind features. To broaden the performance of AWJ machining in terms of precision and 3D machining, a host of accessories have been developed. Representative accessories include:

- A Tilt-A-Jet® dynamically tilts the nozzle up to 9 degrees from its vertical position.[10] It removes the taper from the part while leaving the taper in the scraps.

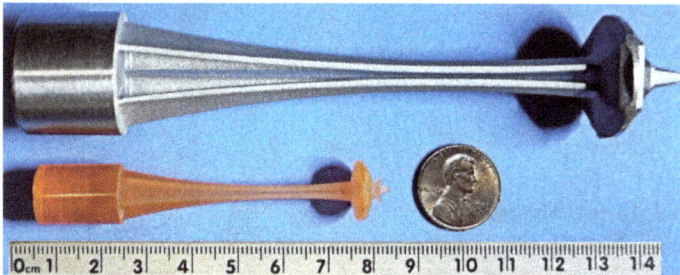

Fig. 4. Space Needle model machined with Rotary Axis (Liu & McNiel, 2010)

[10] http://www.omax.com/waterjet-cutting-accessories/Tilt-A-Jet/61 (8 August 2011)

- A Rotary Axis or indexer rotates a part (Fig. 4) during AWJ machining around what is commonly referred to as the 4th axis.[11] It not only facilitates axisymmetric parts to be machined with AWJs but also enables multimode machining, including turning, facing, parting, drilling, milling, grooving, etching, and roughing.
- An A-Jet™, or articulated jet, tilts the nozzle up to 60 degrees from its vertical position.[12] It is capable of beveling, countersinking, and 3D machining.
- A Collision Sensing Terrain Follower measures and adjusts the standoff between the nozzle tip and the workpiece to ensure that an accurate cut is maintained. Warped or randomly curved surfaces can be cut without the need to program in 3D. The collision sensing feature also protects components from becoming damaged if an obstruction is encountered during cutting.

By combining the Rotary Axis and the A-Jet, complex 3D features can readily be machined.

2.2.2 Forms of waterjets

Waterjets generally take one of three forms: a water-only jet (WJ), an abrasive-waterjet (AWJ), or an abrasive slurry or suspension jets (ASJ). Figure 5 shows drawings of these three jets. On the left is the WJ or the ASJ, depending upon whether the incoming fluid being forced through the small ID orifice is high-pressure water or abrasive slurry. On the right is the AWJ with gravity-fed abrasives entrained into the jet via the Venturi or jet pump effect. The abrasives are accelerated by the high-speed waterjet through the mixing tube.

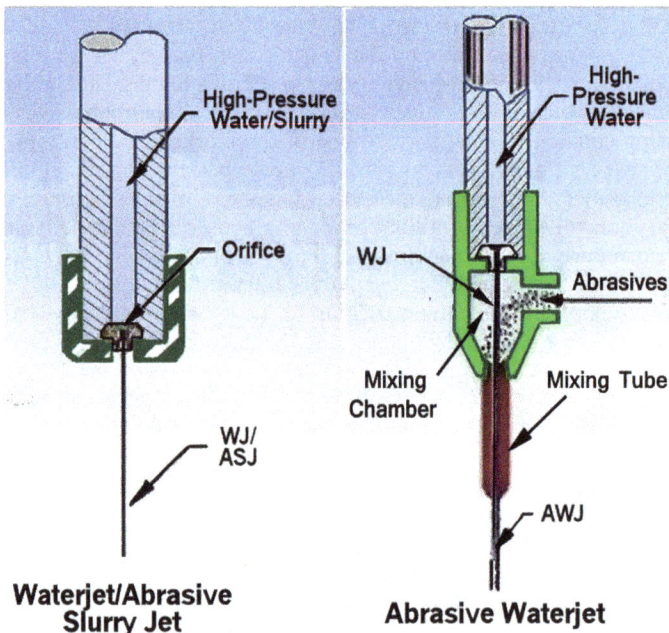

Fig. 5. Three forms of waterjets (Liu, 2009)

[11] http://www.omax.com/accessories-rotary-axis.php (8 August 2011)
[12] http://www.omax.com/waterjet-cutting-accessories/A-Jet/163 (8 August 2011)

For R&D and industrial applications, the majority of waterjet systems are AWJs. Water-only jets find only limited applications in the cutting of very soft materials. In principle, two-phase ASJs have a finer stream diameter, higher abrasive mass flow rate, and faster abrasive speed than do AWJs. As a result, the cutting power of ASJs is potentially up to 5 times greater than that of AWJs at the same operating pressure. Considerable R&D effort has been invested in developing ASJs. However, the high-pressure components, such as orifices, check valves, and seals, through which the high-speed abrasive slurry flows are subject to extremely high wear. The absence of affordable materials with high wear resistance has limited ASJs to pressures around 70 to 140 MPa for industrial applications (Jiang et al., 2005).

2.2.3 Abrasives

The most commonly used abrasive is garnet because of its optimum performance of cutting power versus cost and its lack of toxicity. It is also a good compromise between cutting power and wear on carbide mixing tubes. There are two types of garnet that are generally used: HPX® and HPA®, which are produced from crystalline and alluvial deposits, respectively.[13] HPX garnet grains have a unique structure that causes them to fracture along crystal cleavage lines, producing very sharp edges that enable HPX to outperform its alluvial counterpart. There are other abrasives that are more or less aggressive than garnet.

2.2.4 Speed of water droplets and abrasives

When machining metals, glasses, and ceramics with AWJs, the material is primarily removed by the abrasives, which acquire high speeds through momentum transfer from the ultrahigh-speed waterjet. Therefore, knowing the speed of the abrasives in AWJs is essential for the performance optimization of AWJs. Several methods, such as laser Doppler anemometers or LDVs, laser transit anemometers or LTAs, dual rotating discs, and others, have been used to measure the speed of the waterjet and/or the abrasive particles to understand the mechanism of momentum transfer in the mixing tube in which the abrasives accelerate (Chen & Geskin, 1990; Roth et al., 2005; Stevenson & Hutchings, 1995; Swanson et al., 1987; Isobe et al., 1988). There is a large spread in the test results mainly due to the difficulty in distinguishing the speeds of the water droplets and of the abrasive particles using optical methods.

A dual-disc anemometer (DDA), based on the time-of-flight principle, was found to be most suitable for measuring the water-droplet and/or abrasive speed (Liu et al., 1999). Data discs made of Lexan and aluminum were successfully used to measure water-droplet speeds in WJs and abrasive particle speeds in AWJs. This was achieved by taking advantage of the large differences in the threshold speeds of water droplets and abrasive particles in eroding the two materials.

Figure 6 illustrates typical measurements of water-droplet speeds generated with an AWJ nozzle operating at several pressures from 207 to 345 MPa in the absence of abrasives. The solid curve and the solid circles correspond to the Bernoulli speed, V_B, and the DDA measurements, V_w, with the abrasive feed port of the nozzle closed (i.e., no air entrainment), respectively. The Bernoulli speed is derived from Eq. (2). The close agreement between the two indicates that the WJ moves through the mixing tube with little touching of the

---

[13] http://www.barton.com/static.asp?htmltemplate=waterjet_abrasives.html (8 August 2011)

sidewall. The open circles and dashed curve represent the abrasive speed, V_{wa}, with the feed port open and the corresponding best-fit values.

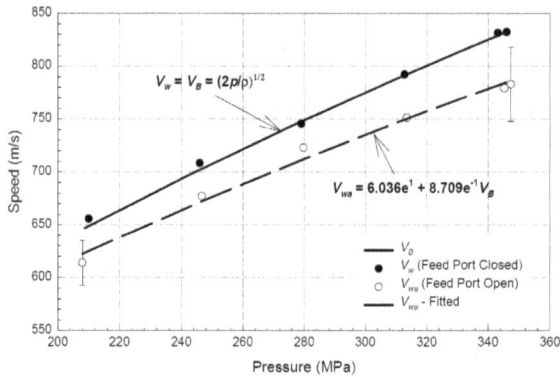

Fig. 6. Water-droplet speed in WJs exiting AWJ nozzle (Liu et al., 1999)

Measurements of abrasive speeds by entraining Barton 220-mesh garnet into the WJ are illustrated in Fig. 7. The measured and best-fit minimum, maximum, and average abrasive speeds are derived for a range of abrasive mass concentrations C_a = 0 to 1.08%.[14] The average abrasive speed at C_a = 0.4% is 300 m/s, about 61% of the water-droplet speed. The decreasing trend in abrasive speed with C_a is evident. The DDA has subsequently been applied to characterize the performance of AWJs (Henning et al., 2011a; Henning et al., 2011b).

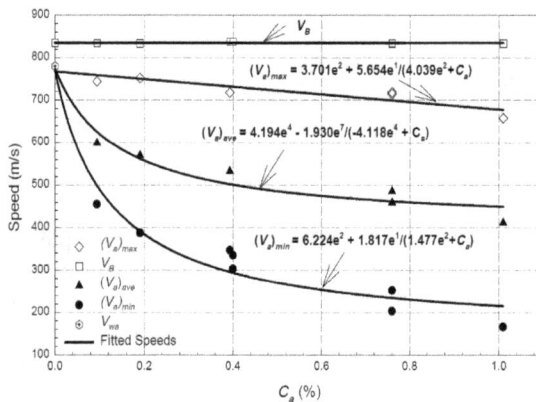

Fig. 7. Abrasive speed in AWJs, p = 345 MPa (Liu et al., 1999)

2.2.5 Control system

Historically, AWJ cutting systems have used traditional CNC control systems employing the familiar machine tool "G-code." G-code controllers were developed to move a rigid cutting

---

[14] C_a is defined as the percentage ratio of the abrasive master flow rate in pounds per minute to that of the water flow rate in gallons per minute.

tool, such as an end mill or mechanical cutter. The feed rate for these tools is generally held constant or varied only in discrete increments for corners and curves. Each time a change in the feed rate is desired, a programming entry must be made.

The AWJ definitely is not a rigid cutting tool; using a constant feed rate will result in severe undercutting or taper on corners and around curves. Moreover, making discrete step changes in the feed rate will also result in an uneven cut where the transition occurs. Changes in the feed rate for corners and curves must be made smoothly and gradually, with the rate of change determined by the type of material being cut, the thickness, the part geometry, and a host of nozzle parameters.

A patented control algorithm "compute first - move later" was developed to compute exactly how the feed rate should vary for a given geometry in a given material to make a precise part (Olsen, 1996). The algorithm actually determines desired variations in the feed rate in very small increments along the tool path to provide an extremely smooth feed rate profile and a very accurate part. Using G-code to convert this desired feed rate profile into actual control instructions for servomotors would require a tremendous amount of programming and controller memory. Instead, the power and memory of the modern PC is used to compute and store the entire tool path and feed rate profile and then directly drive the servomotors that control the X-Y motion. This results in a more precise part that is considerably easier to create than if G-code programming were used.

The advent of personal computing has led to the development of PC-based "smart" software programs for controlling the operations of most modern AWJ systems and a host of accessories for speeding up the cutting while maximizing the precision and quality of cuts. The flexibility of PC programming incorporates the versatility of waterjet technology very well, and the integration of modern PC-based software and hardware takes full advantage of the technological and manufacturing merits of waterjet technology.

One of the advanced software packages used for AWJ machining is the PC-based CAD/CAM.[15] It was particularly designed with "ease of use" in mind to allow operators to focus on the work at hand rather than the intricacies of the AWJ's behavior. The software has a built-in cutting model for common engineering materials that assigns each material a machinability index, as illustrated in Fig. 8. Another important input parameter is the edge or surface finish quality, which is defined in levels from Q1 to Q5, with Q1 representing rough cutting and Q5 representing the best edge quality. Figure 9 illustrates a "five-finger" part to demonstrate the five quality levels as a function of cutting speed. Note that the length of the finger is proportional to the cutting speed or the length of cut. The curvature and amplitude of the striation pattern, which is made of grooves caused by jet fluctuations, increases with increases in the cutting speed.[16] The amplitude of the striation is also proportional to the abrasive size.

To compensate for the AWJ as a flexible abrasive stream, the control algorithm optimally adjusts the cutting speed along various segments of the tool path. As soon as the cutting begins, the nozzle moves slowly along the lead path such that the piercing is complete at the

---

[15] The description of the software package is based on OMAX's Intelli-MAX Software Suite. For detail, refer to (http://www.omax.com/waterjets/intelli-max-software-suite - (8 August 2011)
[16] http://www.micromanufacturing.com/awj.htm (8 August 2011) or
http://oir.omax.com/media/OMAX_JetStream_Simulator.mp4 (8 August 2011)

beginning of the tool path. The nozzle moves relatively fast along straight sections of the tool path and decelerates as corners are approached. Slowing down around corners ensures that there is minimal jet lag as the AWJ cuts the corner. Otherwise, there would be a

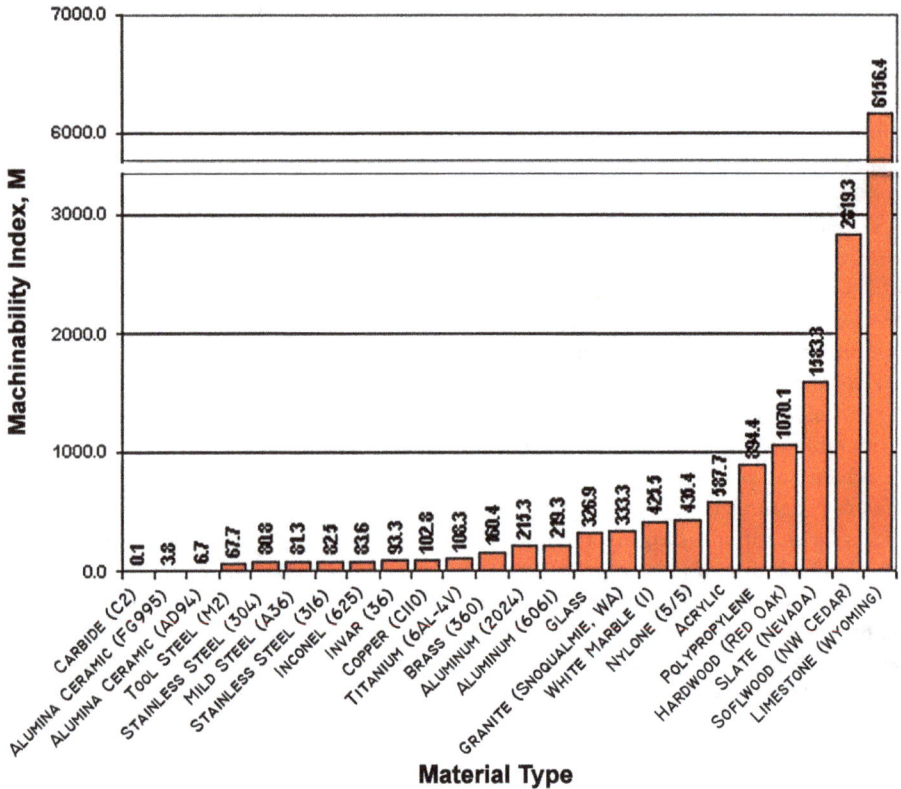

Fig. 8. Machinability of common engineering materials (Liu, 2009)

noticeable taper at the corner. The nozzle speeds up again after it passes the corner and accelerates to its maximum speed along straight segments. Figure 10 shows a color-coded diagram that illustrates the various cutting speeds used along a tool path.

The PC-based CAD is a built-in package that either works as a stand-alone program or allows drawings to be imported directly from other programs. It includes tools that are specific to AWJ machining such as automatic or manual lead in/out tools, tool path generation, collision prediction and correction, surface quality assignment tools, and many others. The PC-based CAM has many special features including the cutting model, six levels of cutting quality, taper compensation, estimate of time required to machine a part, report generation, creation and tracking of multiple home locations, rotating, scaling, flipping, and offsetting, among others. The CAM program also offers several special benefits such as part nesting, low-pressure piercing and cutting for brittle and delicate materials, the resizing of parts, and others.

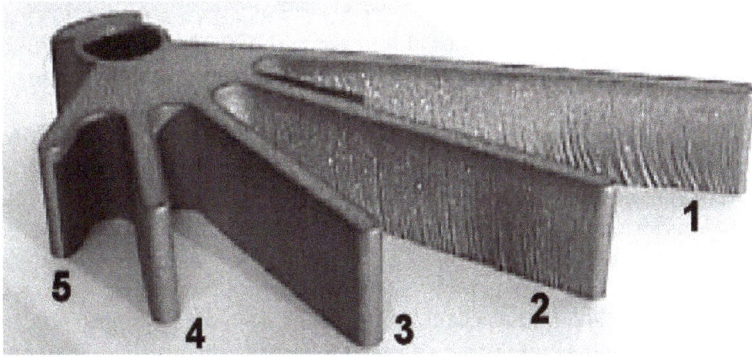

a) Fingers at qualities Q1 through Q5

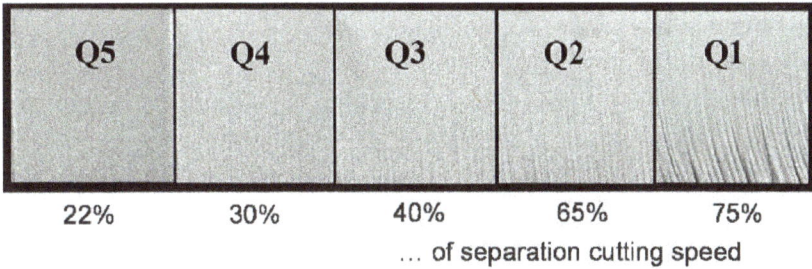

Q5	Q4	Q3	Q2	Q1
22%	30%	40%	65%	75%

... of separation cutting speed

b) Striation patterns for Q1 through Q5

Fig. 9. AWJ-machined five-finger part (Liu et al., 2009)

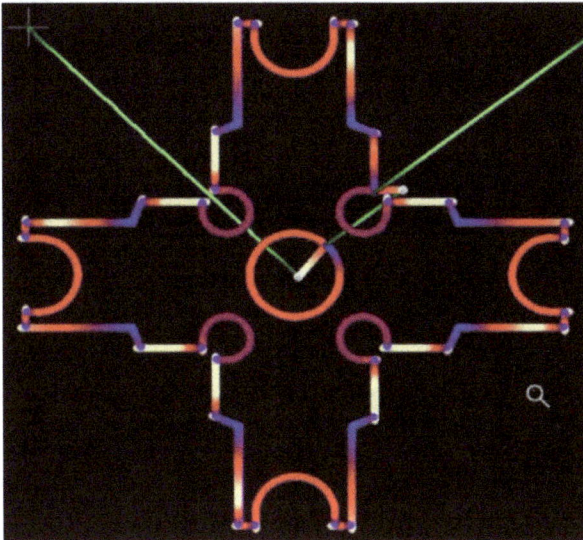

Fig. 10. Cutting speeds along tool path: white & light – fast; blue & dark – slow; green – traverse line (Olsen, 2009)

2.3 Fatigue performance

Current specifications require that AWJ-cut aluminum and titanium parts that will be used in fatigue-critical aerospace structures undergo subsequent processing to alleviate concerns of degradation in fatigue performance. It has been speculated that the striation patterns induced by AWJs (Fig. 9) could be a source of the initiation of micro-cracks under repeated loading. The requirement of a secondary process for AWJ-machined parts greatly negates the merits (cost effectiveness) of waterjet technology. An R&D program was initiated to revisit the fatigue performance of AWJ-machined aircraft aluminum and titanium parts for fatigue-critical applications by incorporating the most recent advances in waterjet technology (Liu et al., 2009a). [17] "Dog-bone" specimens were prepared by using AWJ and CNC machining. Several "low-cost" secondary processes, including dry-grit blasting with 180-grit aluminum oxide and sanding, were applied to remove the visual appearance of the striation patterns on AWJ-machined edges in an attempt to improve fatigue life. Fatigue tests of dog-bone specimens were conducted in the Fatigue and Fracture laboratory at the Pacific Northwest National Laboratory (PNNL).

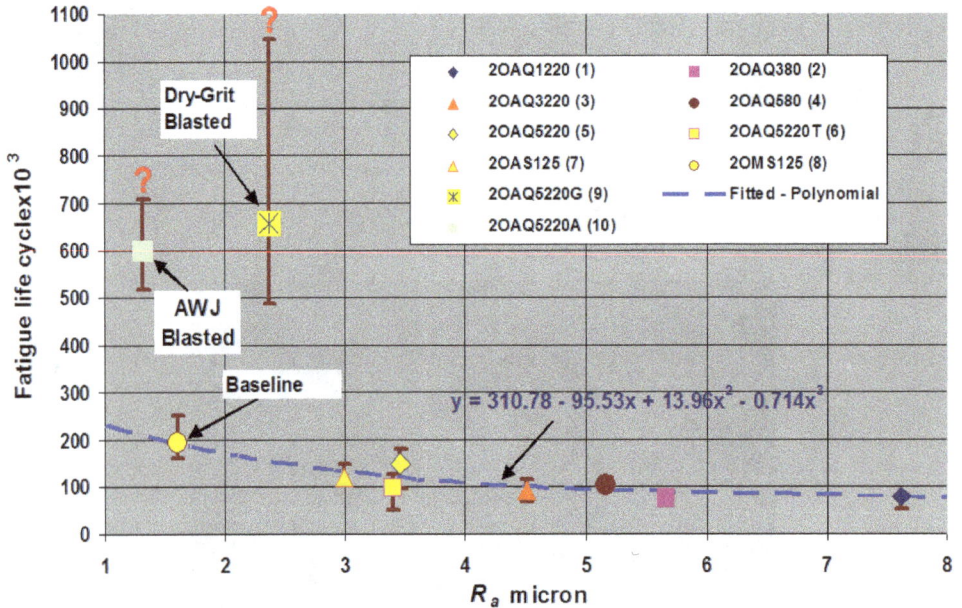

Fig. 11. Fatigue life versus R_a of aircraft aluminum 2024 T3 (Liu et al., 2009b)

Figure 11 illustrates the results of fatigue tests for the aluminum dog-bone specimens. The abscissa and ordinate are the edge surface roughness, R_a, and the fatigue life, respectively. For the AWJ-cut specimens, R_a was measured near the bottom of the edge where the amplitude of the striation is at the maximum. The "error bars" in the figure represent the maximum and minimum fatigue life values from the measurements. Except for the dry-grit

---

[17] This work was a collaboration among OMAX Corporation, Boeing, Pacific Northwest National Laboratory (PNNL), and National Institute of Standards and Technology (NIST).

and AWJ-blasted specimens, Fig. 11 shows that the fatigue life depends mainly on R_a whether the dog-bone specimens were machined with AWJs (both as-cut samples and those with secondary sanding) or conventionally.

On the other hand, the grit-blasting process only reduces the R_a from 3.4 to 2.3 μm, although most specimens did not break at the gage. The question marks (?) in the figure correspond to the number of test cycles at which the test was terminated, whether the specimen broke at the gage or not even though it was not necessarily a failure (at the gage). The "average" fatigue life for the specimens prepared with the combined process is at least 3 or 4 times longer, respectively, than that for specimens machined by a conventional tool or by AWJs (Liu et al., 2009b). Fatigue tests were also conducted on the aircraft titanium dog-bone specimens, and a similar trend of improvement in the fatigue performance was observed (Liu et al., 2011a).

Subsequent measurements at the x-ray diffraction facility of NIST's Center for Neutron Research have demonstrated that the dry-grit and AWJ blasting processes induce residual compressive stresses on the AWJ-machined edges (Liu et al., 2009b). The residual compressive stresses induced by dry-grit and AWJ blasting were responsible for the fatigue performance improvement. The ability to improve fatigue performance further would have a significant impact on many micromachined parts (e.g., orthopedic implants).

3. μAWJ technology

Waterjet machining, a top-down manufacturing process, is accomplished by erosion as individual high-speed water droplets in a WJ or abrasive particles in an AWJ/ASJ impinge onto a workpiece. At the microscopic scale, this erosion process by individual water droplets or abrasive particles is consistent with a micromachining process, especially when the droplet or particle size is at micron and submicron scales. At the macroscopic scale, the size of a machined feature, such as the diameter of a hole or the kerf width of a slot, is proportional to the diameter of the jet stream in which the water droplets and abrasives are confined. Therefore, the waterjet stream diameter governs the size of machined features and must be downsized appropriately for meso-micro machining. Figure 12, modified from the drawing presented by Miller (2005), compares the stream diameters of various types of waterjets with the beam diameters of lasers. The solid and dotted portions of the lines signify the normal and outside ranges of stream/beam diameters, respectively. The outside range is either difficult or impractical to achieve for meso-micro machining.

Most of the advantages of waterjet technology discussed in Section 1 apply equally to AWJ meso-micro machining. Although the technical feasibility of applying waterjet technology for meso-micro machining has been demonstrated since the mid-1990's, μAWJ technology remains in the research and development stage (Miller et al., 1996). For cutting a limited set of soft materials, WJs using micron-size orifices have been applied reliably in production environments. In principle, the two-phase ASJs are more aggressive and inherently have smaller stream diameters than the three-phase AWJs. The lack of suitable engineering materials to fabricate highly wear-resistant check valves and other components exposed to the aggressive ASJ has limited its pressures to between about 70 and 140 MPa (Jiang et al., 2005). Most ASJ systems adopt batch feeding of the slurry just upstream of the orifice in order to isolate the abrasives from the high-pressure pump.

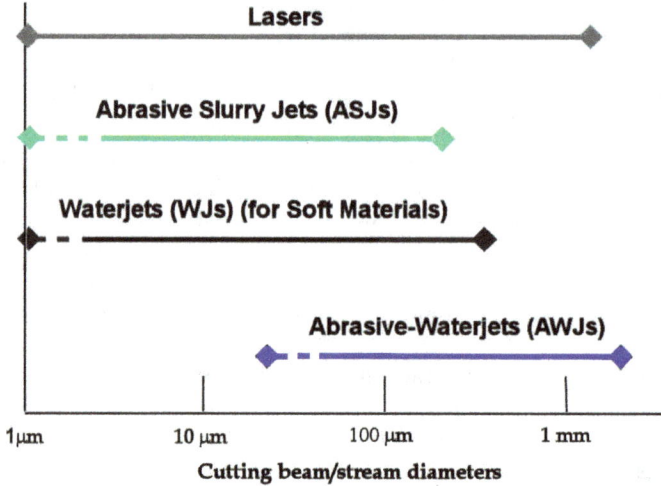

Fig. 12. Comparison of stream/beam diameters of waterjets and lasers (modified from Miller, 2005)

Abrasive-waterjets remain the mainstream of waterjet technology. Recent R&D efforts in further downsizing of AWJ nozzles have shown good promise, as described in Section 1. Most of the issues associated with downsizing AWJ nozzles beyond the current state of the art have been identified. Novel processes have been and are still being developed to meet the challenges to resolving these issues.

3.1 Challenges
Nowadays, the smallest features that can be machined with miniature AWJs are around 200 to 300 μm. Further downsizing of the AWJ presents considerable challenges and difficulties. The supersonic, three-phase AWJ is one of the most complex flow phenomena, particularly because it also involves fluid-fluid, fluid-solid, and solid-solid interactions in a rapidly changing spatial environment. When the ID of the mixing tube is reduced to sizes at which surface tension becomes important, the abrasive slurry has transitioned into a microfluidic flow, leading to additional complexities. In parallel, the size of the abrasive particles must be reduced proportionally with the size of the mixing tube. There are several concerns regarding gravity feeding of fine abrasives, as the flowability of the abrasives deteriorates with particle size distribution. In addition, fine abrasives tend to coagulate or clump together, causing difficulties in achieving consistent and uniform feeding of abrasives to the μAWJ nozzle. Several issues associated with AWJ machining and meso-micro machining are briefly discussed below.

3.1.1 Microfabrication of μAWJ nozzle
The μAWJ nozzle consists of three key components: the orifice, the mixing tube, and the nozzle body, in which the orifice and mixing tube are housed. The optimum ID ratio of the orifice and mixing tube is between 2 and 3. The optimum aspect ratio of the mixing tube (bore length to ID) is about 100 for production AWJ nozzles, which allows adequate acceleration of the entrained abrasives by the high-speed water droplets to produce a focused AWJ stream with

minimal spread at the nozzle exit. Downsizing of the orifice and nozzle body is within the current capability of micromachining technology, as orifices made of sapphire or diamond with IDs of 10 μm and smaller are commercially available. At present, mixing tubes with IDs greater than 200 μm are fabricated by wire EDM. The challenge is to fabricate mixing tubes with IDs less than 200 μm and an adequate aspect ratio.

3.1.2 Microfluidics

As the mixing tube ID is decreased, the capillary effect becomes increasingly important together with the increase in the slurry flow resistance. A meniscus column of water supported by the capillary force will eventually fill the entire bore of the mixing tube. A backsplash of water and abrasives, produced by the front of the AWJ impacting the upper surface of the water column, could reach the interior surface of the nozzle body and the lower portion of the abrasive feed tube, leaving a layer of wet abrasives on these surfaces. After many on-off cycles, buildup of the wet abrasive layers on those surfaces would restrict and eventually block the dry abrasives flowing into the mixing tube. An insufficient quantity of abrasives adversely affects the cut quality. Therefore, optimum nozzle operations to maintain a desired cut quality require the mitigation of the accumulation of wet abrasives on all interior surfaces of the nozzle and the feed tube.

Remedies have been developed, with limited success, to minimize the degree of nozzle clogging by wet abrasives. One of the remedies is the use of vacuum assist and water flushing (Hashish, 2008). The incorporation of such a remedy into μAWJ nozzles is, however, not suitable due to the resultant increase in bulkiness and added complexity in process control.

3.1.3 Feeding of fine abrasives

As a rule of thumb, the optimum size distribution of abrasives for a given nozzle is such that the maximum abrasive size is no larger than one-third of the mixing tube ID to prevent clogging due to the bridging of two large abrasives. For example, for a mixing tube with a 100-μm ID, the abrasive size must be less than about 30 μm. Although fine abrasive or powder flow is a complex phenomenon, because it is affected by so many variables, it is generally accepted that flowability under gravity feed increases with the particle size (Yu et al., 2011; Liu et al., 2008b). Flowability in this sense simply means the ability of a powder to flow. As the abrasives become finer and finer, they eventually cease to flow through the feed tube when being gravity fed. In addition, fine abrasives tend to coagulate or clump together by static electricity or humidity. Abrasives that clump together would have difficulty flowing from the hopper to the mixing tube. As a result, the abrasive flow is unsteady at best, which leads to inconsistent abrasive feeding or even flow stoppage. Inconsistent abrasive flow would result in deterioration of the cut quality.

3.1.4 Design optimization of traverse systems

For an AWJ system with a flexible jet stream, the quality of an AWJ-machined part depends on additional parameters inherent to the characteristics of a waterjet. Note that the jet spreads with the distance it travels. Therefore, the profile of the jet, the abrasive size distribution, the abrasive flow rate, and the uniformity of abrasive feed will all affect the machining precision and cut quality. For example, the kerf width of a slot and/or the minimum diameter of a circle are usually used to define the machining precision. The striation pattern and edge taper are often used as qualifiers for the cut quality. For meso-micro machining, special attention must be devoted to optimization of both the design of the X-Y traverse and the characteristics of the μAWJ. Nowadays, linear traverses can be built with nanometer resolution and position accuracy, but their costs increase exponentially with the resolution. In designing a μAWJ system, one must take into consideration the ultimate machining precision achievable with a μAWJ nozzle. Overdesigning the traverse system for AWJ meso-micro machining would have no real benefit and would only inflate the cost of goods.

3.2 Novel solutions under development

Under support from OMAX's R&D funds and an NSF SBIR Phase I grant, a series of feasibility investigations were conducted regarding the development of μAWJ technology. The success of these investigations, together with the tremendous market potential of μAWJ technology, has led to the award of an NSF SBIR Phase II grant for the development of a μAWJ system prototype for meso-micro machining.

R&D efforts to meet the challenges in developing the μAWJ technology and to resolve various issues associated with nozzle downsizing have led to the development of several novel approaches. Each of the issues described in Section 3.1 has been investigated carefully, and practical solutions have been proposed. Additional solutions are currently being sought.

- A better understanding of the relevant microfluidics has led to the development of novel processes to improve the flowability and uniform feeding of fine abrasives and, thus, mitigate nozzle clogging.
- Efforts are being made to reduce the tolerance stacking error.
- System optimization is being made to develop a prototype of an efficient and cost-effective precision μAWJ machine.
- R&D and beta miniature AWJ nozzles, with and without ancillary devices, have been developed for meso- and micro-scale test cutting.

Success in the above research efforts would facilitate further downsizing of AWJ nozzles with the goal of achieving the capability to machine features around 100 to 50 μm. Another benefit of nozzle downsizing is that multiple small nozzles can be supported by a single pump designed for large nozzles. For the beta nozzles, a multi-nozzle platform has been developed and tested for machining up to four identical parts simultaneously.

Test cuts were made using these nozzles and ancillary devices to fabricate samples from a broad range of engineering materials to demonstrate the versatility of waterjet technology for meso-micro machining. The designs, drawings, and materials of many test samples were furnished by industrial and academic collaborators and customers with specific applications in mind. The finished parts were subsequently returned to the providers for inspection and evaluation. Based on the results of these evaluations, the performance of μAWJ technology for various applications was assessed.

4. AWJ-machined samples and features

In this section, selected machined samples are presented and discussed to demonstrate the versatility of μAWJ technology for meso-micro machining.

4.1 Basic features

The R&D nozzles and ancillary devices discussed above were used to pierce holes and machine slots to demonstrate their ability to machine very small features. Emphasis was made to determine the smallest features that could be machined with these nozzles with and without ancillary devices and also without perfecting the piercing and machining processes. Figure 13 illustrates 3D micrographs of a small hole pierced in a thin stainless steel shim 100 µm thick. The hole was pierced with a 380-µm nozzle and a proprietary ancillary device to reduce the effective diameter of the AWJ.[18] The diameter of the hole on both the entry and exit sides is around 100 µm. As a rule of thumb, the size of AWJ-cut features is slightly larger than the stream diameter of the AWJ or the mixing tube. In other words, the smallest hole that can be pierced with the 380-µm nozzle is approximately 400 µm. The ancillary device has effectively reduced the stream diameter of the AWJ by a factor of 4. The advantage of using such a device is to enable relatively large AWJ nozzles and coarse abrasives to cut parts with features smaller than the diameter of the mixing tube. Note that Fig. 13 represents micrographs of the as-pierced hole. The circularity of the hole could be significantly improved by trepanning after piercing, which would increase the hole diameter slightly.

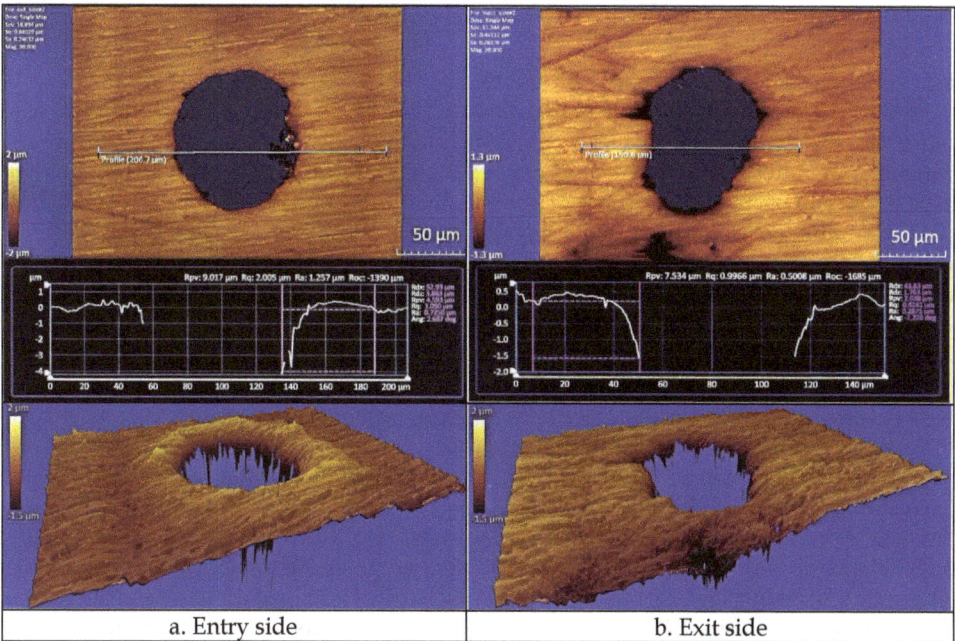

| a. Entry side | b. Exit side |

Fig. 13. Micrographs of holes pierced on 316 stainless steel shim - courtesy of Zygo Corp and Microproducts Breakthrough Institute (Liu et al., 2011b)

[18] The 380mm nozzle is one of the production nozzles. The nozzle size refers to the mixing tube ID which is twice the orifice ID.

Figure 14 illustrates three narrow slots machined in a stainless steel shim 0.25 mm thick. The kerf width of the slots was measured to be approximately 92 µm. These slots were machined with the 380-µm nozzle together with another ancillary device to reduce the effective stream diameter of the nozzle. Note that the kerf width is smaller than the width of the fingerprints impressed on the surface of the shim.[19]

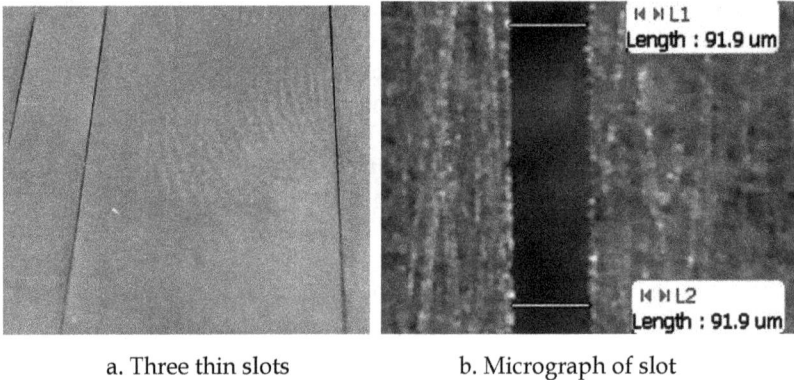

a. Three thin slots   b. Micrograph of slot

Fig. 14. AWJ-machined slots using a novel ancillary device (Liu & Schubert, 2010)

4.2 Components for green energy products

At the Precision Engineering Research Group (PERG) of MIT, a novel concept was conceived to improve the efficiency of small motors/generators by means of surface-mounted armatures.[20] Conventional armatures composed of multiple turns of round wires have a low compaction factor (ratio of volume filled by conducting wires to total allotted conductor volume) because there are gaps among wires resulting in relatively high armature resistance and reduced performance. The concept considered replacing wire-wound armatures with slotted copper tubes of appropriate annular dimensions (Trimble, 2011 - patent pending). The smaller the kerf width of the cuts, the higher the compaction factor and better the performance of the armature become. Researchers at PERG ran into problems in machining narrow slots on copper tubes. Due to the high reflectivity of copper, laser cutting splattered. On the other hand, the tube geometry prohibited the use of wire EDM, while sinking EDMs were too slow and costly.

Abrasive-waterjets were applied cost-effectively to machine large-aspect-ratio microchannels for fuel cells on stainless steel and titanium shims (Liu et al., 2008a). Certain AWJ-cut slot/rib patterns on some materials were fabricated that were otherwise too difficult and/or costly to be machined by conventional tools. By mounting the copper tube on a rotary indexing tool, a 254-µm nozzle successfully machined 16 narrow slots on the copper tube, as illustrated in Fig. 15.[21] Preliminary results demonstrated that µAWJ technology could be an enabling tool for this method of surface armature manufacturing.

---

[19] Since the kerf width of a slot rather than the separation between slots is limited by the stream diameter of the µAWJ, the minimum scale of micromachining is therefore defined by the kerf width.

[20] http://pergatory.mit.edu/ (8 August 2011)

[21] The 254-µm nozzle is an R&D nozzle currently being beta tested.

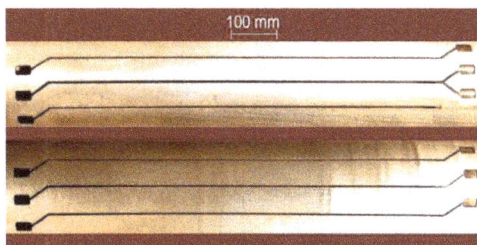

Fig. 15. AWJ-slotted copper tubes (design provided by MIT PERG) (Liu et al., 2011b)

4.3 Biomedical components

Advanced micromachining technology has been favorable for fabricating biomedical devices, which are continually becoming smaller and more intricate in terms of size, shape, and material. μAWJ technology shows great potential for such applications based on the market size and current trends, the urgent need for cost reductions in healthcare, and the nature of biomedical components. For example, mini- and micro-plates for orthopedic implants to repair/reconstruct bone and skull fractures are one of the strong candidates for μAWJ technology (Haerle et al., 2009). Titanium is often used for these plates because of its biocompatibility. While conventional machine tools have difficulty in machining titanium, AWJs cut titanium 34% faster than stainless steel at considerably lower costs (Fig. 8). Note that material toughness does not play an important role in AWJ cutting. Therefore, materials that would normally be cut in an annealed condition may be cut in a hardened condition with insignificant loss of productivity. For emergency operations, in particular, a part could be machined by an AWJ from design to finish in minutes. Another potential benefit is improvements in the fatigue performance of implants via dry-grit blasting (Section 2.3).

Figure 16 illustrates several samples of AWJ-machined mini- and micro-plates made of titanium and stainless steel. The mesh-type mini-plates are made from titanium shim stock 0.34 mm thick. These plates are commonly used in facial and skull repair and reconstruction (Haerle et al., 2009). Machining was carried out at a pressure of 380 MPa with the use of the 254-μm nozzle. The fine-mesh mini-plate (lower right in photo) took about 20 minutes to complete. Optimization of the nozzle performance is expected to reduce the machining time.

Fig. 16. Titanium and stainless steel orthopedic parts. Scale: mm (Liu et al., 2011b)

The same 254-µm nozzle was also applied to cut a flexure to be used as a component of a medical device (Begg, 2011 - patent pending). The material was 6061 T3 aluminum with a thickness of 9.5 mm. The key feature is a narrow bridge, with a target width of 0.25 mm, between two connecting members of the flexure. Figure 17 shows a portion of an AWJ-cut flexure. The two lengths, L1 and L2, shown in the photo are the widths of one of the narrow bridges and one of the connecting members of the flexure. Note that the width of the bridge was measured at 0.31 mm.

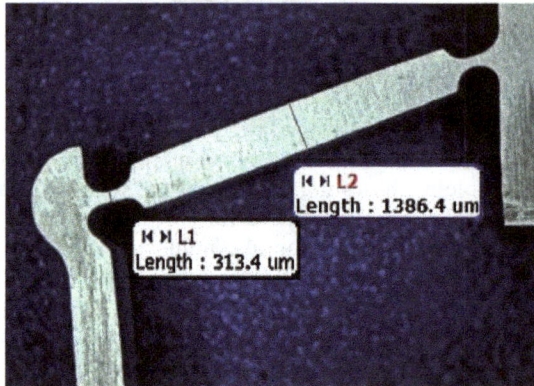

Fig. 17. Small aluminum flexure (MIT PERG) (Liu et al., 2011b)

4.4 Planetary gear set

For machining miniature mechanical components, the 254-µm nozzle was used to cut a set of planetary gears. The set consists of seven gears [a sun gear (9.68 mm OD), a ring gear (19.05 mm OD), and five small planetary gears (3.55 mm OD)], a gear mounting plate, and a gear carrier. Figure 18 illustrates the components of the gear set. Also included are the tool paths of the seven gears corresponding to the screen display of the PC-based CAD program

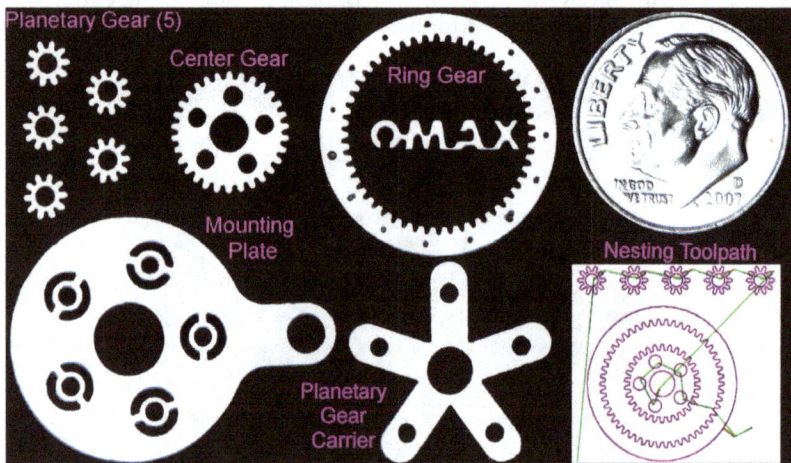

Fig. 18. Components and tool paths of planetary gear (Liu et al., 2011b)

LAYOUT.[22] Note that tool paths with a magenta color represent an edge quality of 3 out of 5, with 5 as the best edge quality. The ring and sun gears were nested to save material. The LAYOUT diagram was then transferred to the CAM program MAKE to machine the planetary gear set from 0.62-mm-thick stainless steel plate using the 254-μm nozzle.[23] Figure 19 illustrates the assembled planetary gear set driven by a micro spur gear head motor (Solarobotics, Model GM14a) powered by a single AAA battery.

Fig. 19. Gear assembly (Liu et al., 2011b)

4.5 Near 3D parts

The 380-μm nozzle was set up together with the Rotary Axis to machine axisymmetric features.[6] The Space Needle models shown in Fig. 4 are examples cut with that setup. Subsequently, a titanium tube with an OD of 6 mm and an ID of 0.6 mm was mounted on the Rotary Axis. Interlocking features were machined on the titanium tube by the nozzle with the Rotary Axis rotating. A steel rod was inserted into the titanium tube serving as a sacrificial

Fig. 20. Titanium interlocking link (Liu et al., 2011b)

---

[22] http://www.omax.com/waterjets/layout-software (8 August 2011)
[23] http://www.omax.com/waterjets/make-software (8 August 2011)

piece to protect the opposite wall of the tube from being damaged by the spent high-speed abrasives. Figure 20 illustrates a photograph of the interlocking link. Since there are no soldering joints on the tube, the link is quite strong as compared with similar ones that are welded together. Also shown in the figure is a magnified view of one of the individual links. Figure 21 illustrates an A-Jet-cut aluminum blisk (OD = 25.5 mm; thickness = 13.1 mm). With the combination of the 380-μm nozzle and the A-Jet capable of tilting the AWJ up to ±60 degrees from the vertical, many complex 3D features can be readily machined.

Fig. 21. A-Jet-machined aluminum blisk. Scale: mm.

The Intelli-ETCH (patent pending) is an advanced utility of the OMAX Intelli-MAX® software that allows a user to recreate images in various materials. These images are created from standard bitmap files (JPG, TIFF, BMP, etc.).[9] By taking the brightness levels of an image and converting those levels into machine speeds, 3D features of the images can be etched onto substrates. Figure 22 illustrates an example of an AWJ-etched lizard on an aluminum substrate. The Intelli-ETCH would have great market potential for controlling μAWJ nozzles for use as versatile jewelry/craft making tools.

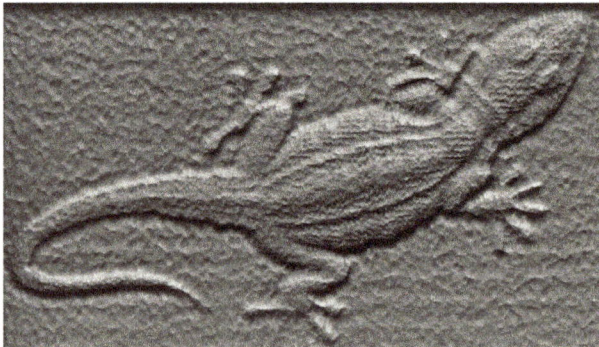

Fig. 22. AWJ-etched lizard on aluminum substrate (Webers et al., 2010)

4.6 Non-metal samples

Several samples made of non-metal materials were machined with the beta and R&D nozzles to demonstrate the material independence of waterjet technology. The materials included various composites and ceramics with machinability indexes ranging from about 700 to 4 (refer to Fig. 8). Figure 23 shows miniature samples machined from various composites using the 254-μm nozzle (Liu et al., 2010a). The material used for each sample is given in the figure subtitle, along with a number in parentheses that is the thickness of the part in millimeters. Details of small features on the order of 100 μm in size remain sharp and crisp. There is no delamination or chipping on the edges. The thickness of the wheel of the smallest bike is about 200 μm. The carbon fiber (dark) and the epoxy (translucent) layers on the wheels are clearly identifiable in Fig. 23e.

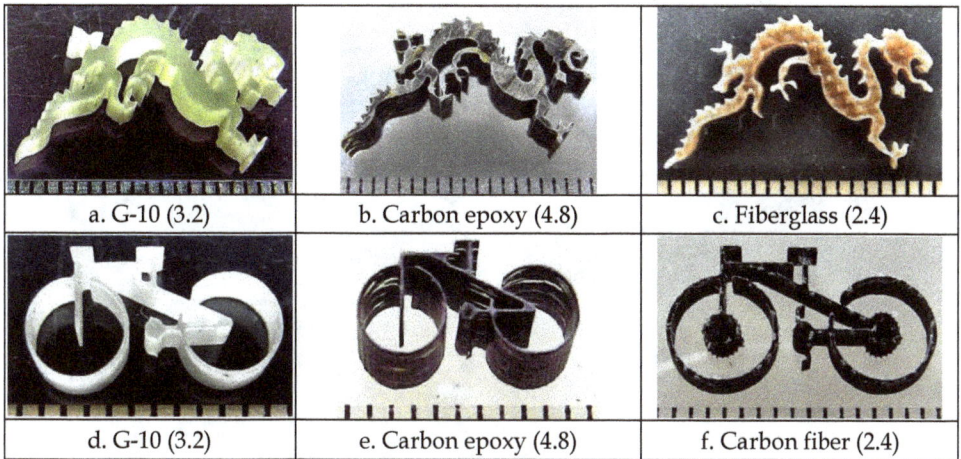

| a. G-10 (3.2) | b. Carbon epoxy (4.8) | c. Fiberglass (2.4) |
| d. G-10 (3.2) | e. Carbon epoxy (4.8) | f. Carbon fiber (2.4) |

Fig. 23. AWJ-machined miniature composite parts. Numbers in parentheses are thickness of part in mm. Scale: 1 mm/div. (Liu et al., 2010a)

Figure 24 illustrates features machined with the 254-μm nozzle in an alumina plate 0.64 mm thick (M ≈ 4). The sharp and crisp edges of all features are evident.

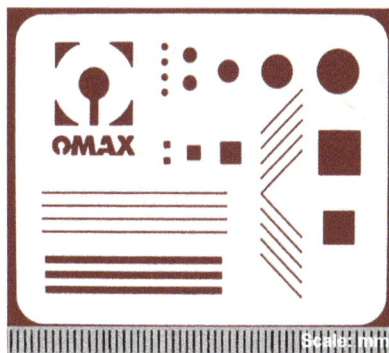

Fig. 24. Features machined with the 254-μm nozzle in alumina thin plate (Liu, 2009)

4.7 Multi-nozzle platform

The downsizing of an AWJ nozzle results in a reduction in the flow rate of the waterjet. Depending on the size of the orifice, the number of nozzles that can be supported by a pump increases accordingly. From Fig. 2, a 22.4-kW pump that is capable of supporting one 360-µm orifice operating at 380 MPa with a water flow rate of 3.4 l/min is capable of supporting four 254-µm nozzles operating at the same pressure. A multi-nozzle platform on which four 254-µm nozzles could be mounted was designed, assembled, and tested, as illustrated in Fig. 25. The platform was subsequently delivered for beta testing at a specialty jewelry manufacturing shop. With the nozzles operating in tandem, four identical parts can be machined simultaneously to boost productivity. Among the advantages of using the 254-µm nozzle together with 320-mesh garnet are that the amplitude of the striation is small and the finished parts are nearly free of burrs.

Fig. 25. Four nozzles mounted on a platform for increased productivity (Liu et al., 2011b)

5. Conclusion

Waterjet technology has inherent technological and manufacturing merits that make it suitable for machining most materials from macro to micro scales. It has been established as one of the most versatile precision machining tools and has proven amenable to micromachining. This technology has emerged as the fastest growing segment of the overall machine tool industry in the last decade.[3]

The smallest features that can be machined with state-of-the-art commercial AWJ systems are limited to greater than 200 µm. Further downsizing of AWJ nozzles for machining features less than 200 µm has met with considerable challenges, as described in Section 3.1. These challenges, which are due to the complexity of the jet flow as the AWJ flow characteristics change into microfluidics, include nozzle clogging by accumulation of wet abrasives, difficulty in the fabrication of mixing tubes with exit orifices less than 200 µm, the degradation in the flowability of fine abrasives, and other relevant issues.

Novel manufacturing and operational processes and ancillary devices have been investigated and developed to meet the above challenges. Miniature beta and R&D nozzles, without the need for vacuum assist and water flushing, have been assembled and tested to machine miniature samples made of various materials for a broad range of applications. Many of the samples with basic features as small as 100 μm were machined to demonstrate the versatility of waterjet technology for low-cost micromanufacturing of components for medical implants/devices and microelectronics, for green energy production systems, and for the post-processing of various micro-nano products.

The advancement and refinement of μAWJ technology continue. Efforts are being made to further downsize μAWJ nozzles for machining features around 100 and 50 μm. The goal is to commercialize a μAWJ system by integrating μAWJ nozzles with a low-cost, low-power, high-pressure pump and a precision small-footprint X-Y traverse. A host of accessories are already available to be downsized for facilitating 3D meso-micro machining.

6. Acknowledgment

This work was supported by an OMAX R&D fund and NSF SBIR Phase I and II Grants #0944229 and #1058278. A part of the work was supported by U. S. Pacific Northwest National Laboratory (PNNL) under Technology Assistance Program (TAP) Agreements: 07-29, 08-02, 09-02, and 10-02. Any opinions, findings, and conclusions or recommendations expressed in this material are those of the authors and do not necessarily reflect the views of the NSF and PNNL. Contributions from research institutes and industrial partners by furnishing sample materials and part drawings and by evaluating AWJ-machined parts are acknowledged. Collaborators include but are not limited to Microproducts Breakthrough Institute (MBI), MIT Precision Engineering Research Group, Ryerson University, and several OMAX's customers and suppliers. The authors would like to thank their colleagues at OMAX for reviewing the article and providing us with constructive feedback.

7. References

Bachelor, G. K. (1967). *An Introduction to Fluid Dynamics*, Cambridge University Press, ISBN 0521663962.

Begg, N. (2011). Blind Transmembrane Puncture Access: Design and Development of a Novel Laparoscopic Trocar and Blade Retraction Mechanism, Master Thesis, Mechanical Engineering Department, MIT, pp. 103.

Chen, W.-L. & Geskin, E. S. (1990). Measurements of the Velocity of Abrasive Waterjet by the Use of Laser Transit Anemometer, *Proceedings of 10th International Symposium on Jet Cutting Technology*, BHRG Fluid Engineering, Amsterdam, Netherlands, October 3-November 2, pp. 23-36, 1990.

Haerle, F.; Champy, M., & Terry, B. (Ed) (2009). *Atlas of Craniomaxillofacial Osterosynthesis: Microplates, Miniplates, and Screws*, 2nd Ed., Thieme, New York, pp. 225.

Hashish, M. (2008). Abrasive-Waterjet Machining of Composites, *Proceedings 2009 Amererican WJTA Conference and Exposition*, Houston, TX, August 18-20.

Henning, A.; Miles, P., Stang, D., (2011a). Efficient Operation of Abrasive Waterjet Cutting in Industrial Applications, *Proceedings of 2011 WJTA-IMCA Conference &* , Houston, Texas, September 19-21.

Henning, A.; Liu, H.-T., & Olsen, C. (2011b). Economic and Technical Efficiency of High Performance Abrasive Waterjet Cutting, to appear in *ASME Journal of Pressure Vessel & Piping*.

Jiang, S.; Popescu, R., Mihai, C., & Tan, K., (2005). High Precision and High Power ASJ Singulations for Semiconductor Manufacturing, *Proceedings of 2005 WJTA Conference and exposition.*, Houston, Texas, August 21–23, Paper 1A-3.

Lamb, H. (1993). *Hydrodynamics* (6th Ed.). Cambridge University Press, ISBN 9780521458689.

Liu, H.-T. (1998). Near-Net Shaping of Optical Surfaces with Abrasive Suspension Jets, *Proceedings of 14th International Conference on Jetting Technology*, Brugge, Belgium, September 21–23, pp. 285-294.

Liu, H.-T. (2010). Waterjet Technology for Machining Fine Features Pertaining to MicroMachining, *Journal of Manufacturing Processes*, Vol.12, No.1, pp. 8-18. (doi:10.1016/ j.jmapro.2010.01.002).

Liu, H.-T. & McNiel, D. (2010). Versatility of Waterjet Technology: from Macro and Micro Machining for Most Materials, *Proceedings of 20th International Conference on Water Jetting*, October 20–22, Graz, Austria.

Liu, H.-T. & Schubert, E. (2010). Piercing and/or Cutting Devices for Abrasive Waterjet Systems and Associated Systems and Methods, U. S. Provisional Patent Application #612348007.US.

Liu, H.-T.; Hovanski, Y., & Dahl, M. E. (2011a). Machining of Aircraft Titanium with Abrasive-Waterjets for Fatigue Critical Applications, to appear in *ASME Journal of Pressure Vessel & Piping*

Liu, H.-T.; Schubert E., and McNiel, D. (2011b). μAWJ Technology for Meso-Micro Machining, *Proceedings of 2011 WJTA-IMCA Conference and Exposition.*, Houston, September 19-21.

Liu, H.-T.; Miles, P., & Veenhuizen, S. D. (1998). CFD and Physical Modeling of UHP AWJ Drilling, *Proceedings of 14th International Conference on Jetting Technology*, Brugge, Belgium, September 21–23, pp. 15-24.

Liu, H.-T.; Hovanski, Y., Dahl, M. E. & Zeng, J. (2009a). Applications of Abrasive-Waterjets for Machining Fatigue-Critical Aerospace Aluminum Parts, *Proceedings of ASME PVP2009 Conference*, Prague, Czech, July 26-30.

Liu, H.-T.; Gnäupel-Herold, T., Hovanski, Y., & Dahl, M. E. (2009b). Fatigue Performance Enhancement of AWJ-Machined Aircraft Aluminum with Dry-Grit Blasting, *Proceedings 2009 American WJTA Conference*, Houston, Texas, August 18-20.

Liu, H.-T.; Hovanski, Y., Caldwell, D. D., & Williford, R. E. (2008a). Low-Cost Manufacturing of Flow Channels with Multi-Nozzle Abrasive-Waterjets: A Feasibility Investigation, *Proceedings of 19th International Conference on Water Jetting*, Nottingham, UK: October, 15-17.

Liu, H.-T.; Schubert, E., McNiel, D., & Soo K. (2010a). Applications of Abrasive-Waterjets for Precision Machining of Composites, *Proceedings* of *SAMPE 2010 Conference and Exhibition*, May 17-20, Seattle, Washington, USA.

Liu, H.-T., Schubert E., and McNiel, D. (2010b) "Development of Micro Abrasive-Waterjet Technology," SciTopics, October, pp. 4 (http://www.scitopics.com/Development_ of_Mirco_Abrasive_Waterjet_Technology.html) (8 August 2011).

Liu, L. X.; Marziano, I., Bentham, A .C., Litster, J. D., White, E. T., & Howes, T. (2008b). Effect of Particle Properties on the Flowability of Ibuprofen Powders, *International Journal of Pharmerceutical*. Oct 1; 362(1-2): pp. 109-17. Epub 2008 Jul 4.

Miller, D. S. (2005). New Abrasive Waterjet Systems to Complete with Lasers, *Proceedings of 2005 WJTA Conference and exposition*, Houston, Texas, August 21-23, Paper 1A-1.

Miller, D.; Claffey, S., & Grove, T. (1996). Technology Package for Abrasive Waterjets, BHR Croup Limited.

Momber, A. W. & Kovacevic, R. (1998). *Principles of Abrasive Water Jet Machining*, Springer-Verlag, Berlin.

Olsen, J. H. (2009). Limits to the Precision of Abrasive Jet Cutting, *Proceedings of the 9th Pacific Rim International Conference on Water Jetting Technology*, Koriyama, Japan, November 20-23, Invited Paper

Olsen, J. H. (1996). Motion Control with Precomputation, *U.S. Patent No. 5,508,596*, April.

Olsen, J. H. (2005). Automated Fluid-Jet Tilt Compensation of Lag and Taper, U. S. Patent No. 6,922,605, July.

Roth, P.; Looser, H., Heiniger, K. C., & Buhler, S. (2005). Determination of Abrasive Particle Velocity Using Laser-Induced Fluorescence and Particle Tracking Methods in Abrasive Waterjets, *Proceedings of 2005 WJTA Conference and exposition.*, Houston, Texas, August 21–23, Paper 3A-2

Smith, R. W.; Hirshberg, M. H., & Manson, S. S. (1967). Fatigue Behavior of Materials under Strain Cycling in Low and Intermediate Life Range, NASA Technical Note D-1574, NASA Lewis Research Center, pp. 25.

Stevenson, A. N. J. & Hutchings, I. M. (1995). Scaling Laws for Particle Velocity in the Gas-Blast Erosion Test, *Wear*, 181-183, pp. 56-62.

Swanson, R. K.; Kilman, M., Cerwin, S., & Tarver, W. (1987). Study of Particle Velocities in Water Driven Abrasive Jet Cutting, *Proceedings of 4th U.S. Water Jet Conference*, ASME, Berkeley, CA, August 26-28, pp. 103-107

Trieb, F. H. (2010). Waterjet Cutting in Austria - History and Actual Status, Keynote Speech, *Proceedings of 20th International Conference on Water Jetting*, Graz, Austria, October 20-22, pp. 3-17.

Trimble, A. Z. (2011). Vibration Energy Harvesting of Wide-Band Stochastic Inputs with Application to an Electro-Magnetic Rotational Energy Harvester, Ph.D. Dissertation, Mechanical Engineering Department, MIT, June, pp. 207.

Yu, W.; Muteki, K., Zhang, L. & Kim, G. (2011). Prediction of Bulk Power Flow Performance Using Comprehensive Particle Size and Particle Shape Distribution, *Journal of Pharmaceutical Sciences.*, Vol.100, No.1, January (also *DOI 10.1002/jps.22254*)

Webers, N.; Olsen, C., Miles, P. & Henning, A. (2010) Etching 3D Patterns with Abrasive Waterjets, *Proceedings of 20th International Conference on Water Jetting*, Graz, Austria, October 20–22, pp. 51-63.

Zeng, J. & Kim, T. J. (1995), Machinability of Engineering Materials in Abrasive Water Jet Machining, *International Journal of Water Jetting Technology*, Vol.2, No.2, pp. 103-110.

11

Integrated MEMS: Opportunities & Challenges

P.J. French and P.M. Sarro
Delft University of Technology,
The Netherlands

1. Introduction

For almost 50 years, silicon sensors and actuators have been on the market. Early devices were simple stand-alone sensors and some had wide commercial success. There have been many examples of success stories for simple silicon sensors, such as the Hall plate and photo-diode. The development of micromachining techniques brought pressure sensors and accelerometers into the market and later the gyroscope. To achieve the mass market the devices had to be cheap and reliable. Integration can potentially reduce the cost of the system so long as the process yield is high enough and the devices can be packaged. The main approaches are; full integration (system-on-a-chip), hybrid (system-in-a-package) or in some cases separate sensors. The last can be the case when the environment is unsuitable for the electronics. The critical issues are reliability and packaging if these devices are to find the applications. This chapter examines the development of the technologies, some of the success stories and the opportunities for integrated Microsystems as well as the potential problems and applications where integration is not the best option.

The field of sensors can be traced back for thousands of years. From the moment that humans needed to augment their own sensors, the era of measurement and instrumentation was born. The Indus Valley civilisation (3000-1500 BC), which is now mainly in Pakistan, developed a standardisation of weight and measures, which led to further developments in instrumentation and sensors. The definition of units and knowing what we are measuring are essential components for sensors. Also if we are to calibrate, we need a reference on which everyone is agreed.

When we think of sensors, we think in terms of 6 signal domains, and in general converting the signal into the electrical domain. The electrical domain is also one of the 6 domains. The signal domain is not always direct, since some sensors use another domain to measure. A thermal flow sensor is such an example, and these devices are known as "tandem sensors". The signal domains are illustrated in Figure 1.

Over the centuries many discoveries led to the potential for sensor development. However, up to the 2nd half of the 20th century sensor technology did not use silicon. Also some effects in silicon were known, this had not led to silicon sensors. The piezoresistive effect was discovered by Kelvin in the 19th century and the effect of stress on crystals was widely studied in the 1930s, but the measurement of piezoresistive coefficients made by Smith in 1954, showed that silicon and germanium could be good options for stress/strain sensors (Smith, 1954). Many other examples can be found of effects which were discovered and a century later found to be applicable in silicon.

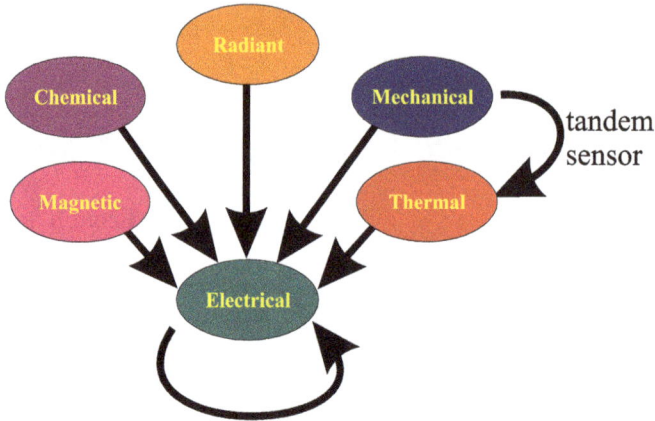

Fig. 1. The six signal domains

An important step towards The beginnings of integrated sensors go back to the first transistor, invented in 1947 by William Shockley, John Bardeen and Walter Brattain, while working at Bell Labs., which was fabricated in germanium. This quickly led to thoughts of integrating more devices into a single piece of semiconductor. In 1949 Werner Jacobi working at Siemens filed a patent for an integrated-circuit-like semiconductor amplifying device (Jacobi, 1949). In 1956 Geoffrey Dummer, in the UK, tried to make a full IC but this attempt was unsuccessful. In 1958 Jack Kilby, from Texas Instruments made the first working IC in germanium (Texas Instruments, 2008). This first device is illustrated in Figure 2. Six months later Robert Noyce, from Fairchild Semiconductor came up with hi own IC in silicon and manage to address a number of practical problems faced by Kilby. From these simple beginnings has come a major industry worth billions. John Bardeen, Walter H. Brattain and William B. Shockley won the Nobel Prize in 1956 and Jack Kilby in 2000.

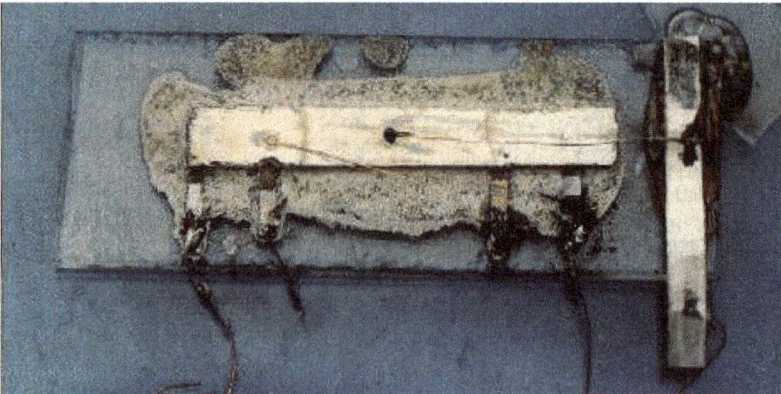

Fig. 2. First working IC

The discovery of sensing effects in silicon and the development of electronic devices in silicon led to many new sensor developments. In the 1950s the idea of p-n junctions for photocells was first investigated (Chapin, 1954). Staying within the radiation domain groups is Philips and Bell Labs. worked in parallel to develop the first CCD devices (Sangster, 1959 & Boyle, 1970).

At Philips, in the Netherlands, work had begun on a silicon pressure sensor and this early micromachined sensor is given in Figure 3 (Gieles, 1968 & 1969). The membrane was made using spark erosion and chemical etching, but the breakthrough was that the whole structure was in one material and therefore thermal mismatches were avoided.

Fig. 3. Early pressure sensor in the early 1960's

Silicon had now been shown to be a material with many effects interesting for sensor development. The work of Gieles showed that the material could be machined. Work from Bean (Bean, 1978) showed the greater opportunities etching silicon with anisotropic etchants, and Petersen (Petersen, 1982) showed the great mechanical properties of silicon. The early days of IC and sensor development were quite separate, but time has shown that these two fields can benefit from each other leading to new devices with greater functionality.

2. Technology

Many of the technologies used in silicon sensors were developed for the IC industry, although the development of micromachining led to a new range of technologies and opportunities for new devices. IC technology is basically a planar technology, whereas micromachining often requires working in 3 dimensions which has presented new challenges, in particular when the two technologies were combined to make smart devices.

2.1 Planar IC technology

The basis of planar technology was developed in the 1940s with the development of a pn-junction, although the major breakthrough was in 1958 with the first IC. This development enabled more and more devices to be integrated into a single piece of material. IC processing can be seen as a series of steps including; patterning, oxidation, doping, etching and deposition. These have been developed over the decades to optimise for the IC requirements and to advance the devices themselves. The following sections will give a brief description of the main steps.

2.1.1 Lithography

Lithography is a basic step carried out a number of times during a process. Basically a resist layer is spun on to the wafer and, after curing, exposed to UV light through a mask. If we use positive resist, this will soften through exposure and negative resist will harden. This can be done using a stepper (which projects the image onto each chip and steps over the wafer) or a contact aligner where the mask is a 1:1 image of the whole wafer. There are also techniques such as e-beam and laser direct write.

2.1.2 Oxidation and deposition
Silicon oxidises very easily. Simply left exposed at room temperature and oxide layer of 15-20Å will be formed. For thicker oxides the wafer is exposed to an oxygen atmosphere at temperatures between 700-1200°C. For thick oxides, moisture is added (wet oxidation) to increase the growth rate.

A number of deposition steps are used in standard processing. The first of these is epitaxy. Epitaxy is the deposition, using chemical vapour deposition (CVD), of a thick silicon layer, usually single crystal, although polycrystalline material can also be deposited in an epi-reactor (Gennissen, 1997). The second group of depositions are low pressure CVD (LPCVD) and plasma enhanced CVD (PECVD). Some examples of LPCVD processes are given in Table 1. PECVD uses similar gasses, but the use of a plasma reduces the temperature at which the gasses break down, which is of particular interest with post-processing, where thermal budget is limited (Table 2). The temperatures for PECVD can be reduced through adjusting other process parameters. These are only examples and there are many other options.

Layer	Gasses	Temperature
Polysilicon	SiH_4	550°C-700°C
Silicon nitride	$SiH_2Cl_2 + NH_3$	750°C-900°C
	$SiH_4 + NH_3$	700°C-800°C
Silicon dioxide undoped	$SiH_4 + O_2$	400°C-500°C
PSG (phosphorus doped)	$SiH_4 + O_2 + PH_3$	400°C-500°C
BSG (boron doped)	$SiH_4 + O_2 + BCl_3$	400°C-500°C
BPSG (phosphorus/boron doped)	$SiH_4 + O_2 + PH_3 + BCl_3$	400°C-500°C
Silicon carbide	$SiH_4 + CH_4$	900°C-1050°C

Table 1. Examples of LPCVD processes.

Layer	Gasses	Temperature
a-Si	SiH_4	400°C
Silicon nitride	$SiH_4 + NH_3 + N_2$	400°C
Silicon dioxide undoped	$SiH_4 + N_2 + N_2O$	400°C
Silicon dioxide, (TEOS)	$TEOS + O_2$	350°C
Oxynitride	$SiH_4 + N_2 + N_2O + NH_3$	400°C
BPSG (phosphorus/boron doped)	$SiH_4 + N_2 + N_2O + PH_3 + B_2H_6$	400°C
	$SiH_4 + CH_4$	
Silicon carbide		400°C

Table 2. Examples of PECVD processes.

The last of the deposition processes is the metallisation, which is usually done by sputtering or evaporation, which is widely used for metals.

2.1.3 Doping
An essential part of making devices is to be able to make p and n type regions. The main dopants are: As, P and Sb for n-type material and B for p-type material. The main techniques to dope silicon are diffusion and implantation. Diffusion is a process where the wafer is exposed to a gas containing the dopant atoms at high temperature. Implantation is a process in which the ions are accelerated towards the wafer at high speed to implant into the material.

2.1.4 Etching
Etching can be divided into two groups, wet and dry. Standard wet etching, in standard IC processing, is used for etching a wide range of materials. However, in recent years, dry etching is also widely used. Commonly used wet etchants are given in Table 3.

Etchant	Target		Target
49% HF	SiO_2	Si etch (85%, 160°C) $126 HNO_3$; $60 H_2O$; $5NH_4F$	Si
5:1 BHF	SiO_2		
Phosphoric acid	SiN	Aluminium etch $16H_3PO_4$; HNO_3; 1Hac; $2H_2O$; (50°C)	Al

Table 3. Commonly used wet etchants.

3. Micromachining technologies

Micromachining technologies moved the planar technology for IC processing into the 3rd dimension. These technologies can be divided into two main groups, bulk micromachining and surface micromachining. In addition there is epi-micromachining which is a variation on the surface micromachining. The following sections give a brief outline of these technologies. A more detailed description can be found in the chapter on micromachining.

3.1 Bulk micromachining
Bulk micromachining can be divided into two main groups: wet and dry. There are also other techniques such as laser drilling and sand blasting. The first to be developed was wet etching. Most wet micromachining processes use anisotropic, such as KOH, TMAH, hydrazine or EDP. These etchants have an etch rate dependant upon the crystal orientation allowing well defined mechanical structures (Bean, 1978). The basic structures made with these etchants are given in Figure 4 with their properties in Table 4.
All of these processes are relatively low temperature and can therefore be used as post-processing after IC processing, although care should be taken to protect the frontside of the wafer during etching.
Bulk micromachining can also be achieved through electrochemical etching in HF. For this etchant there are two distinct structures. The first is micro/nano porous which is usually an isotropic process, or macro-porous which is an anisotropic process. The micro/nano pore structure can be easily removed due to its large surface area to leave free-standing structures (Gennissen, 1995). Porous silicon/silicon carbide can then be used as a sensor material such as humidity or ammonia sensors (O' Halloran, 1998, Connolly 2002).

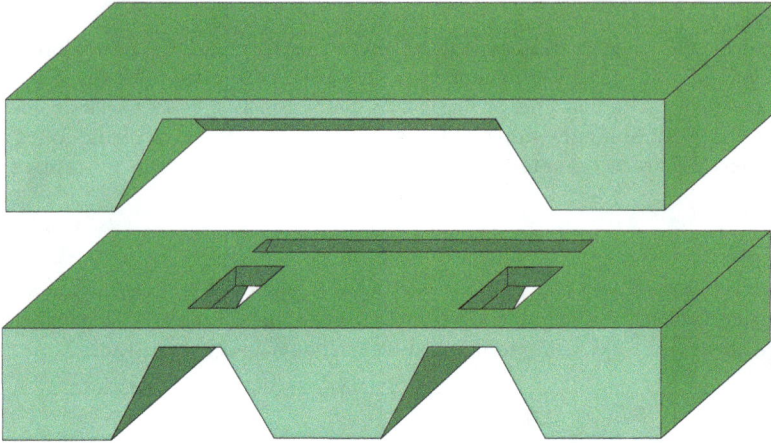

Fig. 4. Basic bulk micromachined structures using wet anisotropic etchants.

Etchant	Mask	Etch rate			Comments
		(100) μm/min (100/(111)	SiO$_2$ [Å/h]	SiN [Å/h]	
Hydrazine	SiO$_2$, SiN Metals	0.5-3 16:1	100	<<100	Toxic, potentially explosive
EDP	Au, Cr, Ag, Ta, SiO$_2$, SiN	0.3-1.5	120	60	Toxic
KOH	SiN, Au	0.5-2, up to 200:1	1700-3600	<10	Not cleanroom compatible
TMAH+ IPA	SiO$_2$, SiN	0.2-1, up to 35:1	<100	<10	Expensive

Table 4. Properties of main anisotropic etchants

The formation of macroporous silicon is usually done using n-type material and illumination from the backside to achieve deep holes with high aspect ratio. The idea was first proposed by (Lehmann 1996) and has been used to make large capacitors (Roozeboom, 2001) and micromachined structure (Ohji 1999). Both of these structures are illustrated in Figure 5.

The macro-porous process usually requires low n-doped material and illumination from the backside, which may not be compatible with the IC process. However, some macro-porous etching has been achieved in p-type material (Ohji, 2000), although the process is more difficult to control.

Deep reactive ion etching (DRIE), addressed some of the limitations of wet etching, although the process is more expensive. Two main processes are cryogenic (Craciun 2001) and Bosch processes (Laemer 1999). The cryogenic process works at about –100°C and uses oxygen to passivation of the sidewall during etching to maintain vertical etching. The Bosch process uses a switching between isotropic etching, passivation and ion bombardment. This results in a rippled sidewall, although recent developments allow faster switching without losing etch-rate, thus significantly reducing the ripples. The etching can be performed from both front and back-side and can be combined with the electronics. In addition to DRIE being used for making 3-D mechanical structures, it has been applied to packaging (Roozeboom 2008).

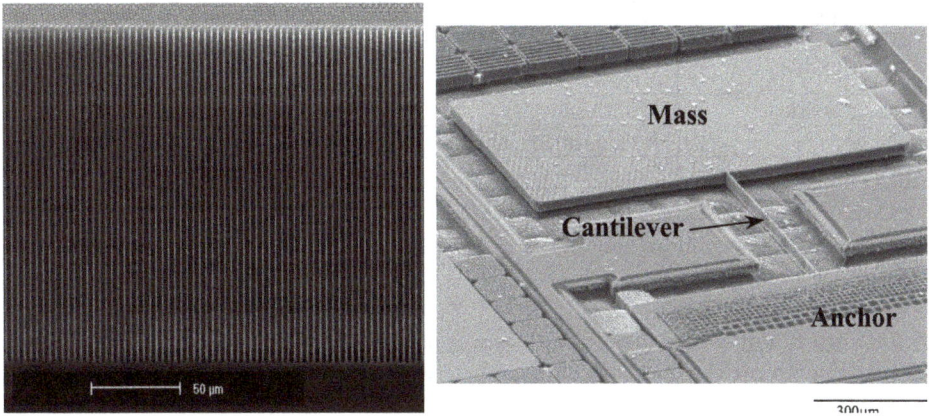

Fig. 5. (left) vertical holes using macro-porous techniques (reproduced with kind permission Fred Roozeboom Philips), and (right) free standing structure (Ohji).

3.2 Surface micromachining

Surface micromachining is quite different from bulk micromachining both in terms of processing steps and dimensions. Basically, this involves the deposition of thin films and selective removal to yield free standing structures. The basic process is given in Figure 6, although this can be augmented with additional sacrificial and mechanical layers.

There are many possible combinations of sacrificial and mechanical layers and a few examples are given in Table 5. It is important with deposited layers to have good stress control, preferably low tensile stress with little or no stress profile (Guckel, 1988, French 1996, 1997, Pakula, 2001). A further important issue when fabricating surface micromachined structures is the release while avoiding stiction. Techniques to achieve this include freeze-drying, super critical drying. Alternatively, vapour-etching has been applied, or dry etching of a sacrificial polymer layer.

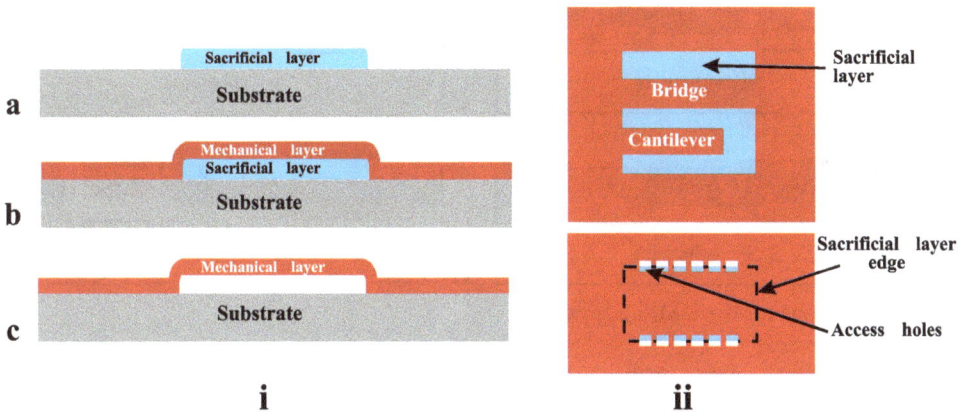

Fig. 6. Basic Surface micromachining process, (ia) deposition and patterning of sacrificial layer, (ib) deposition and patterning of mechanical layer and (ic) sacrificial etching, ii lateral view of typical structures.

Sacrificial layer	Mechanical layer	Sacrificial etchant
Silicon dioxide	Polysilicon, silicon nitride, silicon carbide	HF
Silicon dioxide	Aluminium	Pad etch, 73% HF
Polysilicon	Silicon nitride, silicon carbide	KOH
Polysilicon	Silicon dioxide	TMAH
Resist, polymers	Aluminium, silicon carbide	Acetone, oxygen plasma

Table 5. Examples of combinations of sacrificial and mechanical layers.

3.3 EPI micromachining

Epi-micromachining where the epitaxial layer is used as the mechanical layer. It can be seen as a variation on surface micromachining. There are a number of processes available which are described below.

3.3.1 SIMPLE

The SIMPLE process (Silicon Micromachining by Plasma Etching), forms micromachined structures using a single etch step (Li, 1995). This process makes use of a Cl_2/BCl_3 chemistry which etches low doped material anisotropically and n-type material above a threshold of about $8x10^{19}cm^{-2}$, isotropically. The basic process sequence is shown in Figure 7. The first additional step, to the bipolar process, is a heavily doped buried layer since the bipolar buried layer has a doping level which is too low to be under-etched. This is followed by the formation of the standard bipolar buried layer and epitaxial layer (Figure 7a) followed by an additional deep diffusion where the mechanical structure will be formed (Figure 7b). After this a full standard bipolar process is performed (Figure 7c). The final structure is given in Figure 7d. This shows clearly how the vertical etching continues in the trench during the lateral etching of the buried layer.

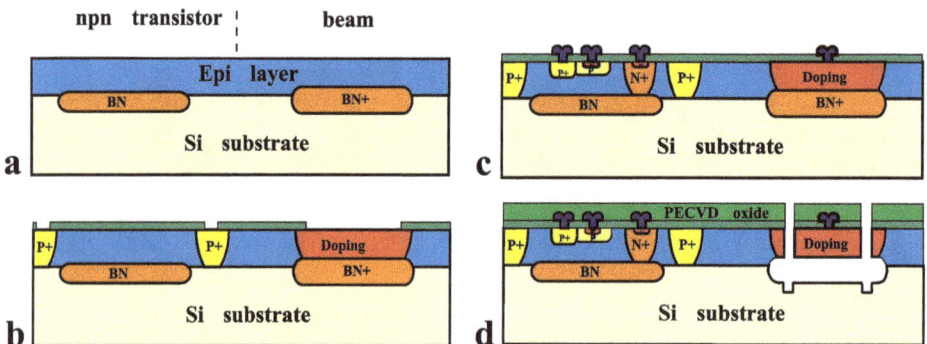

Fig. 7. Main steps for the SIMPLE process.

3.3.2 Silicon on Insulator (SOI)

Silicon-on-insulator wafers are used in standard IC processing, which give a number of opportunities for making mechanical structures. Plasma etching through the epi layer followed by wet etching of the oxide layer is one option. This basic process is given in Figure 8 (Diem, 1995).

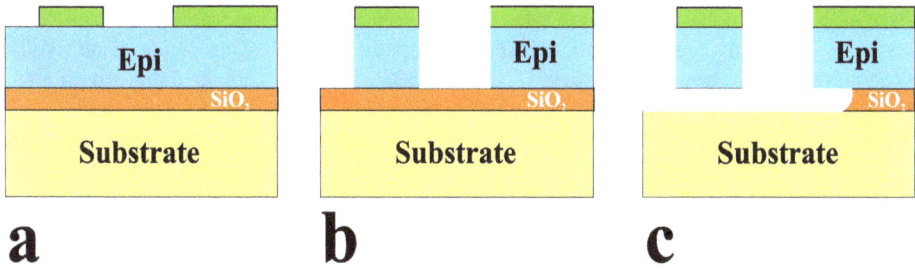

Fig. 8. SIMOX based micromachining

An alternative use of SOI was developed in Twente, The Netherlands, (de Boer, 1995). This process uses different modes of plasma etching to manufacture the free standing structure as shown in Figure 9. First the epi is etched anisotropically, a CHF_3 plasma etch is used to etch the underlying oxide and deposit a fluorocarbon (FC) on the sidewall which protects the sidewall during further etching and also has a low surface tension to reduce sticking. This is followed by a trench floor etch using $SF_6/O_2/CHF_3$. Finally an isotropic RIE etch removes the silicon from under the upper silicon resulting in the free standing structure which is also isolated.

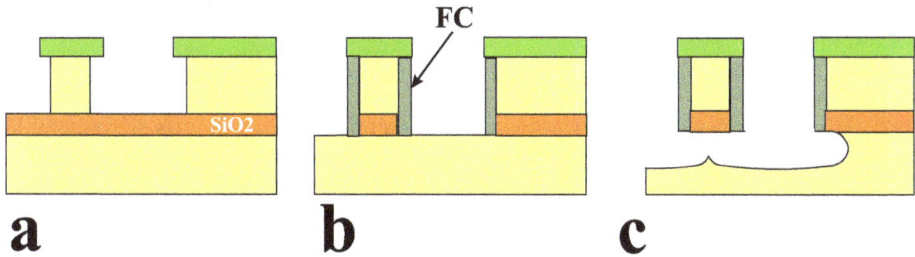

Fig. 9. Basic process steps for making free-standing SOI structures.

3.3.3 Merged Epitaxial lateral Overgrowth (MELO)

The MELO process (Merged Epitaxial lateral Overgrowth) is an extension of selective epitaxial growth SEG. Selective epitaxial growth uses HCl added to the dichlorosilane. The HCl etches the silicon. However, if there is a pattern of bare silicon and silicon dioxide, the silicon deposited on the oxide will have a rough grain like structure, with a large surface area, and will therefore be removed more quickly. Thus, although the growth rate will be lower than a normal deposition, a selective growth can be achieved. Once the silicon layer reaches the level of the oxide, both vertical and lateral growth will occur yielding lateral overgrowth. This basic process is illustrated in Figure 10 (Bartek, 1994).

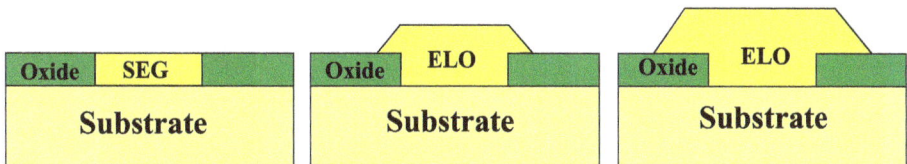

Fig. 10. Basic SEG process extended to ELO

If two of these windows are close enough together they will merge giving the MELO process. As a result we have buried silicon dioxide islands. This lends itself well to micromachining as shown in Figure 11 (Kabir 1993). This process has the advantage of producing single crystal structures, but the disadvantages that beams must be orientated in the <100> direction due to growth mechanisms of silicon (Gennissen, 1999) and since growth continues both laterally and vertically, the lateral dimensions are limited.

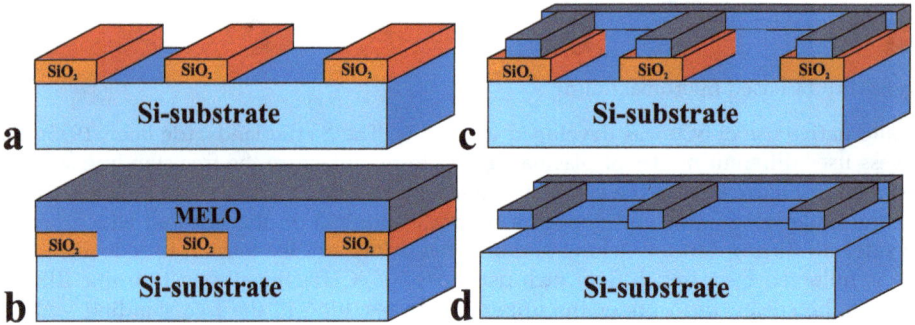

Fig. 11. Basic MELO process.

3.3.4 Sacrificial porous silicon

A process which using silicon as both mechanical and sacrificial layer is the sacrificial porous silicon technique (Lang, 1995, Gennissen, 1995, 1999, Bell, 1996). This process makes use of the fact that, without illumination, p-type material is made porous selectively. The high surface area of the porous materials results in rapid etching, after porous formation, in KOH. This makes the material highly suitable as a sacrificial material. The porous silicon formation rate is highly dependent upon the current density, HF concentration, illumination and the doping in the substrate (Gennissen, 1999). The process sequence is given in Figure 12. In this case the selective etching of p-type material over the n-type epi is used. First a plasma etch is used to etch through the epi-layer to reveal the substrate. The porous layer is then formed and finally removed in KOH at room temperature.

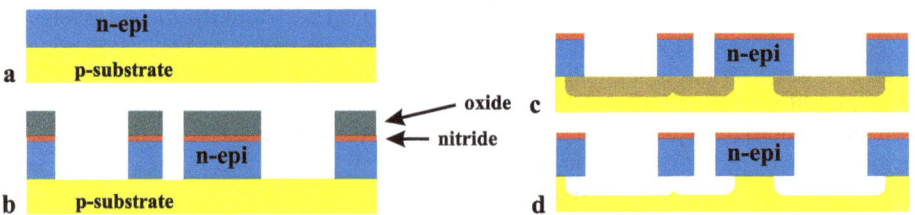

Fig. 12. Basic process steps for sacrificial porous silicon based micromachining

The porous silicon technique is extremely simple and can be applied as a post processing step and it is therefore fully compatible with the electronic circuitry. The only remaining problem is to protect the areas of electronics and metallisation from the HF etchant. One disadvantage of this technique is the added process complexity introduced by the requirement of a backside electrical contact during etching.

3.3.5 Epi-poly

Epi-poly is a polysilicon layers grown in the epitaxial reactor. Although this technique departs from using single crystal silicon as a mechanical material it has greater flexibility in terms of lateral dimensions. Alternatively, the mechanical layers can be formed at the same time as the single crystal epi required for the electronics (Gennissen 1997). The basic process is shown in Figure 13. After the formation of the sacrificial oxide, a polysilicon seed is deposited. A standard epi growth will then form epi-poly on the seed and single crystal where the substrate is bare. The epi growth rate on the polysilicon seed is about 70% of that on the single crystal silicon. Therefore the total thickness of the sacrificial layer and seed can be adjusted to ensure a planar surface after epi growth. A nitride layer is used to stop the polysilicon oxidising during subsequent process, thus minimising intrinsic stress. The mechanical layer is then patterned and released through sacrificial etching as shown in Figure 13d.

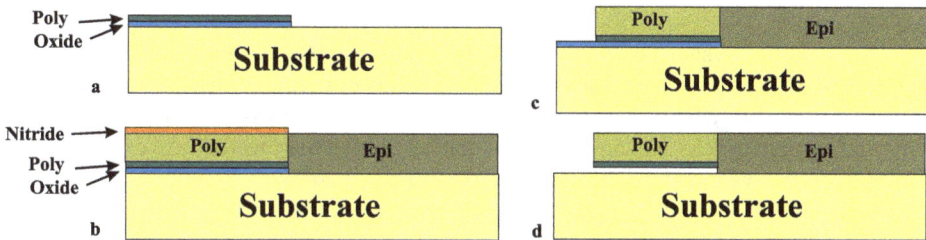

Fig. 13. Basic epi-poly process.

The epi-poly process requires minimal additional processing before epitaxial deposition and no detrimental effect on the electronics' characteristics has been found. The process is therefore fully compatible with the electronics processing.

3.4 Packaging

Packaging for standard ICs has been standardised, but for sensor, many additional challenges can arise and should be considered at an early stage. One of the issues, for example, is that many sensors have to be exposed to the environment. Another issue can be the wire bonding.

3.4.1 Wire bonding

Standard IC processing usually uses wire bonding to make electrical contact from the chip to the outside world. On each chip, metal bond pads are made, and wires (usually gold or aluminium) and made from the pad to the package.

In standard IC, this is all sealed in the package, but if the surface has to be exposed, these wires may not be desired. For example, sensors for catheters may not want bond wires on the front of the chip. In this case through wafer interconnect may be the best option. These options are shown, with the example of a catheter in Figure 14.

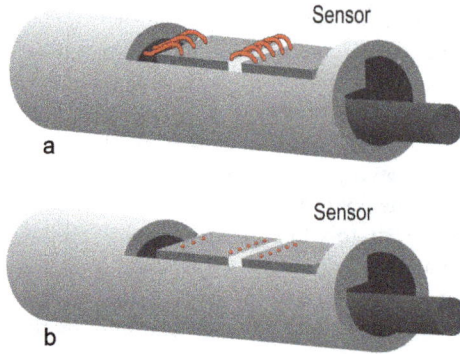

Fig. 14. (a) multi-chip approach using wire bonding, (b) multichip approach using through-wafer-interconnect and a printed circuit board substrate.

3.4.2 Flip-chip

As the field of Microsystems grew, there was more interest in combining different chips in a single package. Wire bonding from chip-to-chip is an option. An alternative is flip-chip. Flip-chip is a process where chips can be mounted. For this process a solder-bump is made on the wafer, after which two chips can be aligned and soldered together. An example of the solder process is given in Figure 15. Adapted from (Fujitsu 2003)

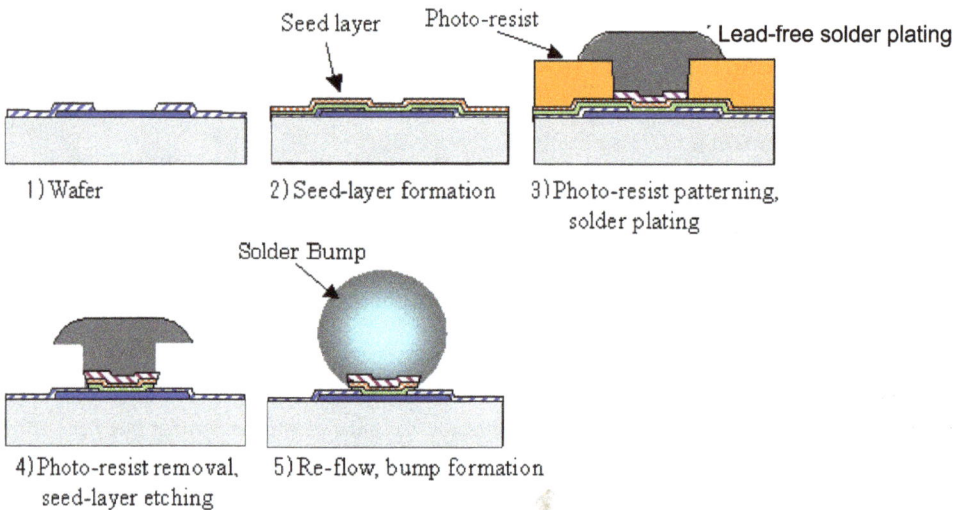

Fig. 15. Example of the Fujitsu solder-bump process.

3.4.3 Multi-chip package

There are a number of options for combining different chips in a single package through flip-chip, through wafer interconnect and simply wire bonding, some of these options , from NXP, The Netherlands, are given in Figure 16 (Roozeboom, 2008-2).

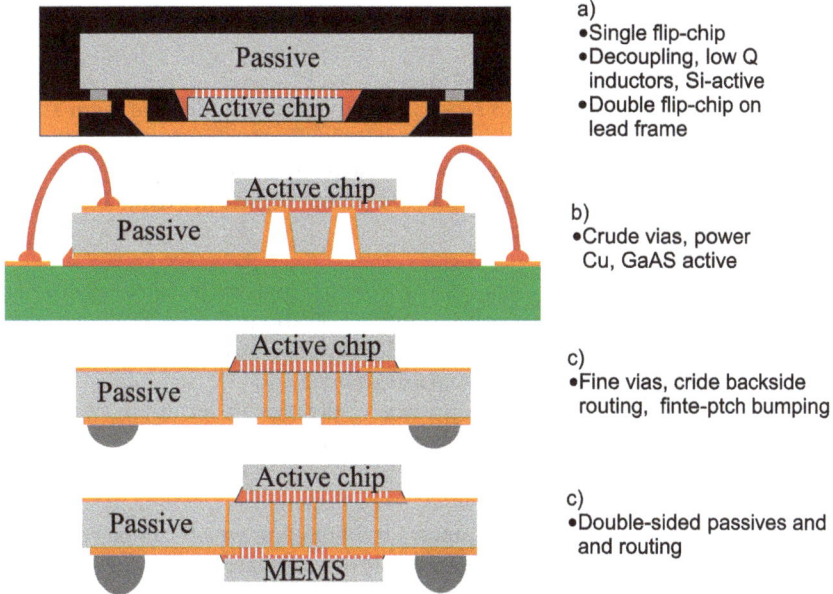

a)
- Single flip-chip
- Decoupling, low Q inductors, Si-active
- Double flip-chip on lead frame

b)
- Crude vias, power Cu, GaAS active

c)
- Fine vias, cride backside routing, finte-ptch bumping

c)
- Double-sided passives and and routing

Fig. 16. Different approaches to systems-in-a-package, based on (Roozeboom, 2008-2).

4. Process integration

When deciding how to integrate it is important to choose how this is incorporated in the process. There are a number of options:

Pre-processing (Smith, 1996 – Gianchandani, 1997)
+ No access to process line necessary
+ No thermal limitations for additional processing
- Thermal considerations for additional layers
- Potential contamination problems

Integrated processing (van Drieenhuizen, 1994)
+ Flexibility
- Require access to the clean line
- Limitations on materials

Using existing layers (Fedder, 1996, Hierold, 1996)
+ Simple
- The layers may not optimised for the sensor application

Post-processing (Bustillo, 1994, Pakula 2004)
+ Flexibility in materials used
+ Fewer contamination problems
- Limited thermal budget

With these options there are a number of issues concerning the integration of the sensor structures with the electronics. In many cases the best option is to separate the sensor from electronics and then combining into a single package. The issues for integration are discussed below.

• COMPATIBILITY WITH THE CLEANROOM

Materials or processing steps for the sensor may not be compatible with the cleanroom and therefore cause contamination.

• COMPATIBILITY OF THE PROCESSING

Both ICs and mechanical structures can be very sensitive to any additional thermal process.

• YIELD

A major issue in industrial application is the yield. Standard IC processing has been developed to have a high yield. Any additional processing can potentially reduce yield, which will increase your overall cost.

• APPLICATION ISSUES

High temperatures operations may not benefit from integration. In some applications, the integrated option may be too expensive. In other applications the environment where the sensor has to work, may not be suitable for the electronics. However, if the above issues can be addressed there can be great benefits from integration.

Both surface and bulk micromachining present challenges in terms of integration. The following sections describe a number of integration options.

4.1 Bulk micromachining integration

As shown above there are a number of process technologies, which can be combined with electronics. Bulk micromachining is, in general, low temperature and does not yield any thermal limitations to the electronics. In some DRIE processes charging of the gate oxide can occur, but adjustment of the process can avoid this. With wet etching the main issue is masking/protecting the front side.

Using wet anisotropic etching we require a masking layer on the back of the wafer, which can be a PECVD layer and etching is performed through the wafer to produce, for example, membranes for pressure sensors, or mass/spring structures for accelerometers. One of the issues is how to define the thickness of the membrane. This can be done is several ways (Palik 1982): time stop, p+ etchstop (Gianchandani, 1991), electrochemical etchstop (Kloeck, 1989) and galvanic etchstop (Ashruf 1998 and Connolly 2003).

4.2 Surface micromachining integration

Surface micromachining presents more challenges. Most LPCVD processes use to high a temperature to be suitable as post-processing. However, PECVD processes and metal layers do have a much lower temperature, and can therefore, in principle, be added after IC processing. These options are discussed further below.

4.2.1 Pre-processing

Although described as pre-processing there are usually some processing steps also required after the standard processing. In the case of pre-processing the major consideration is to ensure that the mechanical properties are not detrimentally affected by the standard processing. Polysilicon has been used for pre-processing and this basic structure is given in Figure 17 (Smith 1996, Gianchandani 1997).

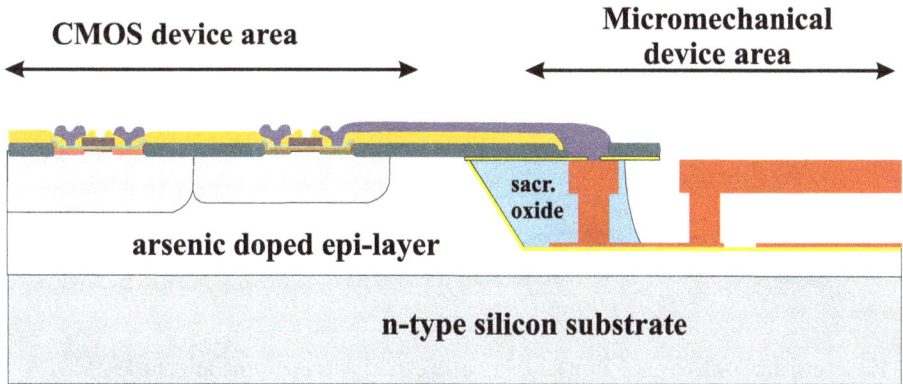

Fig. 17. Structure fabricated using pre-processing combined with CMOS.

4.2.2 Integrated processing

With the integrated process option the wafers are removed from the standard line and after the addition of micromachining steps return to the standard line. The position of the additional process steps is extremely important. In some cases the additional depositions are added after the main thermal processing but before the aluminium. Depending on the sensitivity of the electronics devices to thermal budget, a maximum thermal budget for the micromachining is determined. An example of an integrated process combining polysilicon-based surface micromachining with bipolar electronics is given in Figure 18 (van Drieenhuizen, 1994).

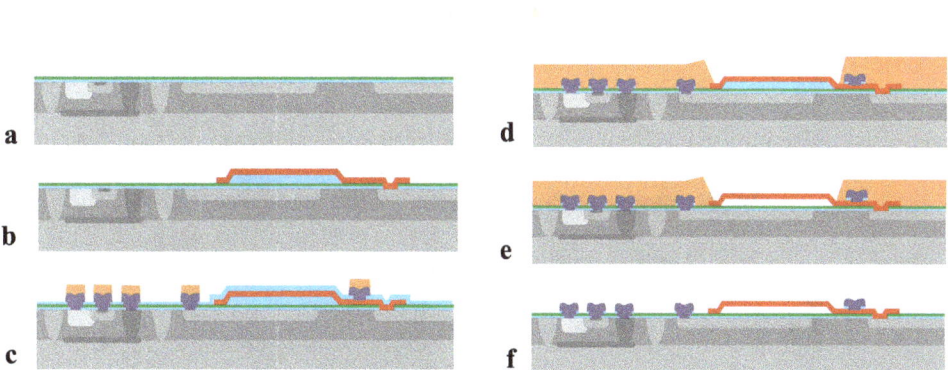

Fig. 18. Polysilicon based integrated surface micromachining process

A similar process has been developed by Fischer et. al. (Fischer, 1996) using an aluminium gate CMOS process. The basic process is shown in Figure 19. The interesting feature of this process is that the deposition of the micromechanical structures is performed before the gate oxidation.

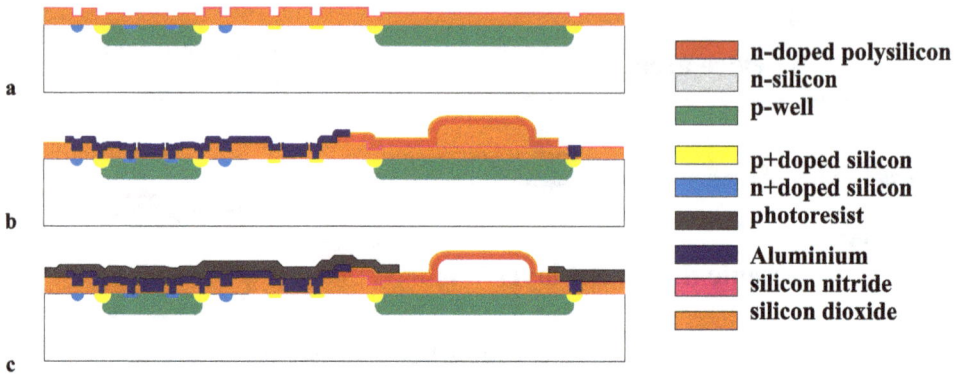

Fig. 19. Compatible surface micromachining process: a) After completion of the CMOS process (before gate oxidation) and capping silicon nitride; b) formation of the sacrificial and mechanical layers followed by gate oxidation and aluminium deposition; c) formation of the resist protection mask and sacrificial etching.

4.2.3 Post-processing

In the post processing option wafers go through the complete standard process. After standard processing either existing layers can be used or additional layers can be added. Both approaches can be found in the literature. Two examples can be seen in Figure 20. The first, Figure 20a uses the gate poly as the mechanical material (Hierold 1996) and the second, Figure 20b, (Fedder, 1996) a combination of oxide and metal used in standard processes.

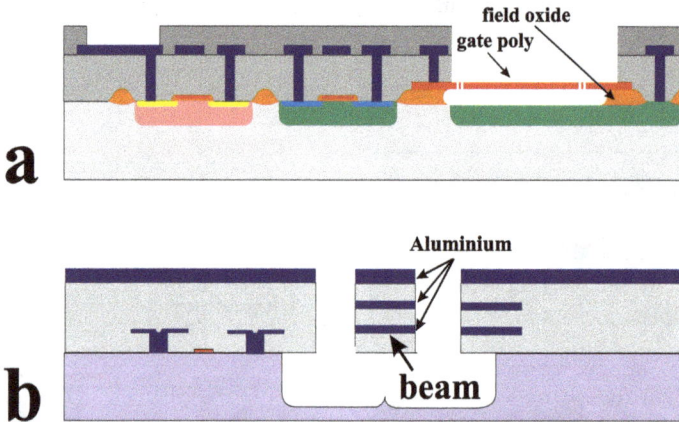

Fig. 20. Post processing micromachining with CMOS (a) using the gate poly and (b) using aluminium and oxide.

An alternative approach is to use one of the aluminium layers as a sacrificial layer, and protecting the remaining aluminium layers with oxide (Westberg, 1996). The resulting structure is shown in Figure 21. These approaches have the advantage of simplicity, but these layers are not optimised for their mechanical properties. The approach is through careful design.

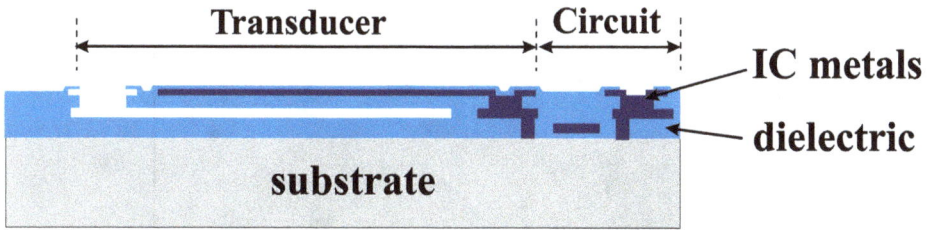

Fig. 21. Micromachining integrated into a CMOS process

One of the limitations of post-processing is the thermal budget. In particular, if aluminium is used for metallization, the maximum temperature that can be used is about 400ºC. An alternative is to use another metal. Figure 22 shows an example where tungsten is used as the metal (Bustillo, 1994).

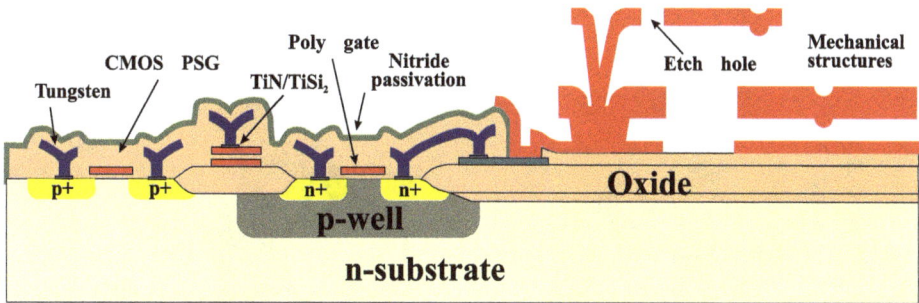

Fig. 22. Post processing micromachining using Tungsten metallisation for the electronics.

Alternatively, low temperature deposition materials, such as PECVD layers can be used. One PECVD layer with good mechanical properties is SiC (Pakula, 2004). With these low temperatures layers such as polyimides (Bagolini, 2002) can be used as sacrificial layer.

4.3 System-in-a-package
In the development of IC the packaging options were wire bonding in a package. This is still used today, but there are many other options. In many applications, the best option is to combine the sensors and the electronics in a single package, but not on a single chip. In this case the sensor can be optimised, without compromising the electronics. Furthermore, the larger production volume of the electronics chip can help to keep costs down. An example of this is given in Figure 23.

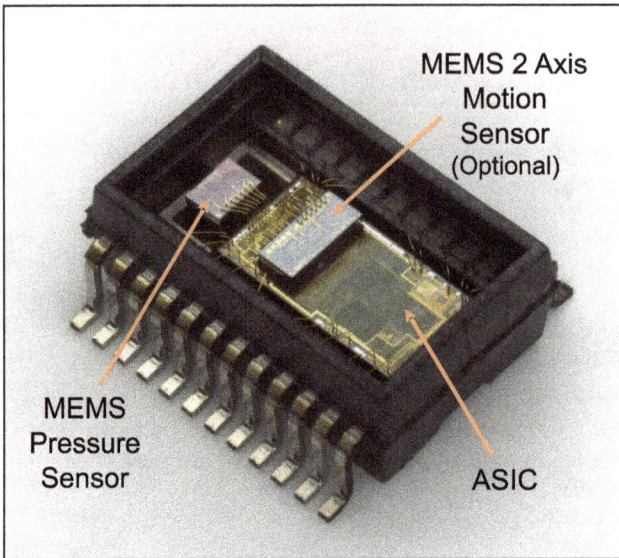

Fig. 23. Tyre Pressure Sensing System developed by LV Sensors integrated ASIC, pressure sensor and 2-axis acceleration sensor (reproduced with kind permission Janusz Bryzek).

The package can then be built into a complete system, as shown in Figure 24.

Fig. 24. Complete system fitted with the valve and a wireless communication system (reproduced with kind permission Janusz Bryzek).

4.4 Integrated sensor

One of the problems in developing a simple electronic compass based on the Hall effect was the offset. This was solved by the development of the "Spinning" Hall plate (Munter, 1990, Bellakom 1994). This technique can reduce the offset by a factor of 1000 to make it practical to be used as a compass. However, in order to make full use of this advantage, the devices should be integrated with a low-offset amplifier. Since the Hall plate only requires a thin diffused layer, this can be integrated with both bipolar and CMOS. This has since become a commercial product developed by Xensor Integration, in Delft. A photograph of this device is given in Figure 25.

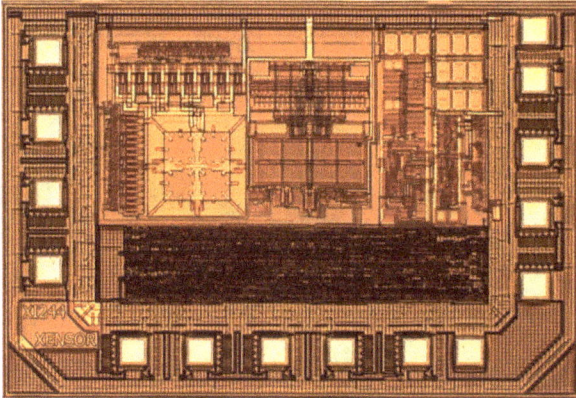

Fig. 25. Single Chip CMOS compatible Compass sensor based on hall technology. The chip incorporates a low-offset Hall sensor with ultra low-noise low offset amplifier, a sigma delta ADC and an SPI digital output. (Courtesy of Xensor Integration, The Netherlands).

Another sensor system which can benefit from integration is the temperature sensor. In this case the sensor can be a simple p-n junction or a transistor and the accuracy can be achieved through the read-out electronics. Furthermore, the electronics can be used to reduce the costs of calibration. A chip photograph of this device is given in Figure 26 (Bakker 1998, 2000, Pertijs 2005).

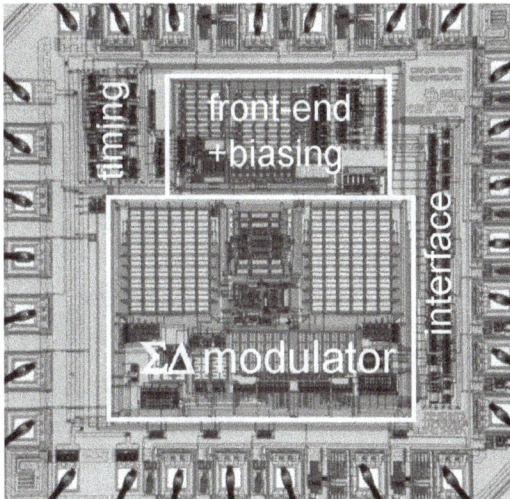

Fig. 26. Chip photograph and block diagram of a smart temperature sensor chip with bus interface.

A further example of sensor integrated is given in Figure 27. The basic sensor, was developed in the 1990s, and is shown on left. The thermopiles used for measuring temperature differences use polysilicon and aluminium. On the right is the integrated version, which contains all power management, read-out and A-D conversion. Although the chip is more expensive, the total system is much cheaper and more efficient.

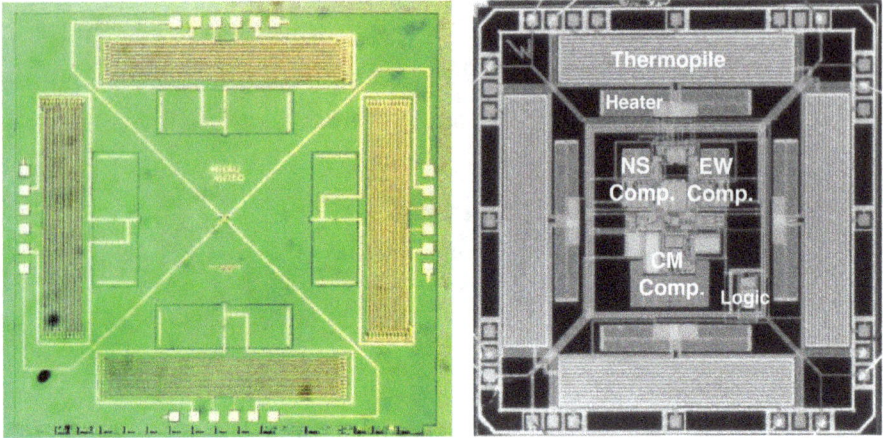

Fig. 27. Two wind-sensors (left) non integrated and (right) fully integrated (Makinwa, 2002).

The examples given above are where using standard processing. There have been many successes by integrating micromachining with electronics on a single chip. An important breakthrough for silicon sensors was with the introduction of integrated silicon accelerometers. Such an example is given in Figure 28. These devices could incorporate self-test and read-out electronics to make a complete system.

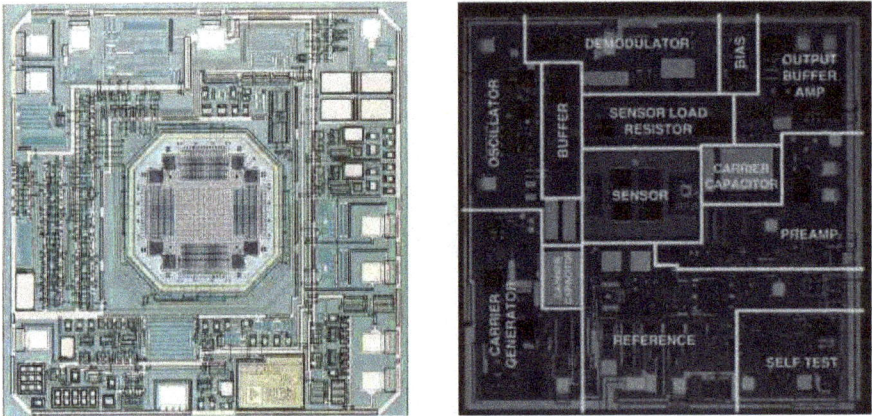

Fig. 28. Analog Devices integrated accelerometer, (left) a 2-D accelerometer, (right) a 1-D device with indication of the different circuitry (Copyright Analog Devices, Inc. All rights reserved).

A further example of a fully integrated device is the TI mirror array which, known at the digital light processor (DLP). This device makes use of the multiple aluminium layers to create the mechanical structure and also the mirrors. This is an excellent example of integrated devices. For this application the reflectivity of the aluminium is essential (Hornbeck, 1996). This device uses three layers of aluminium-based films for the mirror and its suspension system. The build up of the device is illustrated in Figure 29.

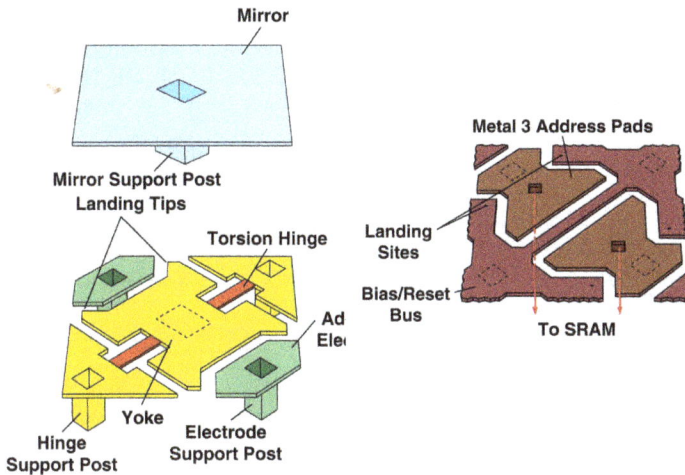

Fig. 29. Schematic of the TI digital display device (from http://www.dlp.com/).

5. Conclusions

There is no simple answer whether or not to integrate, for all applications. In many applications, the best option is system-in-a-package, in others it is better to move the electronics further away to protect it from the environment (in particular with high temperature applications). Another issue which may work against integration is potential increase in costs, through the more complex processing. In these cases the multi-chip approach is used. There are, however, examples where great benefits can be gained from integrating the sensor with the electronics, and a number of examples are given in this chapter. There is no simple answer to all applications and each case needs to be examined in its own right to decide the best path.

6. Acknowledgments

The author would like to thank the many colleagues who have provided pictures included in this chapter. Also the Technology Foundation STW, in The Netherlands, for their funding of work in Delft and their general support for Applied Research in The Netherlands. Last, but not least, colleagues in DIMES for their collaboration over the years.

7. References

Ashruf, CMA, French, PJ, Bressers, PMMC, Sarro, PM and Kelly, JJ, (1998), *A new contactless electrochemical etch-stop based on gold/silicon/TMAH galvanic cell*, Sensors & Actuators Vol A66 pp 284-291.

Bagolini, A, Pakula, L, Schotes, TLM, Pham, HTM, French, PJ and Sarro, PM, (2002) *Polyimide sacrificial layer and novel materials for post-processing surface micromachining*, J. Micromech. Microeng., Vol 12, pp 390-394.

Bakker, A and Huijsing, JH., (1998). *CMOS smart temperature sensor with uncalibrated 1oC accuracy*, Proceedings ProRISC, 26–27 November 1998, Mierlo, The Netherlands, pp. 19–22.

Bakker, A, (2000) *High-accuracy CMOS smart temperature sensors*, PhD thesis, TU Delft, The Netherlands, 2000, 132 p. ISBN: 90-901-3643-6.

Bartek, M. Gennissen, PTJ, Sarro, PM, French, PJ and Wolffenbuttel, RF, (1994) *An integrated silicon colour sensor using selective epitaxial growth* Sensors and Actuators, Vol A41-42, pp 123-128.

Bean, KE., (1978) *Anisotropic etching of silicon*, IEEE Trans Electron Devices, Vol ED-25, pp 1185-1193

T.E. Bell, TE, Gennissen, PTJ, de Munter, D and Kuhl, M (1996) *Porous silicon as a sacrificial material*, J. Micromech. Microeng., Vol 6, pp 361-369.

Bellekom, AA., and Munter, PJA, (1994) *Offset reduction in spnning-current Hall plates*, Sensors Mater., Vol 5, pp 253-263.

Boyle, WS. & Smith, GE. (1970) *Charge coupled semiconductor devices*, Bell Syst. Techn. J., Vol 49 pp 587- 593

Bustillo, JM, Fedder, GK, Nguyen CT-C and Howe, RT, (1994) *Process technology for the modular integration of CMOS and polysilicon microstructures*, Microsystem Technology, Vol 1, pp 30-41.

Chapin, DM., Fuller, CS. and Pearson, GL. (1954), *A new silicon p-n junction photocell for converting solar radiation into electrical power*, J. Appl. Phys. Vol 25, pp 676-677.

Connolly, EJ, O'Halloran, GM, Pham, HTM, Sarro, PM, French,PJ (2002). *Comparison of porous silicon, porous polysilicon and porous silicon carbide as materials for humidity sensing applications*, Sensors and Actuators Vol A 99 pp 25-30.

Connolly, EJ, Sakarya, S, French, PJ, Xia, XH and Kelly, JJ (2003) *A practical galvanic etch-stop in KOH using sodium hypochlorite*, Proceedings IEEE MEMS 2003, Kyoto, Japan, January 2003, pp 566-569.

Craciun, G, Blauw, M, van der Drift E and French, PJ, (2001) *Aspect ratio and crystallographic orientation dependence in deep dry silicon etching at cyrogenic temperatures*, Proceedings Transducers 01, Munich Germany, June 2001,

de Boer, M, Jansen, H and Elwenspoek, M, (1995), *Black silicon V: A study of the fabricating of moveable structures for micro electromechanical systems*, Proceedings Transducers 95, Stockholm, Sweden, pp 565-568.

Diem, B, Rey, P, Renard, S, Viollet Bosson, S, Bono, H, Michel, F, Delaye, MT and Delapierre, G, (1995) SOI 'SIMOX' from bulk to surface micromachining, a new age for silicon sensors and actuators, Sensors and Actuators, VolA46-47, pp 8-26

Fedder, GK, Santhanan, S, Read, ML, Eagle, SC, Guillou, DF, Lu, MS-C and Carley, LR, (1996) *Laminated high-aspect ratio microstructures in a conventional CMOS process*, Proceedings MEMS 96, San Diego, USA, Feb 1996, pp 13-18.

Fischer, M, Nägele, M, Eichner, D, Schöllhorn, C and Strobel, R (1996*) Integration of surface micromachined polysilicon mirrors and a standard CMOS process*, Sensors and Actuators, Vol A52, pp 140-144.

French, PJ, van Drieënhuizen, BP. Poenar, D, Goosen, JFL, Sarro, PM and Wolffenbuttel RF, (1996), *The development of a low-stress polysilicon process compatible with standard device processing*, IEEE JMEMS, Vol 5, (1996), pp 187-196.

French, PJ, Sarro, PM, Mallée, R, Fakkeldij, EJM and Wolffenbuttel, RF, (1997), *Optimisation of a low stress silicon nitride process for surface micromachining applications*, Sensors and Actuators, Vol A58, pp 149-157.

Fujitsu (2003), *Fujitsu Achieves Breakthrough in Ultrafine-Pitch Solder Bumping*, http://www.fujitsu.com/global/news/pr/archives/month/2003/20031215-02.html accessed August 2011

Hierold, C, Hilderbrandt, A, Näher, U, Scheiter, T, Mensching, B, Steger, M and Tielert, R, (1996) A pure CMOS surface micromachined integrated accelerometer, Proceedings MEMS 96, San Diego, USA 1996, pp 174-179.

Hornbeck, LJ, (1996) *Digital light processing and MEMS: An overview*, Proc. IEEE/LEOS 1996 Summer Topical Meetings, Aug. 7–8, 1996

Gennissen, PTJ, French, PJ, de Munter DPA, Bell, TE, Kaneko, H, Sarro, PM (1995), *Porous silicon micromachining techniques for acceleration fabrication* Proceeding ESSDERC 95, Den Haag, The Netherlands, 25-27 September 1995, pp 593-596.

Gennissen, PTJ, Bartek, M., Sarro, PM. and French, PJ. (1997) *Bipolar compatible epitaxial poly for smart sensor-stress minimisation and applications*, Sensors and Actuators, VolA62, pp 636-645.

Gennissen, PTJ, (1999) *Micromachining techniques using layers grown in an epitaxial reactor*, PhD Thesis, Delft University of Technology, 1999. Delft University Press. ISBN 90-407-1843-1.

Gianchandani, Y and Najafi, K, (1991) A *bulk dissolved wafer process for microelectromechanical systems*, IEDM Techn. Digest 1991, 757-760

Gianchandani, YB, Shinn, M and Najafi, K, (1997), *Impact of long high temperature anneals on residual stress in polysilicon*, Proceedings Transducers'97, Chicago, USA, June 1997, pp 623-624.

Gieles, ACM, (1968) Integrated miniature pressure sensor, *Dutch patent application, no 6817089*

Gieles, ACM, (1969) *Subminiature silicon pressure transducer*, Digest IEEE ISSCC, Philadelphia, 1969, pp. 108-109.

Guckel, H, Burns, DW, Visser, CCG, Tillmans, HAC, Deroo, D (1988), *Fine-grained polysilicon films with built-in tensile strain*, IEEE Trans. Electron. Dev. 35 pp 800–801.

Hierold, C, Hilderbrandt, A, Näher, U, Scheiter, T, Mensching, B, Steger, M and Tielert, R, (1996) *A pure CMOS surface micromachined integrated accelerometer*, Proceedings MEMS 96, San Diego, USA 1996, pp 174-179.

Jacobi, W. (1952), *Halbleiterverstärker*, DE patent 833366 priority filing on April 14, 1949, published on May 15, 1952.

Kabir, AE, Neudeck, GW and Hancock, JA (1993) *Merged epitaxial lateral overgrowth (MELO) of silicon and its applications in fabricating single crystal silicon surface micromachining structures*, Proceedings Techcon 93, Atlanta, GA, USA.

Kloeck, B, Collins, SD, de Rooij, NF and Smith, RL, (1989), *Study of Electrochemical Etch-Stop for High-Precision Thickness Control of Silicon Membranes*, IEEE Trans. Electron Dev., Vol 36, pp 663-669.

Laemer, F, Schilp, A, Funk, K, Offenberg, M, *Bosch deep silicon etching: improving uniformity and etch rate for advanced MEMS applications*, Proc. IEEE MEMS 1999 Conf., Orlando, FL, USA (1999).

Lang, W, Steiner, P, Sandmaier, (1995), H, *Porous silicon: a novel material for microsystems*, Sensors and Actuators Vol.A51 pp 31-36.

Lehman, V, (1996) *Porous silicon- a new material for MEMS*, Proc. IEEE MEMS Workshop '96, San Diego, USA pp.1-6.

Li, YX, French, PJ, Sarro, PM and Wolffenbuttel, RF, (1995) *Fabrication of a single crystalline capacitive lateral accelerometer using micromachining based on single step plasma etching*, Proceedings MEMS 95, Amsterdam Jan-Feb 1995, pp 398-403.

Makinwa, KAA and Huijsing, JH, (2002) *A smart CMOS wind sensor"*, Solid-State Circuits Conference, Digest of Technical Papers. ISSCC. 2002 IEEE International Volume 2, Issue , Page(s):352 – 544

Munter, PJA, (1990), *A low-offset spinning-current Hall plate*, Sensors & Actuators, Vol A21-A23, pp 743-746.

O'Halloran, GM, Groeneweg, J, Sarro, PM and French, PJ, *Porous silicon membrane for humidity sensing applications*, Proceedings Eurosensors 98, Southampton, UK, 13-16 September 1998, pp 901-904.

Ohji, H, Trimp, PJ and French, PJ (1999) *Fabrication of free standing structures using a single step electrochemical etching in hydrofluoric acid*, Sensors and Actuators, . Vol A73 pp. 95-100

Ohji, H, French, PJ and Tsutsumi, K, (2000). *Fabrication of mechanical structures in p-type silicon using electrochemical etching*, Sensors and Actuators Vol A82 pp 254 – 258

Pakula, L, Pham HTM and French, PJ, (2001) *Novel material for low temperature post-processing micromachining*, Proceedings MME, Cork, Ireland, September 2001, pp 70-73

Pakula, LS, Yang, H, Pham, HTM, French, PJ and Sarro, PM, (2004) *Fabrication of a CMOS compatible pressure sensor for harsh environments*, J. Micromech. Microeng., Vol 14, pp 1478-1483.

Palik, ED, Faust, Jr., JW, Gray, HF and Greene, RF, (1982) *Study of the etch-stop mechanism in silicon*, J. Electrochem. Soc. 137 (1982) 2051-2059

Pertijs, MAP, (2005). *Precision Temperature Sensors in CMOS technology* promotor: prof. dr. ir. J.H. Huijsing, 330p. ISBN-10: 90-9020097-5; ISBN-13: 978-90-9020097-2

Petersen, KE., (1982) *Silicon as a mechanical material*, Poc. IEEE, VOL. 70, pp 420-457.

Roozeboom, F, Elfrink, R, Rijks, TGSM, Verhoeven, J, Kemmeren A, and van den Meerakker, J, (2001), Proc. 34th Int. Symp. on Microelectronics (IMAPS 2001), Baltimore, Oct. 9-11, 2001, pp. 477-483- Int. J. Microcircuits and Electronic Packaging, 24 (3) (2001) pp. 182-196

Roozeboom, F, Blauw, MA, Lamy, Y, van Grunsven, E, Dekkers, W, Verhoeven, JF, van den Heuvel, F, van der Drift, E, Kessels WMM and van de Sanden, MCM. (2008) *Deep Reactive Ion Etching of Through-Silicon Vias*, in 'Handbook of 3-D Integration: Technology and Applications of 3D Integrated Circuits', (P. Garrou, C. Bower and P. Ramm, eds.), Wiley-VCH Verlag, Weinheim, pp. 47-91.

Roozeboom, F, Klootwijk, J, Dekkers, W, Lamy, Y, van Grunsven, E and Kim, H, (2008-2) *System-in-package integration of passives using 3D through-silicon vias*, Solid State Technology, vol 51, No. 5

Sangster, LFJ & Teer, K. (1969) *Bucket-brigade electronics - new possibilities for delay, time-axis conversion and scanning*, IEEE J. Solid-State Circuits, SC-4 pp 131-136.

Smith, CS. (1954). *Piezoresistive effect in Germanium and Silicon*, Physical Review. Vol 94, (1954) pp 42-49

Smith, JH, Montague, S, Sniegowski, JJ, Murray, JR, Manginell, RP and McWhorter, PJ, (1996), *Characterisation of the embedded micromachined device approach to the monolithic integration of MEMS with CMOS*, Proceedings SPIE Micromachining and Microfabrication Process Technology II, Austin, Texas, USA, October 1996, vol 2879, pp 306-314.

Texas Instruments. (2008) . *The chip that jack built*, HTML accessed 10 August 2011

van Drieёnhuizen, BP, Goosen, JFL, French, PJ, Li, XL, Poenar, D and Wolffenbuttel, RF, (1994), *Surface micromachined module compatible with BiFET electronic processing*, Proceedings Eurosensors 94, Toulouse, France, September 1994, p 108

Westberg, D, Paul, O, Andersson, GI and Baltes, H (1996) *Surface micromachining by sacrificial aluminium etching*, J. Micromech. Microeng., Vol 6, pp 376-384.

12

Electrochemical Spark
Micromachining Process

Anjali Vishwas Kulkarni
Centre for Mechatronics, Indian Institute of Technology Kanpur,
India

1. Introduction

Electrochemical spark micromachining process (ECSMM) is a process suitable for micromachining of electrically non-conducting materials. Besides the classic semiconductor technology, there are various methods and processes for micromachining such as Reactive Ion Etching (RIE) (Rodriguez et al., 2003), femto-second pulse laser radiation (Hantovsky et al., 2006), chemical etching and plasma-enhanced chemical vapor deposition (Claire, 2004)], spark assisted chemical engraving (Fasico and Wuthrich, 2004) and micro-stereolithography (Rajaraman, 2006) in practice. Use of photoresist as sacrificial layer to realize micro-channels in micro fluidic systems is discussed in (Coraci, 2005). All these methods are expensive as they need the vacuum, clean environment and mostly involve in between multi processing steps to arrive at the final microchannel machining results. There is a need of an innovative process which is cost effective and straight forward without employing intermediate processing steps. One such process thought of and being researched is electrochemical spark micromachining (ECSMM) process. The ECSMM process is a stand alone process unlike others and it does not demand on intermediate processing steps such as: masking, pattern transfer, passivation, sample preparation etc. The use of separate coolants is also not required in performing the micromachining by ECSMM.

Micromachining needs are forcing reconsideration of electrochemical techniques as a viable solution (Marc Madau, 1997). Another similar process termed as spark assisted chemical engraving (SACE) (Wuthrich et al., 1999) has been employed for the micromachining of glass. ECSMM is a strong candidate for microfabrication utilizing the best of electrochemical machining (ECM) and electro discharge machining (EDM) together. Applications of ECS for microfabrication can be in the field of aeronautics, mechanical, electrical engineering and similar others. It can successfully process silicon (Kulkarni et. al., 2010a), molybdenum (Kulkarni et. al., 2011c), tantalum (Kulkarni et. al., 2011a), quartz (Deepshikha, 2007; Kulkarni et. al., 2011a), glass ((Kulkarni et al., 2011a, 2011b); Wuthrich et al. 1999)), alumina (Jain et al., 1999), advanced ceramics (Sorkhel et al., 1996) and many other materials.

The chapter discusses the details of the experimental set-up developed in the next section. The procedure for micromachining using the developed set-up is outlined next. The experimental scheme to perform machining on glass pellets (cover slips used in biological applications) is presented. Discussion of the micro machined samples is presented. This discussion is based on various on line and post process measurements performed. The qualitative material removal mechanism is presented based on the results and discussions.

2. Experimental set-up

A functional set-up of the ECSMM process is designed, developed and fabricated as shown in Figure 1 (Kulkarni et. al., 2011b). The main components of the ECS set-up are as follows and are described in the following sub sections:

1.   Machining Chamber
2.   Power Supply System
3.   Exhaust System
4.   Control PC

2.1 Machining chamber

The machining chamber houses X-Y table, Z axis assembly, tool feed and tool holder assembly and ECS cell. X, Y, Z and tool feed stages are motorized.

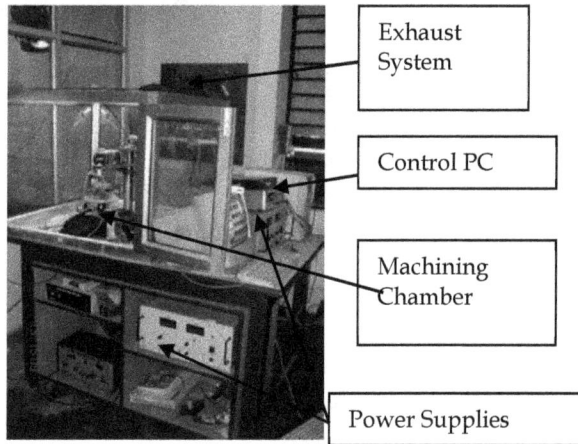

Fig. 1. Photograph of experimental set-up (Kulkarni et al., 2011b)

2.1.1 X-Y table

X-Y table has resolution of 2 µm in X and Y directions and traverse of 100 mm in X as well as Y directions. The guide ways use non-recirculating balls as rolling elements. The mechanical drive is a ground lead screw of 400 µm pitch made of aluminium alloy. Rotation to the X and Y screws is provided by separate stepper motors. The table is mounted on a chrome plated MS plate. Chrome plating protects the plate from corrosion. The MS plate has mounting tapped holes on a 25 mm grid to mount the ECS cell. Bellows are provided to protect the motors and lead screws from the electrolyte splashes and fumes produced.

2.1.2 Z axis assembly

The Z axis is automated to move up or down to maintain a constant work piece-tool gap. The worm and worm wheel with a gear ratio of 1:38 transmit the power to a lead screw of 200 µm pitch. All the parts are fabricated with stainless steel and brass to resist corrosion due to acidic environment. It has positioning accuracy of 50 µm and maximum vertical travel of 80 mm.

2.1.3 Tool feed and tool holder assembly
Tool feed assembly is mounted on Z axis assembly. A glass tool holder is designed and developed. This tool holder provided the tool insulation and hence reduction in the stray currents. This glass tool holder is used to hold the tool wire in place. A fixture made of Perspex material is designed and fabricated to hold the tool holder on Z assembly. Cu wire of 200 µm diameter is used as a cathode (tool).

2.1.4 ECS cell
It is a rectangular box of 10 cm x 8 cm x 6 cm dimensions made up of Perspex material. It is mounted on X-Y table. It houses separate fixture arrangement for graphite anode and work piece holder. It is filled with the electrolyte. The electrolyte level is maintained at 1mm above the flat surface of the work piece. Electrolyte used is NaOH in varied concentration in the range of 14-20 %.

2.2 Power supply system
DC regulated power supplies of different ratings are used for driving stepper motors, machining supply and control circuitry. Use of separate power supply ensures the noise free operation.

2.3 Exhaust system
Proper exhaust system is designed and provided to take away the electrolyte fumes generated during the spark process inside the machining chamber. A small DC operated fan is placed in the machining chamber where the fumes are generated. These are carried away by a hose pipe and thrown away from the room with an exhaust fan.

2.4 Control PC
Stepper motors used for driving X, Y, Z and tool feed are all interfaced to motion controller card installed in PC. Precise control and drive of the machine is achieved with NI 7834 PCI card and NI 7604 drive board interfaced to a computer. Contouring functions in LabVIEW platform are used to carve different shapes of the micro channels [Kulkarni et al., 2008].

3. Experimental procedures

The supply voltage, electrolyte concentration and table speed are the control parameters. Pilot experiments are performed to determine the optimum window of these operating parameters.

It is observed that sparking occurs at supply voltage of 30 V and above. Glass samples break above 50 V supply voltage. Hence the working supply voltage range chosen is 40 V – 50 V.

Use of base solution is preferred over the acidic electrolyte. It was observed that in the acidic environment the surface roughness increases. The fumes formed of acidic solutions during the electrochemical sparking process are harmful. During the pilot experiments it was observed that machining takes place in diluted sodium hydroxide (NaOH) solution as electrolyte. The concentration window was decided upon by performing many experiments to arrive at a permissible concentration range. It was observed that machining does not take place below 14% concentration of NaOH. Above 20 % concentration of NaOH, the machined surface roughness is notable. Hence 14% -20% concentration range for NaOH electrolyte is arrived at. Moreover use of low concentration of NaOH as electrolyte makes the ECSMM process as a 'green process'. Level of electrolyte is maintained at 1 mm above the work piece surface in the ECS cell.

The table speed is chosen ranging between 12.5 µm/s – 25 µm/s. It is such that the traverse is not too slow to dig the micro channel and not too fast to miss the micro machining in that region.

Micro channels are formed using the ECSMM process on microscopic glass pellets using platinum wire as a tool of 500 µm diameter. Pellets are of 180 µm thickness, 18 mm diameter circles in size. Length of the tool protruding out of the tool holder is 4 mm. The gap between the cathode tool electrode tip and the work piece surface is maintained at around 20 µm using the tool feed device mounted on Z-axis. The distance between the tool and the anode is 40 mm. Figure 2 shows the photograph of the electrolytic cell with the spark visible at tool tip and electrolyte interface. Graphite anode is seen in the cell. It is a non consuming electrode.

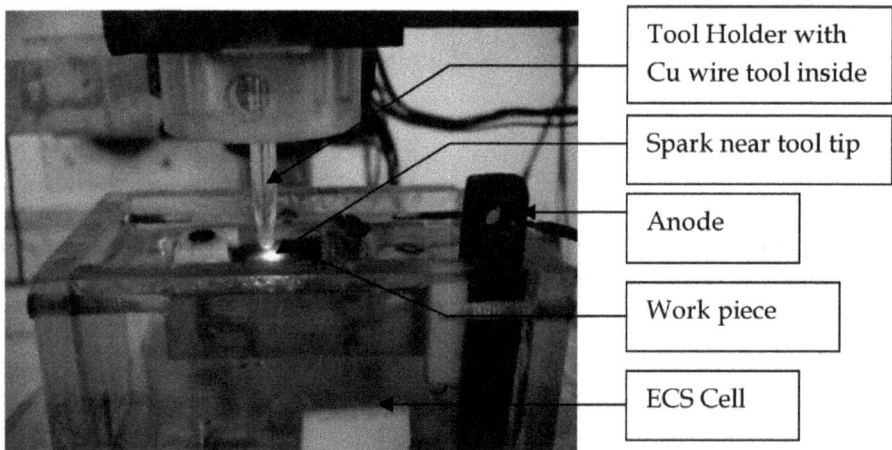

Fig. 2. Photograph of the ECSMM cell with graphite anode, tool and work piece. The spark is visible near the tool tip (Kulkarni et al., 2011b).

Experiments are conducted with Voltage, Electrolyte Concentration and Table Speed as the control variables. The experiments are conducted in accordance with the central composite design scheme developed by the software 'Design Expert 07' to study the response surface. The range of the control variables chosen is as shown below:

- Factor 1 (V_s): Supply voltage ranging between 40 V - 50 V
- Factor 2 (EC): Electrolyte Concentration (NaOH) ranging between 14% - 20%
- Factor 3 (TS): Work piece Table Speed ranging between 12.5 µm/s – 25 µm/s

The design resulted in total of twenty one experiments, out of these twenty one experiments, six central experiments were performed at 45 V supply voltage, 17 % electrolyte concentration and 18.75 µm/s table speed as the values for the control variables.

The responses measured are: average process current (I), width of microchannel (W) and depth of microchannel (D) formed using ECSMM. The scheme of the experiments is as shown in Table 1. Columns 2-4 list V_s, EC, and TS respectively. Columns 5-7 give average current, width, and depth of the microchannels respectively as the responses measured post process.

	Control Variables			Responses			Comments
R #	V$_s$ (V)	EC (%)	TS (µm/s)	I (A)	W (µm)	D (µm)	Channel Type
1	45	17	18.75	0.05	760.5	-	Through
2	40	14	25	0.025	520	-	Through
3	50	14	12.5	0.105	450	-	Through
4	45	17	18.75	0.105	580	-	Through
5	45	17	18.75	0.12	790.5	-	Through
6	45	17	30	0.09	1030	-	Through
7	36.6	17	18.75	0.06	421.5	81.5	Blind
8	40	20	25	0.08	870	*	Blind
9	50	20	12.5	0.015	1110	-	Through
10	50	20	25	0.0933	1090	-	Through
11	40	20	12.5	0.205	720.5	*	Blind
12	45	17	18.75	0.115	970.5	97.5	Blind
13	45	17	18.75	0.966	1030	97.5	Blind
14	40	14	12.5	0.025	585	-	Through
15	50	14	25	0.0733	600	-	Through
16	45	17	18.75	0.5	855	97.5	Blind
17	53.4	17	18.75	2.493	610	81.5	Blind
18	45	11.9	18.75	0.6	480	77.25	Blind
19	45	11.9	18.75	0.05	560	-	Through
20	45	22	18.75	0.08	485.5	123.6	Blind
21	45	17	8.45	0.16	810	-	Through

(* could not be measured)

Table 1. Experimental parameters and responses

3.1 On line measurements
The average process current is measured with the help of a digital multimeter. Besides this average current, the time varying process current is measured on line by digital storage oscilloscope. For this purpose the 'resistive shunt method' is used. In this a 1 Ω resistance is connected in series with cathode and ground of the power supply to the ECS cell. The time varying voltage across this resistance is the direct measure of the time varying process current. The wave forms are saved on the control PC via RS 232 connectivity module of the oscilloscope (Hameg 1008). The on-time, off-time and the frequency of the sparks occurring are measured. These parameters are otherwise theoretically estimated. The occurrences of these pulses directly indicate the correlation between the presences of the sparks during the process. The analysis of these current pulses will be helpful to devise the electrical model of the process.

3.2 Post process measurements
A set-up is developed to measure the depth of the microchannels at various points with the resolution of 10 µm. The dial gauge used for this purpose is mounted on the Z Axis of a standard machine to achieve these measurements.

To study the surface topography and width measurement, SEM analysis of the micro channels is performed. SEM at different and higher magnification is performed to get the insight into the surface topography due to this process. SEM at increasing magnification clearly shows the imprints and development of how the material is removed from the work piece surface.

The results based on the above studies are presented in the following section.

4. Results and discussions

Measurements of on line average current, and post measurements of width and depth of microchannels are presented in Table 1 in column 5-7. Theses are discussed in details in the following sub sections. To study the surface topography and width measurement, SEM analysis of the microchannels is performed. Following section describes the microstructure analysis of the microchannels.

4.1 Microstructure Analysis by SEM

Detailed SEM is performed for the samples of central experiments (at 45 V, 17% electrolyte concentration, and 18.75 μm/s table speed) to study the effect of sparking on the microstructure. SEM is performed at successive higher magnification to visualize the surface closely. Figure 3 shows the photograph of three microchannels (forming an inverted 'C' section) carved at 45 V, 17% electrolyte concentration and 18.75 μm/s table speed. These are formed using X traverse through 2500 pulses in carving channel 1, Y traverse in carving channel 2 and then negative X traverse of X-Y table in carving the third channel, i.e. channel 3. The length of each section in the C type micro channel is about 5000 μm. The average width of channel 2 is around 535 μm and the average depth is around 370 μm.

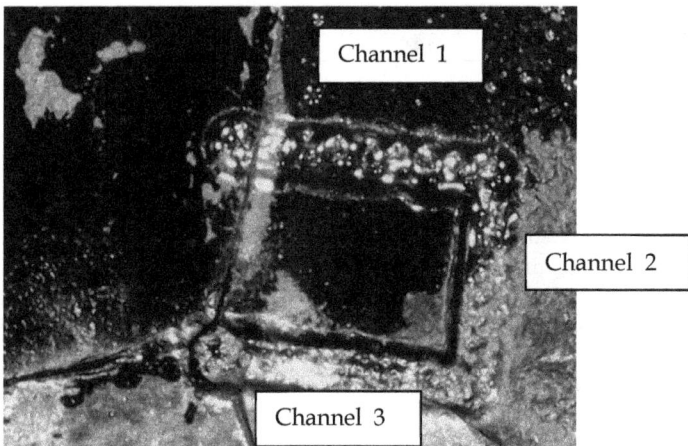

Fig. 3. USB Photograph of the channels carved at 45 V, 17% electrolyte concentration and 18.75 μm/s table speed. The approximate length of each channel is 5000 μm, average width is 535 μm, and depth is 370 μm.

Figure 4 gives the microstructure of the micromachined glass surface at 162 X magnification carved at 37 V, 17% electrolyte concentration and 18.75 μm/s table speed. From SEM picture it is obvious that a shallow microchannel is obtained. This may be due to the lower supply voltage of 37 V. The width at two different regions of microchannel is 267 μm and 410.3 μm. Thus, average width of microchannel is 338.65 μm.

Figure 5 gives the microstructure of the micromachined coverslip surface at 881X magnification.The microstructure clearly shows the removal of material along the path of tool movement. Valleys and ridges are clearly visible which are due to melting and the layer by layer material removal in the spark affected region. A piece of material is seen which got re solidified and remained there.

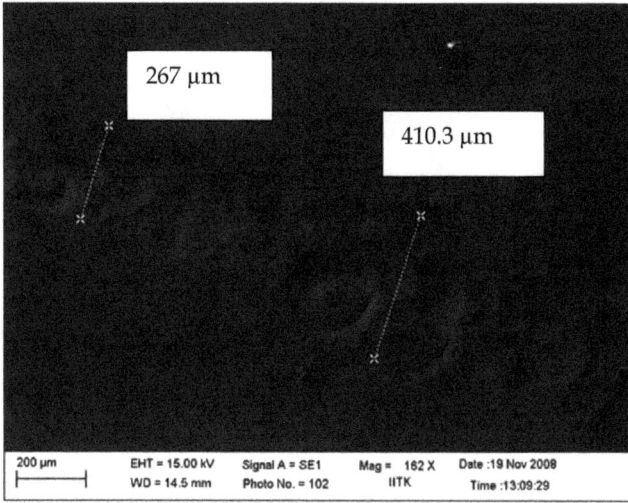

Fig. 4. SEM image (162X) showing width at two places along the microchannel, the average width of the channel is around 338.65 μm at 37 V, 17% electrolyte concentration and 18.75 μm/s table speed.

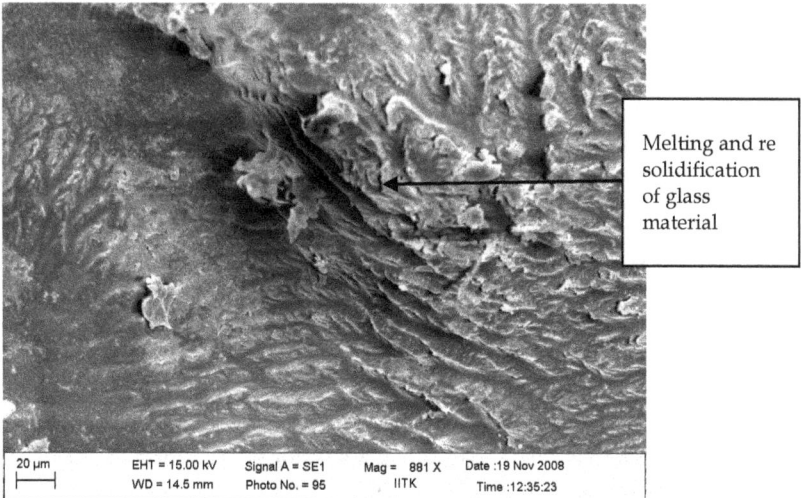

Fig. 5. SEM image of pellet (881X) treated at 45 V, 12% electrolyte concentration and 18.75 μm/s table speed.

Figure 6 gives the microstructure of the microchannel at 4500X magnification. The tearing off of the material is seen. The region shows the melting and solidification of the workpiece material. The thickness of the smallest layer at the corner is around 7.8 μm.

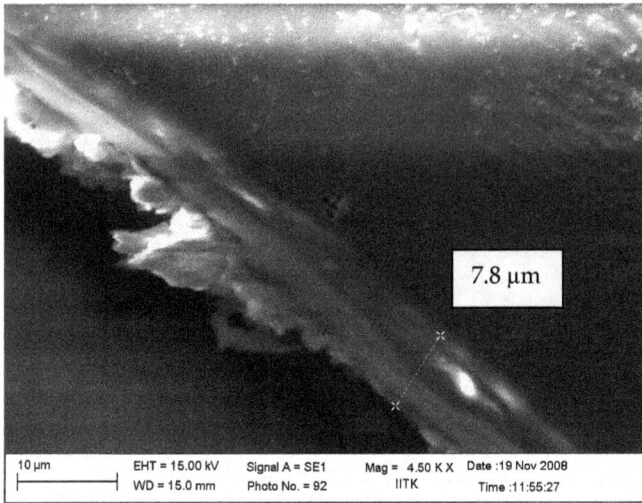

Fig. 6. SEM image (4500X) giving thickness of layer at the corner around 7.8 µm at 45 V, 17% electrolyte concentration and 18.75 µm/s table speed.

4.2 Current analysis
4.2.1 On line average process current
Column 5 of Table 1 gives the values of the average process current measured on line. The electro chemical action causing the migration of ions, and electrons contributes to the average current. The average current value as seen in Table 1, column 5 is ranging from 0.0125 – 0.9 A. Occasionally it has shoot up to 2.493 A for R#17 for 53.4 V supply voltage.

The interesting process phenomenon is not obvious from only recording the average current. The processes' time varying nature can be only revealed by studying the transient current. The transient current waveforms reveal the process complexity and help in understanding the holistic and time changing phenomena during the single entire current cycle of ECSMM process, as explained in the next section. The actual machining is occurring during the very short time of the instantaneous current pulses carrying high energy density.

4.2.2 On line, transient process current
The time varying current is measured with the help of a digital storage oscilloscope as mentioned in section 3.1. The snap shots of the stored waveform are presented in Figures 7 a and b. In Figure 7a it can be noted that there are many spikes during a time of 10 ms duration corresponding to 1 division of oscilloscope window. Each pulse represents a spark occurrence. The average process current can be seen at a level of 0.1 A. A pulse of height of 0.3 A can be seen of time period greater than 20 ms. A second pulse of instantaneous current value more than 0.4 A can be seen after a period of around 15 ms. It's time period is about 6ms. Many short duration (<1ms) pulses can be seen in between these two remarkable pulses. These pulses show the stochastic nature of the spark formation process.

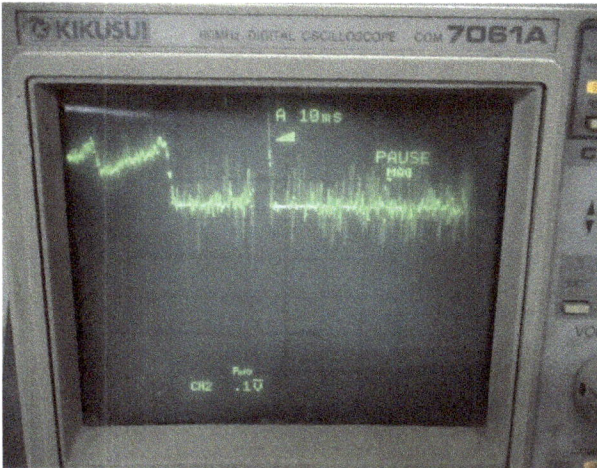

Fig. 7 a. Snap shot of online time varying ECSMM process current during glass pellet micromachining.

In another waveform, Figure 7 b, two complete current pulses and two halfway current pulses can be seen. These are of different time durations, ranging from 0.1 ms to 0.3 ms time period. Hence the resulting frequency is variable and is ranging from 2.5 kHz to 5 kHz. The sparking frequency depends on many factors such as size of bubble formed, bubble growth time, time of its survival, etc. The size of the spark or discharge depends on the instantaneous current value. It is clear from these time varying current pulses that sparks of different energy strike the work piece surface resulting in softening, melting, and / or vaporizing of the work piece material.

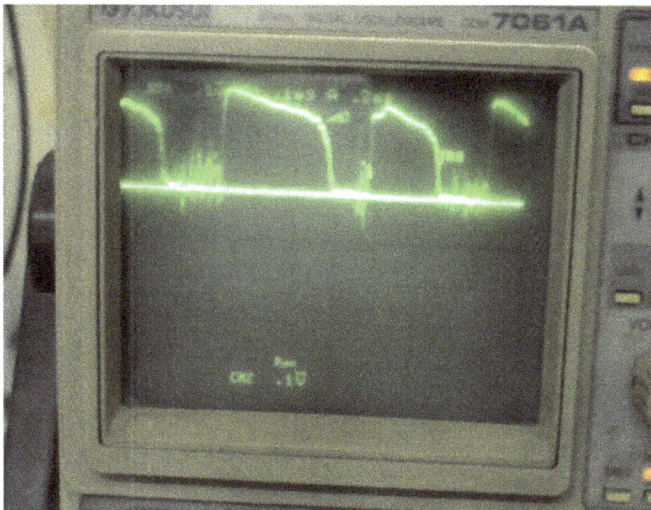

Fig. 7 b. Snap shot of online time varying ECSMM process current during glass pellet micromachining.

The spark energy can be estimated by taking V_s = 45 V, $i_{instantaneous}$ = 0.2 A and time = 0.2 ms. The instantaneous spark energy with a striking area of diameter less than the tool diameter, i.e. 200 μm, is of the order of 500 kJ/m^2.

4.3 Width of microchannels

Columns 6 of Table 1 gives post process measured values of the width of the microchannels using the dial gauge. The last column summarizes the type of the microchannel formed. It says whether a channel is a through channel or a blind microchannel achieved. The rows corresponding to the successfully achieved microchannels are shown in bold face. The depth of the microchannels in Run # 8 and # 11 could not be measured for some reasons. In case of the through channels the depth of the microchannel achieved is more than 180 μm. Either the higher machining time or the lower travel speed or the smaller gap due to local irregularities is responsible for through machining to occur.

The minimum width achieved is 421.5μm for Run # 7 for 36.6 V, 17 % electrolyte concentration and 18.75 μm/s table speed. The maximum width achieved is 1110 μm for Run # 9 for 50 V, 20 % electrolyte concentration and 12.5 μm/s table speed. That means for higher voltage, higher electrolyte concentration and lower table speed combination of parameters, the width achieved is higher.

4.4 Depth of microchannels

The minimum depth achieved is 77.25 μm for Run # 18 for 45 V, 11.9 % electrolyte concentration and 18.75 μm/sec table speed. The maximum depth achieved is 123.6 μm for Run # 20 for 45 V, 22 % electrolyte concentration and 18.75 μm/sec table speed. Higher electrolyte concentration results in higher depth. Microchannels of the width between 400 – 1100 μm are achieved. The depth achieved is 75 -120 μm.

For other experiments, through machining has been occurred where the depth of cut is more than the thickness of the work piece. This may be partly due to the gap adjustment between the tool and the work piece surface. It is a crucial operation to maintain the gap at or above 20 μm without the closed loop control. This calls for a close loop control for maintaining the gap between the tool and the work piece surface.

A novel technique to measure the depth of these microchannels is devised and discussed in Kulkarni et. al., (2010b) and Kulkarni et. al., (2010c).

Parametric models pertaining to the average current, width and depth of the microchannels' are presented elsewhere.

Section 5 describes the systematic description on understanding the ECSMM process mechanism in view of the transient current.

5. Understanding the process mechanism

The material removal mechanism in ECSMM is complex as it is revealed by the SEM and current pulses analysis in the previous sections. This is primarily due to the non-thermal nature of these sparks. In the existent literature the spark energy is considered to be of thermal nature and thermal analysis and material removal are considered to be due to this thermal source (Jain et al.,1999; Basak & Ghosh, 1992). Experimentally it has been found that the spark is a non thermal discharge. This has been confirmed (Kulkarni et al, 2009) while making an attempt to measure the spark temperature by a pyrometer. Pyrometer failed to measure the temperature as the radiation is a non thermal type. Instead it is a discharge process similar to that of the breakdown of the hydrogen gas bubble isolating the tool tip from the surrounding electrolyte.

Secondly, electro chemical systems are known to exhibit complex non-linear behavior. These nonlinearities arise due to electro hydro dynamism, ionic reactions, bubble generation, their growth and their breakdown phenomena. The overall process seems to be discrete in nature though the supply voltage is DC. Positive as well as negative spikes are also observed in the current waveforms. The electrochemical kinetics includes negative faradic impedance in the electrolyte solution. There are many intermittent, small amplitude current spikes, of smaller duration. These seem to be representing the partial sparks due to the break down of the small hydrogen bubbles. The partial discharge is due to the total isolation of single or many such bubbles completely isolating the electrolyte contact. On the other place the total or complete sparking is that occurring due to the complete isolation of the cathode tip from the electrolyte surface. This can be understood by the pictorial representation as in Figure 8 a and b. In Figure 8 a, there is a local isolation of the tool tip from the surrounding electrolyte due to a small hydrogen bubble. This causes an instantaneous sparking across the bubble, resulting in a small amplitude current spike. Where as, in Figure 8 b, the tool tip is surrounded by a single larger bubble. Many small sized bubbles coalesce in a single larger bubble resulting in complete isolation of the tool tip from the electrolyte. The sparking resulting due to this kind of total isolation will result in the intensive sparking manifesting the large amplitude current spikes. This kind of behavior is reflected in the nature of the on line time varying current pulses studied. It was observed that the frequency of sparking

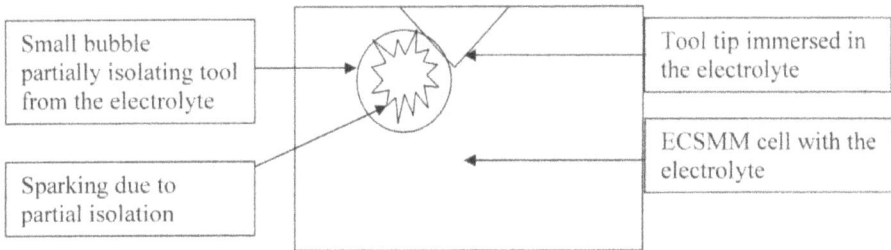

Fig. 8a. Partial sparking due to local isolation by small bubble resulting in a low energy spark

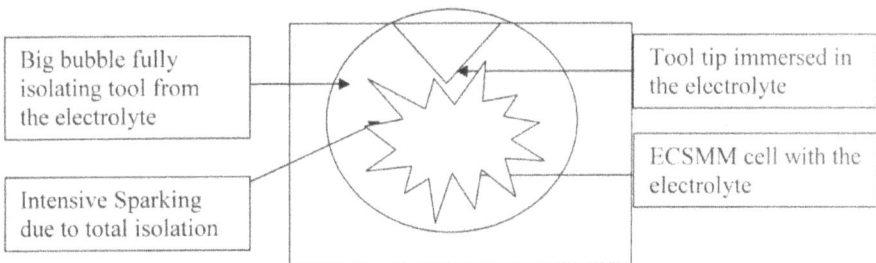

Fig. 8b. Intensive sparking due to complete blanketing of the tool tip by large sized bubble resulting in high energy spark

(oscillations) varies with varying supply voltage (Kulkarni, 2000). The sparking frequency is high (in the tens of MHZ range) and it lowers (in the few hundreds of kHz range) for higher supply voltage. This supports the possibilities of the many partial sparks due to breakdown of the single isolated hydrogen gas bubbles. Theses partial sparks or discharges are of less current value and hence having less energy. These may not result in material removal from the workpiece. These may die out before reaching the workpiece surface.

5.1 Intermediate processes and their interrelation

Thus ECSMM process comprises of many intermediate processes such as electro chemical action causing the migration of ions, followed by the nucleate pool boiling of hydrogen gas bubble due to immense local heating of tool tip immersed in electrolyte. The gas bubble growth dynamics is a complicated phenomenon. It is changing the isolation between the cathode tip from the electrolyte and hence creating a varying electric field. This varying electric field in turn affects the bubble growth dynamics.

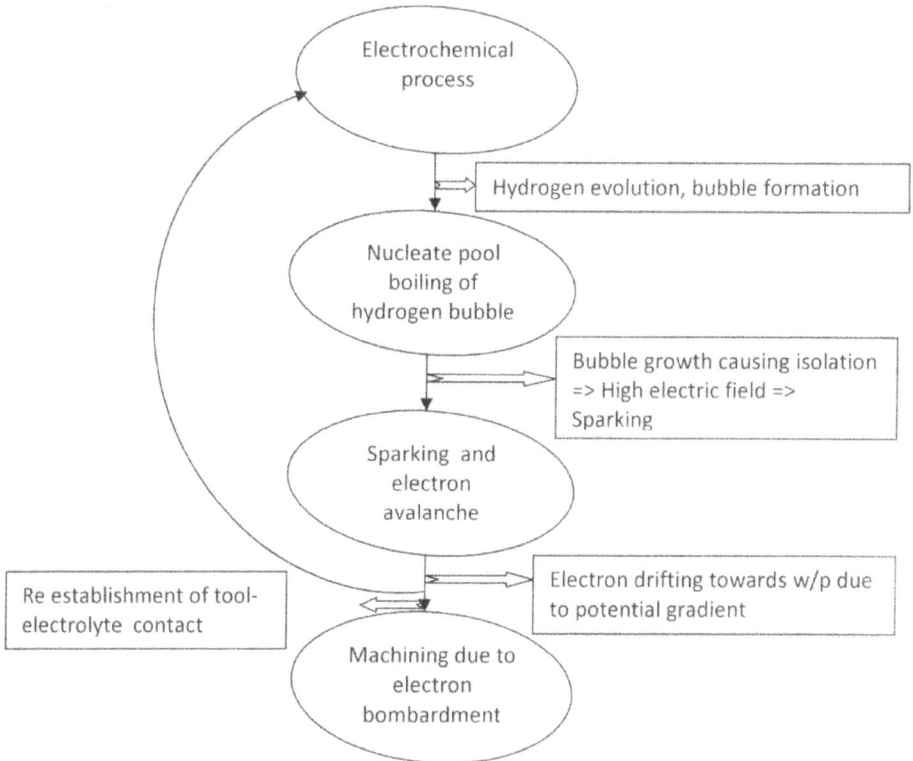

Fig. 9. Operational flow of ECSMM process showing intermediate processes.

It starts with the electron generation, these in turn generating the secondary electrons and hence causing the electron avalanche. These energetic electrons get drifted away from cathode (tool) to the work piece very quickly due to the high potential gradient getting generated within the tool – work piece gap because of the hydrogen bubble isolating the electrolyte, as described. These drifted electrons bombard on the work piece surface. A large current spike is seen as a result of electron flow from cathode to work piece as actually seen during the transient current measurements. The bombardment of electrons on the work piece surface results in intense heating and hence metal removal takes place.

The overall material mechanism of the ECSM process can be understood in the light of the electrochemistry, heat transfer, ionization theory and electrical response of the system. The operational flow of the overall process is as shown in Figure 9. Each involved intermediate process and the cross relation with other sub processes is illustrated further in section 5.1.1-5.1.4.

5.1.1 Electrochemical process

When the supply to the electrolyte cell is applied in the proper polarity, (i.e. positive terminal connected to anode and negative terminal to cathode) electrochemical action starts. electrons move from the cathode–electrolyte interface, and go to the solution. At the anode-electrolyte interface, equal number of electrons are discharged from the solution to the anode. Electrochemical reactions that occur at the electrode–electrolyte interface continuously supply electrons from cathode to solution and solution to anode. The type of reaction depends on the characteristics of electrodes, electrolyte and applied voltage. This is called as the 'migration' state of the ECSMM process.

a.   **Reactions at anode and electrolyte interface:**

The electrochemical reactions at anode–electrolyte interface cause generation of oxygen gas. Dissolution of anode does not occur as the anode material used is graphite which is non consumable.

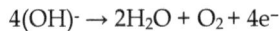

$$4(OH)^- \rightarrow 2H_2O + O_2 + 4e^-$$

b.   **Reactions at cathode and electrolyte interface:**

Following electrochemical reactions take place at cathode–electrolyte interface, and cause evolution of hydrogen gas.

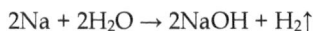

$$Cu^{2+} + 2e^- \rightarrow Cu$$

$$2H^+ + 2e^- \rightarrow H_2\uparrow$$

$$Na^+ + e^- \rightarrow Na$$

$$2Na + 2H_2O \rightarrow 2NaOH + H_2\uparrow$$

$$2H_2O + 2e^- \rightarrow H_2\uparrow + 2OH^-$$

Hydrogen gas evolves at the cathode, subsequently forming an isolating film, as depicted in Figure 8 a, or b, which leads to sparking across the bubbles between the cathode and electrolyte interface.

c. Reduction of electrolyte in the bulk:
It is given by:

$$NaOH \rightarrow Na+ + OH-$$

These liberated positive ions move towards cathode and negative ions move towards anode. In the external circuit, electrons move towards the cathode–electrolyte interface, and go to the solution. At the anode–electrolyte interface equal numbers of electrons are discharged from the solution to the anode. Electrochemical reactions that occur at the electrode–electrolyte interface continuously supply electrons from cathode to solution and solution to anode. This ionic and electronic current is the average current of the order of 100 – 200 mA.

5.1.2 Nucleate pool boiling of hydrogen bubble
The tip of the cathode gets heated up this causes nucleate pool boiling of hydrogen bubble that leads to development and formation of isolation vapor chamber of H_2 gas. The heat transfer controlled growth model applies and the radius of the bubble as a function of time can be found by the corresponding equations. According to this model, the vapor bubble starts growing till it reaches its departure diameter, reaching which the bubble gets detached from the lower surface of the tool. An isolating film of hydrogen gas bubble covers the cathode tip portion in the electrolyte, abruptly a large dynamic resistance is present and the current through the circuit becomes almost zero. At the same moment, a high electric field of the order of 10^7 V/m gets applied. This high electric field causes the bubble discharge, sparking takes place. This leads to generation of energetic electros. These electrons generate secondary electrons. These get drifted towards the workpiece surface due to potential gradient.

5.1.3 Sparking and electron avalanche
The high electric field causes spark within the gas bubble isolating the tip. The spark should occur between the tip of the tool and the inner surface of the electrolyte. At the instant when spark occurs, an avalanche of electrons caused by ionization flow towards work piece kept around 20 μm distance away from the tool tip. This avalanche of electron is manifested as the current pulses of short duration and high amplitude as seen in Figure 7a,b. As the potential gradient after varied varied, these electrons drift through the sparking channel towards the work piece surface. Experimentally it is found that this time to reach the electron avalanche to the work piece depends on the separation distance. This time is longer for work piece kept at a 500 μm distance (Kulkarni et al., 2002) than that kept at 20 μm. This fact supports that it is a drifting phenomena of the electron avalanche.

5.1.4 Material removal
The bombardment of electrons on the glass work piece surface results in intense heating and hence material removal takes place. Atoms of the parent material get dislodged and material removal takes place. There are partial sparks occurring, theses may not be having enough energy to cause the material removal. These hamper the efficiency of the process and also affect the surface finish.

At the same instant, the bubble geometry gets disturbed the contact between the tool and the electrolyte reestablishes. Electrochemical reaction takes over, bubble gets built up and the cycle keeps repeating itself. This makes the process discrete and repetitive.

All these intermediate processes described in sections 5.1 through 5.1.4 are correlated with the transient current pulses as observed in Figures 7 a and b. Figure 10 presents this correlation pictorially. The figure is self explanatory illustrating the time events during the ECSMM process w.r.t current.

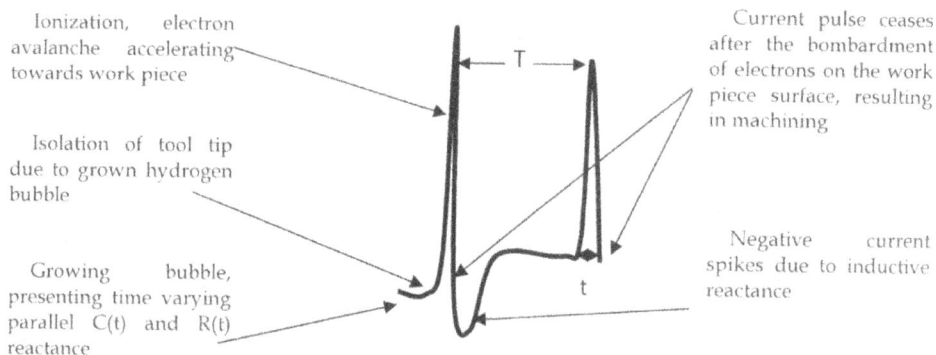

T: Time between two sparks, i.e. time required for the bubble growth till isolation of tool tip from electrolyte (T ranges between few hundreds of μs to few tens of ms)
t: Time required to reach the electron avalanche to the work piece surface
(t ranges between tens of μs to few hundreds of μs)
Sparking frequency $f_{sparking} = 1/(T+t)$
($f_{sparking}$ ranges between few hundred hertz to few tens of kHz)

Fig. 10. Part of an entire transient, instantaneous current pulse illustrating various time events during the ECSMM process w.r.t current

6. Concluding remarks

ECSMM process is found to be suitable for production of micro channels on glass pellets. The width of the micro channels achieved is in the range of 400 – 1100 μm. The depth achieved is in the range of 75 -120 μm. The time required to form these micro channels of 5mm length is about 5000 μm. SEM analysis shows that the micro machined surface is produced by melting and vaporization. The current pulses show the stochastic nature of the spark formation process.

The material removal mechanism is complex. It involves various intermediate processes such as: electrochemical reactions followed by nucleate pool boiling, followed by breakdown of hydrogen bubbles, generating the electrons, these electrons drifting towards the workpiece and causing the material removal. The process starts all over again by electrochemical reactions once the bubbles are burst due to sparking. And re establishment of contact between tool electrode – electrolyte takes place.

Close control for gap adjustment is must. Research efforts must be made to reduce the low energy sparks due to partial isolation to enhance the efficiency of the process and surface finish.

7. Acknowledgements

I am indebted to Prof. V K Jain for his immense guidance and support throughout my academic life at IIT Kanpur. I am thankful to Prof. K A Misra for his guidance in carrying out the work. Financial support for this work from Department of Science and Technology, Government of India, New Delhi, is gratefully acknowledged (Grant no. SR/S3/MERC-079/2004). Thanks are due to the staff at Manufacturing Science Lab and Centre for Mechatronics, at IIT, Kanpur. Ms. Shivani Saxena and Mr. Ankur Bajpai, Research Associates in the project, helped in carrying out the experiments. Their help is duly acknowledged. Thanks are also due to the staff at Glass Blowing section of IIT, Kanpur.

8. References

Basak, I. Ghosh, A. (1992). Mechanism of Material Removal in Electrochemical Discharge Machining: A Theoretic Model and Experimental Verification. J. Mater. Process. Technology, 71, 350–359

Basak, I., and Ghosh, A. (1996). Mechanism of Spark Generation During Electrochemical Discharge Machining: A Theoretical Model and Eexperimental Investigation. Jr. of Materials Processing Technology, 62 46-53

Bhattacharyya, B., Doloi, B. N., and Sorkhel, S. K. (1999). Experimental Investigation Into Electrochemical Discharge Machining of Non conductive Ceramic Material. Journal of Materials Processing Technology, 95, 145-154

Claire, L.C., Dumais, P., Blanchetiere, C., Ledderhof, C.J., and Noad, J.P., (2004). Micro channel arrays in borophsphosilicate Glass for Photonic Device and optical sensor applications, *Tokyo konfarensu Koen Yoshishu* L1351C, 294authors name missing

Coraci, A., Podarul, C., Maneal, E., Ciuciumis, A. and Corici, O. (2005). New technological surface micro fabrication methods used to obtain microchannels based systems onto various substrates', *Semiconductor Conference CAS 2005 Proceedings*, Vol. 1, pp.249–252

Crichton, I.M. and McGough, J.A. (1985). Studies of the discharge mechanisms in electrochemical arc machining', *J. of Appl. Electrochemistry*, Vol. l15, pp.113–119

Deepshikha, P. (2007). Generation of microchannels in ceramics (quartz) using electrochemical spark machining, MTech thesis, IIT Kanpur

Fascio, V., Wüthrich, R. and Bleuler, H. (2004). Spark assisted chemical engraving in the light of electrochemistry, *Electrochimica Acta*, Vol. 49, pp.3997–4003

Han, M-S., Min, B-K. and Lee, S.J. (2008). Modeling gas film formation in electrochemical discharge machining processes using a side-insulated electrode', *J. Micromech. Microeng.*, doi: 10.1088/0960-1317/18/4/045019

Hnatovsky, C., Taylor, R.S., Simova, E., Rajeev, P.P., Rayner, D.M., Bhardwaj, V.R. and Corkum, P.B. (2006). Fabrication of micro channel in glass using focused femto second laser radiation and selective chemical etching', *Applied Physics*, Vol. 84, Nos. 1–2, pp.47–61

Jain, V.K., Dixit, P.M. and Pandey, P.M. (1999). On the analysis of electro chemical spark machining process', *Int. J. of Machine Tools and Manufacture*, Vol. 39, pp.165–186

Kulkarni, A. V., Jain, V.K. and Misra, K.A. (2011c). Application of Electrochemical Spark Process for Micromachining of Molybdenum, ICETME 2011, Thapar University, Patiala, Mr J S Saini, Mr Satish Kumar, Mr Devender Kumar, Eds., pp. 410-415.

Kulkarni, A. V., Jain, V.K. and Misra, K.A. (2011b). Electrochemical Spark Micromachining: Present Scenario, *IJAT* vol. 5, no. 1, pp. 52-59.

Kulkarni, A.V., Jain, V.K. and Misra, K.A. (2011a). Electrochemical spark micromachining (microchannels and microholes) of metals and non-metals, *Int. J. Manufacturing Technology and Management*, vol. 22, no. 2, 107-123.

Kulkarni A. V., Jain V. K., and Misra K. A., (2010c). Development of a Novel Technique to Measure Depth of Micro-channels: A Practical Approach for Surface Metrology, Proc. of the ICAME 2010, R. Venkat Rao, Ed, pp. 1008-1012.

Kulkarni A. V., Jain V. K., and Misra K. A., (2010b). Traveling Down the Microchannels: Fabrication and Analysis, AIM 2010, 978-1-4244-8030-2/10 ©2010 **IEEE,** pp. 1186-1190.

Kulkarni, A. V., V. K. Jain, V.K. and Misra, K.A. (2010a). Simultaneous Microchannel Formation and Copper Deposition on Silicon along with Surface Treatment, IEEM 2010 IEEE, DOI: 10.1109/IEEM.2010.5674509, pp 571-574.

Kulkarni, A. V. (2009). Systematic analysis of electrochemical discharge process, Int. J. Machining and Machinability of Materials, 6, ¾, pp 194-211.

Kulkarni, A. V., Jain, V. K., Misra, K. A. and Saxena P., (2008). Complex Shaped Micro-channel Fabrication using Electrochemical Spark, Proc. Of the 2nd International and 23rd AIMTDR Conf. Shanmugam and Ramesh Babu, Eds, pp. 653-658.

Kulkarni, A. V. Sharan and G.K. Lal, (2002). An Experimental Study of Discharge Mechanism in Electrochemical Discharge Machining, International Journal of Machine Tools and Manufacture, Vol. 42, Issue 10, pp. 1121-1127.

Kulkarni, A. V. (2000). An experimental study of discharge mechanism in ECDM, M.Tech. Thesis, IIT Kanpur, Kanpur, India.

Marc Madou, (1997). Fundamentals of micro fabrication, *CRC Press*

Rajaraman, S., Noh, H-S., Hesketh, P.J. and Gottfried, D.S. (2006) 'Rapid, low cost micro fabrication technologies toward realization of devices for electrophoretic manipulation', *Sensors and Actuators B*, Vol. 114, pp.392–401

Rodriguez, I., Spicar-Mihalic, P., Kuyper, C.L., Fiorini, G.S. and Chiu, D.T. (2003) 'Rapid prototyping of glass materials', *Analytica Chimica Acta*, Vol. 496, pp.205–215.

Sorkhel, S.K., Bhattacharyya, B., Mitra, S. and Doloi, B. (1996) 'Development of electrochemical discharge machining technology for machining of advanced ceramics', *International Conference on Agile Manufacturing*, pp.98–103

Wuthrich, R., Fascio, V., Viquerat, D. and Langen, H. (1999) 'In situ measurement and micromachining of glass', *Int. Symposium on Micromechatronic and Human Science*, pp.185–191

13

Modeling and Simulation of MEMS Components: Challenges and Possible Solutions

Idris Ahmed Ali

Dean, College of Engineering & Applied Sciences
Al Ghurair University, Academic City, Dubai,
UAE

1. Introduction

Micro-electro-mechanical systems (MEMS) represent a very important class of systems having applications ranging from small embedded sensors and actuators, passive components in RF and microwave fields, and micro-mirrors in the optical range. The importance of MEMS stems from their many advantages, among which are, their small compact size amendable to integration with other components, low loss and parameter variability.

From structural point of view, each MEMS component is, by itself, a very small electromechanical system of heterogeneous structure composed of materials with different chemical composition (dielectric substrate, metal alloys and conducting wire) and different physical (electrical, thermal, mechanical) properties. Moreover, MEMS components may represent static systems or they may contain some moving parts, such as in variable capacitor, moving membranes and cantilevers. The dimensional scale of the different parts of MEMS components may vary from very small (microns or even nanometers) in one dimension, such as thickness of a plate, to comparatively large of few hundred microns in other dimensions, thus resulting in large aspect-ratios.

When MEMS components are put into oration, they constitute systems, in which electrical, thermal, mechanical, and other physical phenomena take place and interact with each other. From mathematical modeling and simulation point of view, this calls for multi-physics treatment, in which coupled systems of differential equations of different combinations of electromagnetic, mechanical, fluid, heat transfer and/or transport equations, are formulated then solved depending on the type of boundary conditions imposed by MEMS component under investigation.

Mathematical modeling and simulation has been used in all fields and disciplines of engineering for decades, for theoretical characterization of devices and systems before manufacturing, or even before prototyping, for a number of reasons among which are reduction in manufacturing cost and time. However, the heterogeneous nature of MEMS structures, coupled with multi-physics phenomena that take place during their operation, makes modeling and simulation of MEMS components, a complex and challenging task.

The main objectives of this chapter are to outline the nature of MEMS componets, from both the structural and physical points of view and identify the difficulties that these heterogeneous structures impose on modeling and simulation of these systems and to give a comprehensive account of different modern modeling methods and techniques used to overcome MEMS modeling and simulation difficulties, with illustrative examples. From here onwards, the word 'components' will be suppressed and refer to MEMS component just as MEMS for brevity.

The chapter starts with some basic definitions, then shed light on the use and advantages of mathematical modeling and simulation in MEMS design and manufacturing, investigates the nature of MEMS and the main challenges facing researchers in their modeling and simulation. It presents systematic approach to MEMS modeling and simulation, gives a survey coverage of different methods and techniques used for simulation of MEMS, such as finite-differences time-domain (FDTD), finite-element (FE) and their different variants proved to be successful in their applications to such microstructures. A detailed illustrative MEMS modeling and simulation example is given and other examples developed by other active research workers, are cited. The chapter concludes with future outlook and new trends in MEMS modeling and simulation.

2. Background

Over the last few years, extensive research work in modeling and simulation of MEMS took place, and a large number of techniques appeared in the literature to tackle this complicated modeling problem, each having its own advantages and limitations (Bushyager et al., 2003; Lynn Khine & Moorthi, 2006; Khine A. et. al., 2008; Yong Zhu & Espinosa, 2003; Shanmugavalli et. al. 2006; Heung-Shik Lee et. al., 2008; Fengyan and Vaughn, 2009; Chiao & Liwei, 2000).

Bushyager and his coworkers considered modeling electro-statically actuated RF passive components, such as parallel–plate capacitors and tuners (Bushyagers & Tenzeris, 2002; Bushyager et .al 2003; Bushyager, Mc Garvey & Tenzer 2000). They used finite-difference time-domain technique (FDTD) or its variant MRTD, with adaptive gridding to tackle the problem of large aspect ratio and moving boundary conditions (see section 4).

Fengyan and Vaughn (2009) developed an on-line tool for simulating micro scale electro-thermal actuators (ETA), using both distributed and lumped analysis. The model allows the user to input the electrical Current of ETA to compute both temperature distribution and the displacement of the ETA. Finite–Element method was used to find temperature distribution due to joule heating and the average temperature across the beam is used to find the displacement by lumped analysis. The Model takes into account both properties' variation with temperature as well as radiation effects at high temperature. The main characteristics of their model is the use of computationally expensive distributed analysis for modeling electro thermal phenomenon and the computationally efficient lumped analysis for the modeling the thermo mechanical phenomenon thus improving the overall computational efficiency, accessibility and ease of use. The last two properties (accessibility and ease of use) are due to on-line availability of model for remote users.

Chiao & Liwei (2000) Considered self buckling behavior of micromechanical clamped-clamed micro beams under resistive heating using both analytical and finite-difference technique. Their model consisted of electro-thermal part in which electric current flowing through the beam gives rise to Joule Heating effects, and thermo-mechanical part which deals with the mechanical buckling of the beam due to thermal expansion. Results was verified by measurements with good agreement. They considered the effected of residual stress and found that when compressive residual is considered in the analysis the critical current causing bulking of the beams, decreased where as tensile residual stress hinders the actuation of the beams. They also considered the effect of process variation such as the width and the thickness of the beam, and found that they affect the performance of the beams such as the current–deflection curves, however variation expansion and Young's modulus with temperature, were not considered and called for characterization of these properties' variation in order to improve the accuracy of MEMS Models.

Most, if not all, of these attempts relied on numerical (computational), rather than analytical techniques, due to ability of numerical methods to model structures with arbitrarily complex geometrical shapes. Despite the advantages of numerical techniques researchers were faced with a number of difficulties when trying to use commercial simulation packages and adapt traditional numerical methods, such as finite-difference, time-domain (FDTD) and finite-element (FE), used in modeling structures at macro-scale. The main difficulties and challenges are highlighted in section 4.

3. Modeling & simulating definition & advantages

3.1 Definitions

Before we go into further details of this chapter definition of some of the main concepts is in order. In the following are given, definitions of the main keywords of this chapter; modeling and simulation.

Modeling: Modeling, and a model, has different meanings depending on the context or field of application. In engineering science and technology, Modeling refers to mathematical representation of a physical phenomenon, system or device. Usually mathematical models can take many forms such as dynamic system models, statistical models, game models, differential equation…etc. But for the purpose of this chapter, we will be mainly concerned with mathematical models that are represented by differential equations.

Simulation: Similar to modeling, simulation also has different meanings depending on the context and type of application and it can take many forms. However for the purpose of this chapter, we will concentrate on computer simulation or computational modeling, which is defined as: a computer program, or package, that attempts to simulate or imitate an abstract model of a particular system or device. Computer simulations have become very much related and integrated with mathematical modeling, and usually modeling and simulation are taken as being one discipline that can be used to explore and gain insight into the new technology and predict and estimate behavior of complex systems and devices that are too much complicated for analytical solutions.

Modeling and Simulation generally have iterative nature. A model is first developed then simulated to gain some understanding. The Model is then revised, and simulated again and this process goes on until an adequate level of understanding is developed for the system/device under consideration.

3.2 Need for MEMS modeling and simulation

There are many reasons why we need modeling and simulation for MEMS, among which are:

a. Due to small dimensions of MEMS, direct experimentation for determination of some physical properties of MEMS is difficult, and measurement errors occur when dealing with these micro-level systems.

b. Time reduction: Designers need simulation tools that allow them try "what If "experiments in hours instead of days and weeks, thus reducing time to market.

c. Production cost reduction: Modeling and simulations are needed in order to study the behavior of the design under different experimental conditions and different level of parameters, before prototyping, thus reducing production cost.

d. MEMS are usually embedded within other systems or packaged with other micro-machined components and systems. Therefore modeling and simulation is needed at macro or system level as well as micro-level.

e. Modeling and simulation can render fast design cycle that allows extensive scoping for more and accurate decision making.

f. Modeling and simulation allows better understanding of the device/system operation and gives scope for optimization of its operation.

g. Modeling and simulation enable designers and system developers to see and further investigate systems behavior which could have not been discovered otherwise.

4. MEMS modeling challenges

The difficulties and challenges that face MEMS modeling and simulation experts are due to the very nature of MEMS and can be related to the following MEMS features: Physical principle of operation, geometrical structure, miniaturization, packaging, manufacturing and processing of MEMS, and environmental conditions. These are detailed in the following:

a. Physical Principles:

As mentioned earlier, MEMS are characterized by interaction of many physics domains in a single device or system. This is pictorially represented in Fig. 2. Interaction of many physical phenomena in MEMS needs dealing with different types of equations, each governing a certain physical phenomenon e.g. electrical, mechanical or thermal. Moreover these governing equations are always coupled sometimes strongly and sometimes weakly, thus calling for solution of coupled system of equation a phenomenon called multi- physics approach. Depending on the number of physics domains involved, we can further classify these multidomain systems as: Double-Physics (interaction of two physics domain, Triple-Physics (interaction of three physics domains), Quadruple-Physics (interaction of four physics domains) and so on. In section 7.2 , we show specific modeling and simulation examples illustrating these different physics categories.

b. MEMS geometrical structure:

One of the main obstacles facing accurate modeling of micro-systems is proper definition (construction) of the system geometry, one of the main first requirements in any modeling and simulation process. This is due to different deformations and irregularities during MEMS operation, such as in the case of tuners and microbeams that contain moving parts. In order to tackle this problem, Peyrou David et.al, 2004, proposed a reverse engineering technique, whereby the model is first built using the real shape of the device, then a virtual model is made from the deformed shape, using different software packages. This method was applied for simulation of RF-MEMS capacitive switch and electrical contact resistance of RF MEMS, with satisfactory results (Peyrou David et.al, 2004).

c. Large Aspect Ratio:

Many MEMS, such as metallic sheets on top of large substrate (e.g. capacive stub tuners), have very small dimension in one direction (thickness) compared to relatively large dimension (length) in other direction, thus leading to large aspect ratio. Large aspect ratio, in turn lead to creation of large number of computations cells thus leading to long computation time (Ali, I. Kabula M. & Hartnagel H. L., 2010).

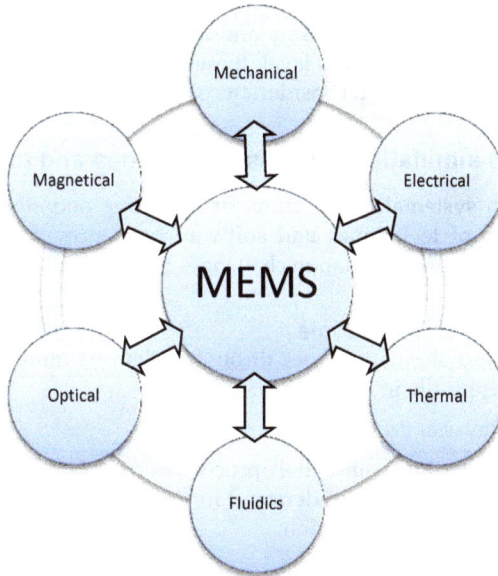

Fig. 1. Multiphysics nature of MEMS

d. Geometry change due to moving parts:
Some MEMS components, such as electrostatic and thermal actuators, contain moving parts, which lead to moving boundary conditions.Commercially available traditional frequency domain simulators are not adequate to handle these features, due to many approximations that are more to simplify modeling and time-domain techniques, are more appropriate (Bushyagers & Tenzeris. 2002; Bushyager et. al., 2003). However even the well-known time-domain techniques such as FDTD, are not adequate as they stand and some sort of modification are needed to deal with large grid size and moving boundary conditions. These are solved by using adaptive gridding in both space and time, which are included in FDTD technique with variable gridding capability or the use of multi resolution time domain technique known as MRTD for short.

e. Miniaturization:
Due to the very small dimensions of MEMS many physical phenomena such as surface tension from humidity, become significant at micro-level. Downscaling of the dimensions of MEMS structures, which are usually three-dimensional makes the scale of these effects (force, displacement ... etc.) also significantly small which, in turn lead to accumulation of numerical errors.

f. Packaging:
In many cases MEMS are assembled and packaged with other electronic components and devices, and this calls for proper modeling and simulation at systems level taking into consideration all input/output feedback etc.

g.   Manufacturing and processing:
Manufacturing and process uncertainties which result from the fact that the dimensions and properties used in simulation of the MEMS cannot be exactly produced during fabrication.
h.   Environmental Conditions:
MEMS are sometimes needed to work at different environmental conditions, such as sensors on the surface of the aircraft wings or sensors and actuators in nuclear power plants or under sea. Being tiny components at micro level, these systems are much affected by these external conditions. This calls for careful considerations of these conditions when modeling.

5. MEMS modeling and simulation process, techniques and tools

In this section, we show a systematic procedure for modeling and simulation process of MEMS, and give a survey of techniques and software tools proved to be successful in tackling many of the modeling and simulation challenges illustrated in the last section.

5.1 Modeling and simulation process cycle
The process of modeling and simulation goes through a defined number of steps, briefly described and depicted graphically in Fig.1 below:

Step 1: Description of the Physical Problem:

The first step in the modeling and simulation process is the physical description of the problem under investigation in order to understand its geometrical structure and the entire physical phenomenon involved in its operation.

Step 2: formulation of governing equations:

Depending on the physical principles governing operation of the device/system, all equations and mathematical expressions governing this operation are setup together with appropriate boundary and/or initial conditions.

Step 3: approximation of governing equations:

At this step, some approximations that facilitate solution with adequate effort and/or resources are made. If more than one equation is involved, the degree of coupling between different types of equations is determined at this step.
Steps 2 and 3 are usually very much interrelated, and the process of approximation and formation of governing equations are done iteratively in one step.

Step 4: Method(s) of solution:

Here all adequate methods of solutions of the governing equations are explored, and the most appropriate one identified. Appropriateness here calls for proper consideration of all available computation resources (both hardware and software), and selection of the most efficient one in terms of computational resources.

Step 5: Solution of the governing equation(s):

The method(s) identified in step 4 is applied to find the solution of the system of equations under specified boundary/initial conditions.

Step 6: Verification of Results:

After finding the general solution in step 5, some simplified standard cases which can be verified by experimental measurements or for which analytical or known solutions are

```
┌─────────────────────────┐
│ 1. Description of the    │
│    physical problem      │
└─────────────────────────┘
            ⇓
┌─────────────────────────┐
│ 2. Develop mathematical  │◄──────────────┐
│    model                 │               │
└─────────────────────────┘               │
            ⇓                              │
┌─────────────────────────┐               │
│ 3. Make possible         │               │
│    approximation         │               │
└─────────────────────────┘               │
            ⇓                              │
┌─────────────────────────┐               │
│ 4. Select solution       │               │
│    algorithm and solve   │               │
└─────────────────────────┘               │
            ⇓                              │
┌─────────────────────────┐               │
│ 5. Interpret results and │               │
│    check                 │               │
└─────────────────────────┘               │
            ⇓                              │
         ◇ Results              No   ┌─────────────────────┐
         ◇ OK?    ────────────────►  │ 6. Make necessary   │
         ◇                           │    modification     │
            ⇓ Yes                    └─────────────────────┘
┌─────────────────────────┐
│ 7. Use the Model         │
└─────────────────────────┘
```

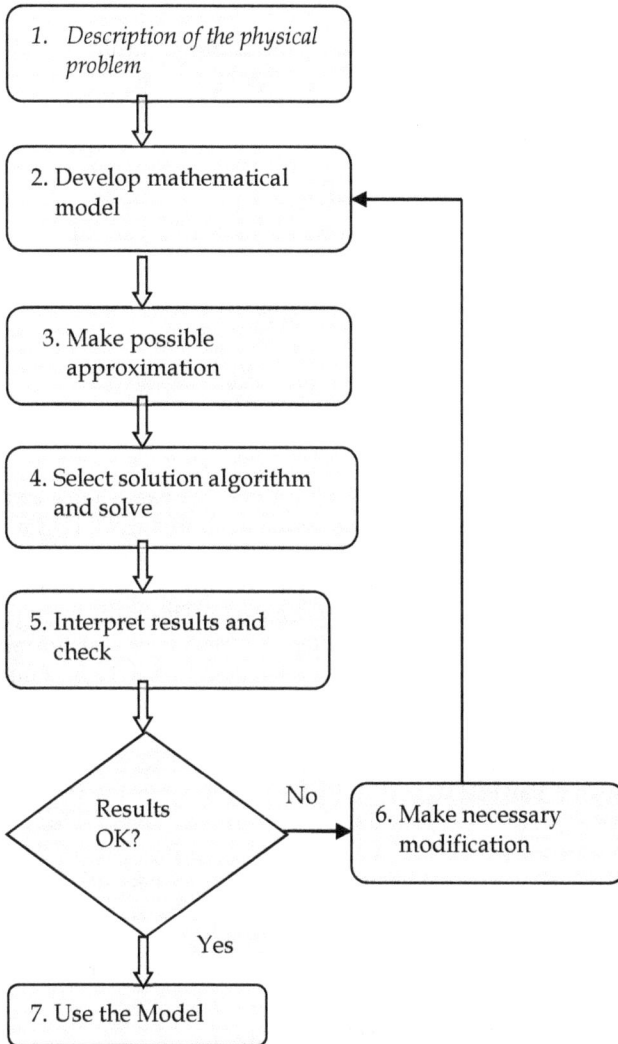

Fig. 2. The main Steps of Modeling and Simulation process (details of each step are given in the main text)

identified for comparison reasons. If adequate solution that agree with measurements or known solution is obtained, move to the following step or otherwise go back to check method of solution or even the model itself. This checking process goes on until adequate solution is obtained (see decision block in the flowchart of Fig.2.

Step 7: Use model for exploration and design optimization:

This is the last step at which you make use of your efforts in modeling and simulation. Here the model is used for analysis of varied situation under different conditions i.e. change of input parameters and study of what-if conditions.

5.2 MEMS modeling techniques

There are a number of modeling techniques, with different facilities and capabilities, for modeling and simulation of MEMS at both component and system levels. However, we will concentrate on two mostly used techniques; the finite difference time domain (FDTD) and Finite Elements techniques and their variants.

5.2.1 Finite – Difference – Time Domain (FDTD)

FDTD technique is direct method of Solution of Maxwell's curt equation (Taflove, A. & Hagness, S., 2001):

$$\nabla \times \bar{E} = -\mu \frac{\partial H}{\partial t} \tag{1}$$

$$\nabla \times H = \sigma \bar{E} + \epsilon \frac{\partial E}{\partial t} \tag{2}$$

where E and H are the vector electric and magnetic fields, μ, ϵ and σ are the respectively, permeability, permittivity and conductivity of the medium.

FDTD method is a full-wave time stepping procedure that uses simple central – Difference approximation to evaluate the space and time derivatives. FDTD method has been successfully used for simulation of complex structure at macro scale, however one of the most constraints of FDTD is the requirement that cell size be equal to or smaller than the smallest feature of the device or system being simulated. This leads to large grid size when simulating structures consisting of several different elements or structure with large aspect ratios, such as MEMS circuits. The large grid size leads, in turn to large computing time.

In order to overcome this difficulty, a number of modification and / or addition have been proposed, among which are the method of sub gridding, and the use of non-uniform grid (Shanmugavalli et. al., 2006; Bushyager et. al., 2001). The first, subgridding method subdivides sections of the FDTD grid, thus creating finer grids in areas of large field variation or interfaces between areas of different properties. This method can be used when small complex devices are embedded in a coarse grid, such as small passive compounds (compactors' or inductors) connected by micro strip lines or when MEMS are packaged. Here a subroutine is needed to interface grids of different sizes. In the non uniform grid technique, each direction of the structure under investigation is treated separately, thus allowing each axis to be divided into a number of sections each having a different cell size.

5.2.2 Multiresolution Time Domain (MRTD) technique

MRTD employs multiresolution principles to discretize Maxwell's equations in wavelet expansions, which provide a set of functions with adaptive resolution. Higher resolution functions can be added and subtracted during simulation. Accordingly MRTD has built-in gridding capability in both time and time and space

1. MRTD can model changes in structure over time because it is a time-domain simulator for example: variable parallel-plate capacitor.
2. Variable grid capability and ability to arbitrarily place metals in structures allow for modeling of a structure that changes configuration during simulation.
3. Adaptive (variable) gridding allows modeling of MEMS in packages such as variable capacitor connected to large feed lines .In this case the surrounding structure can be modeled with a comparatively coarser grid, while the MEMS is modeled with a fine grid (higher resolution).

When MRTD is applied to Maxwell's Curl equations (1 and 2 above), the orthogonality of the wavelets provides an efficient discretization that results in an explicit time marching scheme similar to that of FDTD technique. Due to this property and the characteristics of MRTD stated in 1-3 above, MRTD has been successfully applied by Bushyager and his co-workers to a number of modeling and simulation of RF MEMS structures, such as parallel-plate capacitor (Bushyager & Tentzeris 2001), MEMS tunable capacitors (Bushyager et. al. 2002), analysis of MEMS embedded components in multilayer packages (Tentzeria et. al. 2003), for intracell modeling of metal/dielectric interfaces (Bushyager & Tentzeris, 2003), for static MEMS capacitor (Bushyager, McGarvey & Tentzeris, 2001).

5.2.3 Finite Element Technique

The **finite element method (FEM)** is a numerical technique for finding approximate solutions of partial differential equations (PDE) and integral equations. The solution approach is based either on eliminating the differential equation completely (steady state problems), or rendering the PDE into an approximating system of ordinary differential equations, which are then numerically integrated using standard techniques such as Euler's method, Runge-Kutta, etc. Historically, finite element method originated from the need for solving complex material and structural problems by civil and mechanical engineers.

The first step in finite-difference analysis is to divide the structure under consideration into small homogeneous pieces or elements with corners called nodes. The nodes define the boundaries of each element and the entire collection of elements defines the analysis region or the geometry of the structure. These elements can be small where detailed geometric properties are required and much larger elsewhere. A simple variation of the field is assumed in each element and the main goal of FEA is to determine field quantities at each node. In general, FEA techniques solve for the unknown field quantities by minimizing an energy functional using variational techniques. The energy functional is an expression describing all the energy associated with the structure under consideration. For instance, this functional for 3-dimentional time-harmonic electromagnetic fields E and H is represented as:

$$F = \int_v \frac{\mu|H|^2}{2} + \frac{\epsilon|E|^2}{2} + \frac{JE}{2j\omega} \, dv \qquad (3)$$

Where the first two terms represent energy stored in the magnetic and electric fields and the third is the energy dissipated or supplied by the conduction currents.

Eliminating H using relation between E and H, and setting derivatives of this functional with respect to E to zero (for minimization), an equation of the form f(J,E)=0 is found. Appling K[th] order approximation of F at each of the N nodes and enforcing boundary conditions results in a system of equations of the form:

$$\begin{bmatrix} J_1 \\ \vdots \\ J_n \end{bmatrix} = \begin{bmatrix} y_{11} & \cdots & y_{1n} \\ \vdots & \ddots & \vdots \\ y_{n1} & \cdots & y_{nn} \end{bmatrix} \begin{bmatrix} E_1 \\ \vdots \\ E_n \end{bmatrix} \qquad (4)$$

Here J_i and y_{ij} (i=1,2,..n and j=1,2,..,n) are known quantities and the electric fields E at each node are the unknowns to be determined. This needs inversion of the y matrix and the larger the number of nodes, the larger the size of the matrix and hence more computational requirements. For example a mesh of a 3-D cube with 1000 nodes (10 nodes per side) and 3 degree of freedom creates a 3000 by 3000 matrix. Doubling the number of nodes per side gives a system of 24000 by 24000 matrix. Therefore it is necessary to as fewer as possible number of nodes is preferable from the point of view of computer resources. On the other hand the more the number of nodes the more is the result accuracy. Usually a compromise is made between required accuracy and computational resources. For proper grasp of this useful technique, the reader is referred to any of the many text books on finite element analysis (see for instance Silvester & Ferrari (1990)).

5.3 MEMS simulation software tools

During the last decade, and specially in the period 2002-2010, there were a lot of research activities and developmental in the area of modeling and simulation tools dedicated to MEMS, and a number of such software simulation tools moved from research centers to the commercial world. These tools can be classified broadly as: *system level modeling and simulation tools* and *detailed device–level tools known as field solvers*. Technique used in type of tools, have been in use at macro level for some time and have gained maturity. the micro level , however . There tools are still going some modification due to the nature of MEMS, which comprises many physical domains such as; mechanical, thermal, electrical and optical domains. In the following sections, we give a brief overview of these techniques and tools.

5.3.1 System level tools

These tools predict the main behavior of the MEMS, using a set of ordinary differential equations and non linear functions at block diagram level in terms of input and output of each block. These tools are usually required by MEMS Designers for accurate prediction of the overall performance of these systems when connected or operated with other circuit components and for proper optimization (Lynn Khine & Moorthi P., Hine A. et. al. 2008). In addition they enable the designer predict the performance of the device under consideration for a set of defined input parameters, it allows him/her to study the effect of variation of one or a certain set of parameters relative to other parameters on the overall performance of the device, thus leading to optimized design.

5.3.2 Field solvers

The majority of these solvers are based on the well-known numerical techniques of finite elements (FE) or finite–difference time–domain, methods briefly out lined in the lost section, with only few tools based on analytical or asymptotic solution of the governing equation. These methods are virtual and can analyze complex geometric shapes.

Finite-Elements–based tools

The majority of MEMS software tools now available in the market are based on the well – established FE technique. These include ANSYS (COMSOL), ConvectorWare, Intellisuite, ABAQUS, and DELEN, to mention only some example. In the following we give a brief account of their mostly used tools in the field of MEMS analysis and design.
a.   ConventorWare

This is a fully integrated MEMS Design environment, the latest version of which is convectorWare2010. It runs on PCS and work Stations based on sun Solaris and windows
This package consists of four main parts; Designer, Architect, Analyzer and Integrator. Designer is concerned with design specification and modeling of the MEMS Structure including two-dimensional layout, process emulation and finite elements meshing. It

includes a full- featured layout editor, foundry access kits , 2D and 3 D import/export capabilities, mesh generator and a library of 3D models of standard MEMS packages. It is tightly integrated with the other two modules ARCHITECT and ANALYZER, but it can also be used as stand- alone package to be integrated with other third-party FEA tools.

i. ARCHITECT 3D: This module creates schematic models of MEMS designs using MEMS component Library, perform fast system level simulations with finite elements accuracy and significantly reduce product development time by mean of fast simulation speeds.

ii. ANALYZER: This Module provides a single unified interface to a complete suite of 3D field solver such as coupled Electro-Mechanics, Piezo–Electric or Microfluuidics. With this module the designer can perform: parametric studies to optimize design, Incorporate packaging effects such as ambient temperature and peruse, Predict and validate experimental measurements, Investigate manufacturing effects such as residual stress.

iii. INTEGRATOR: This module extracts reduced-order models of physics effects for use in ARCHITECT. For Example it can be used to add gas damping effects on a ARCHITECT model. Similarity it can extract mechanical stiffness of structural components such as tethers that can be modeled as a linear or non linear mechanical springs a electro statistic forces between electrodes and combs that can be modeled as electrostatic springs.

b. IntelliSuite:

This is an integrated design environment specifically developed to link the entire MEMS organization together. It enables designers to manage their MEMS product throughout it life cycle. Intellisuite consists of a number of advanced tools that work together, each covering a certain MEMS development stage. The main components of Intellisuite are:

i. Synble: that allows the designer to capture MEMS at a schematic level, explore his/her design, then optimize and synthesize. It has a powerful schematic editor specifically developed for MEMS and other multiphysics modeling.

ii. Synthesis: these are sophisticated algorithms that automatically convert schematics into mask layouts or a meshed structure for multiphysics analysis.

iii. Blueprint: a physical design tool that includes advanced layouts, cross section exploration and automated mask to hex mesh.

iv. CleanRoom: used for creation of process flow and mask set before entering the physical clean room thus allowing visual prototype before actual fabrication.

v. Fastfield: a collection of multiphysics tools featuring coupled electrostatic, electromagnetic and electromechanic engines.

vi. Extraction: collection of techniques that capture electromechanical and damping behavior into compact models.

c. COMSOL Multi-physics

COMSOL Multiphysics is a simulation software specifically designed to tackle multi physics problem, usually encountered all disciplines in Engineering Design starting from the definition of geometry, through meshing and material property definition and ending with solving and visualization of results. It is based on solution of the partial differential equations that given different physical phenomenon be it electromagnetic, heat transfer, structural mechanics, magnetostatics, simultaneously or sequentially solved using finite element analysis.

By its very nature of structure it is particularly appropriate module for MEMS, modeling and simulation, and in addition , it has a specialized module for MEMS, called MEMS Module that simulates standard MEMS problems, such as piezoelectric and thin film damping problems.

Finite difference time domain tools

a.   XFDTD Package:
This is based on finite difference time domain technique (FDTD), for the solution of vector Maxwell's equations developed by REMCOM Inc. It is an easy to use package, but it does not have the capability of dealing with multidomain (multiphysics) problems.

5.3.3 Other tools

In addition to the above mentioned main MEMS Simulation software tools, there are other tools most important of which is the hybrid approach in which both system-level and field solvers in circuit simulation environment. They rely on a new class of parameterized behavioral method that can automatically access relational databases that contain field software solution results. These models have been successfully used for analyzing packaging effects of inertial sensors and design of PZR pressure sensors (Lorenz G., Greiner, K., & Breit S. 2006).

6. Selected MEMS-based Modelling Examples

Modeling and simulation can not only be acquired by reading and studying different methods and techniques, but through actual involvement in building models, simulating and practicing. This is similar to driving a car. Car driving can be learned through reading instructors notes, books and articles about car driving, but can only mastered by actually driving a car.

In this section, we illustrate the process of modeling and simulation described in section 2.2, by going through an actual case carried out by the author and his coworkers, and highlight some few successful cases done by other research workers.

6.1 Detailed illustrative example on modeling and simulation process

Micromachined DC and high frequency power sensors

In order to show how the modeling and simultion process described in section 3.3 is implemented in practice, modeling and simulation of micromachined power sensors at both the dc and high frequencies (microwave and millimeter waves) is presented. The sensor fabrication techniques is based on bulk and surface micromachining of AlGaAs/GaAs heterostructures, and the sensor concept is based on transduction principle that converts the RF into thermal power, which is then measured by thermoelectric means (Mutamba et al., 2001; Ali, I. Kabula M. & Hartnagel H., 2010)).

Modeling and simulation is based on the solution of the coupled electromagnetic field and heat transfer equations, using hybrid finite-element/finite-difference techniques. The model can be used as an effective tool for adjustment of design parameters, such as geometric configuration of the sensor resistive element, and arrangement of thermocouples around the resistor for optimum sensitivity and noise figure.

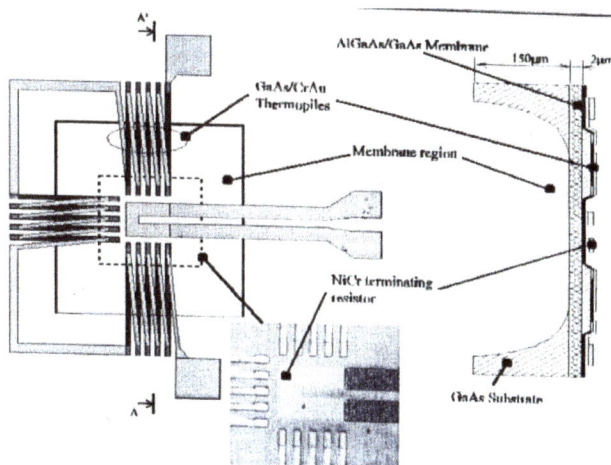

Fig. 3. The structure of the power sensor. left-hand : top view, right-hand-side sectional view along A-A'.

Step 1: Description of the Physical Problem:

A sectional view of the configuration of the sensor to be modeled is depicted in Fig. 3. The sensor is composed of a thermally isolated thin (2 μm) AsGaAs/GaAs membrane region with a terminating resistor, in which heat generated by DC or RF power is dissipated and converted into heat. A high-thermally resistive membrane region is obtained by selective etching of the GaAs against AlGaAs. This helps to increase temperature gradient between the resistor region and the rest of the chip, thus leading to high sensitivity. Temperature increase in the resistor region is sensed by a set of Gs/Au-Cr thermoelements, whose dc output is proportional to the input RF power. Detailed description of technical realization of this sensor is given by Mutamba et al., (2001).

Steps 2,3: The Governing Equations and Approximations:

Here we consider both the thermal and the electric models. The mathematical model is required to be simple enough to be handled by known methods which demand reasonable time and cost, and at the same time give an adequate description of the physical problem under investigation.

Thermal model

A simplified pictorial view of the sensor is shown in Fig. 3 for the purpose of thermal modeling. Here the thermocouple wires are not shown, and heat generated by the NiCr resistor, is assumed to be distributed throughout the sensor structure mainly by conduction. Due to axial symmetry of the sensor structure about it longitudinal axis, we only consider half of its geometry as shown in Fig.4. We select Cartesian coordinates (x,y,z) with its origin at the sensor input and the direction of power transmission along the positive z-direction.

In order to obtain a simple and manageable model, the following simplifying assumptions, are made:

i.   The presence of the thermocouples is ignored as the metallic part (gold) is very small compared to other dimensions.

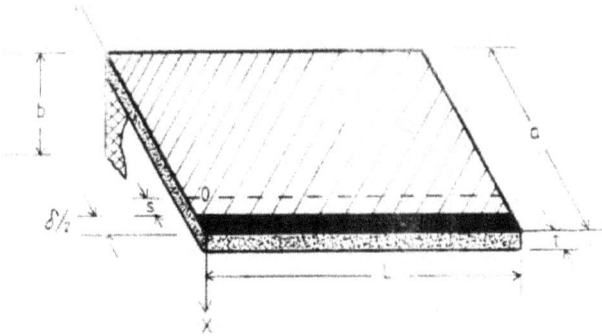

Fig. 4. Simplified half-section used for modeling.

ii. As a first approximation, heat distribution is assumed to be two-dimensional (in Y-Z plane). This assumption can be justified due to the presence of the thermally isolated thin membrane whose area dominates the horizontal dimensions of the sensor.

iii. Constant thermal properties. Although thermal properties, especially thermal conductivity of GaAs vary with temperature, these properties were assumed to be constant due to the expected moderate temperature rise.

iv. Radiation losses are ignored as the sensor were to work in a very confined region and the expected temperature rise is limited.

The equation governing heat flow as a result of microwave or dc heating is the well-known heat conduction equation, also known as Fourier equation:

$$\frac{\partial T}{\partial t} = \alpha \nabla^2 T + Q / \rho C \tag{5}$$

where T is the temperature, α is thermal diffusivity , Q heat generation term, ρ is the density and C is the specific heat of the medium .

Thermal boundary conditions

These conditions can be obtained from the prevailing thermal conditions at the boundaries and the interface between different material layers of the sensor.

i. Convective heat transfer at boundaries y=0 and y=a.

ii. Specified temperature at boundaries z=0 and z=L.

iii. At the interface between different sensor material layers: assuming perfect thermal contact leads to the continuity of heat flux and temperature at these interfaces between layers.

iv. Assume initially that the sensor is at a constant temperature To.

The heat source term, Q, in equation (5), is determined by solving either of two different sets of equations, depending on the operation mode of the sensor (dc or high frequency) as shown below.

Electric model

Here we consider the case of AC and DC case separately, because of their different governing equations.

DC Operation Mode:

In the case of the dc operation mode, the electrostatic potential equation is used:

$$-\nabla(\sigma\nabla V) = q_i \tag{7}$$

where V is the electrical potential, σ is the electrical conductivity, and q_i is the current source. In the case of constant electrical conductivity, equation (7) reduces to the simple Poisson's equation:

$$\nabla^2 V = q_i / \sigma \tag{8}$$

Boundary conditions assumed for solution of equation (8) were that a constant voltage +V applied at one arm and –V at the other arm of the CPS With the symmetry condition along the longitudinal axis of the sensor, +V was assumed at one arm at input of the sensor and zero voltage at the end of the resistor (see Fig. 3). Having found V from equation (8), the heat generation term, Q, is obtained as:

$$Q = \sigma |E^2| \tag{9}$$

where E is the electric field intensity, given by:

$$\bar{E} = \nabla V \tag{10}$$

High Frequency Mode:

In the case of high frequency operation mode, the heat generation term, Q, is obtained from the solution of Vector Maxwell's equations:

$$\nabla \times \bar{E} = -\mu \frac{\partial \bar{H}}{\partial t} \tag{11}$$

$$\nabla \times \bar{H} = \sigma \bar{E} + \varepsilon \frac{\partial \bar{E}}{\partial t} \tag{12}$$

in which $\bar{E}$ and $\bar{H}$ are electric and the magnetic field vectors, and ε, μ ,σ are respectively the permittivity, the permeability and the conductivity of the material (medium) through which electromagnetic wave propagation takes place.

Steps 4 & 5: Solution of Governing Equations:

If the simplifying assumption (i) – (iv) are used, equation (1) with boundary conditions ii), iii) and iv) can be solved analytically using the method of separation of variables. However, in the more general case of three-dimensional form of equation (1) with boundary conditions (i) to (iv), the versatile numerical methods of finite difference in time-domain or finite elements are more appropriate, since they can deal with any sensor structure. Investigation of available software packages revealed that XFDTD package, based on finite difference time domain technique, is most appropriate for the solution of vector Maxwell's equations and FEMLAB, a package based on FEM for solution of the heat equation. As the two packages (XFDTD and FEMLAB) are based on different solution techniques using different simulation tools, it was necessary to have an interface that links the two packages. This was achieved by a special script file written using MATLAB built-in functions. Material properties used in the simulation are shown in table 1.

Material	Thermal conductivity (W/m K)	Specific heat (J/kg K)	Density (kg/m³)	Relative permittivity	Electrical conductivity (mho)
GaAs	44	334	5360	12.8	$5x10^{-4}$
NiCr	22	450	8300	-	$9.1x10^5$
Au	315	130	1928	-	$4.5x10^7$
Al/GaAs	23.7	445	3968	-	$5x10^{-4}$

Table 1. Electrical and thermal properties used in the simulation.

Step 6: Verification of Solution:

Due to the small dimensions of the sensor structure, it was difficult to determine temperature distribution by direct temperature measurements. Therefore, a technique based on thermal imaging was used. The top surface of the test structures were coated with a thin film of liquid crystal (R35CW 0.7 from Hallcrest Inc, UK), that changes color with the changes in the sensor surface temperature. This change in temperature was monitored by a CCD camera connected to a microscope and a personal computer was used to store the recorded shots for later analysis. For dc operation mode, a stable current source was used, and both the input voltage and current were monitored for accurate determination of input power. For RF mode of operation, the current source was replaced by an RF probe that connected directly to the input pads of the CPS (see Fig.3).

In order to show how closely the simulated results resemble the actually expected temperature distribution, we compare simulated current density distribution in Fig.5 with the an experimental shots of the sensor while it was burning shown in Figs 6,7; one with thermocouple (Fig. 6) and the other without thermocouples (Fig. 7). The experimental results were obtained by increasing the input power level at small increments until the resistor was destroyed at an input power level of about 80 mW. The accumulation of current density around the inner corner of the resistor in the simulated result (Fig. 5), explains the destruction of the resistor at one of its sharp inner corners. furthermore experimental results show that the degree of destruction is more severe (as illustrated by the size of the elliptical shape surrounding the resister) when the thermocouples are removed (Fig. 7). This can be attributed to spreading of heat away from the resistive termination by thermocouples when they present, thus decreasing level of destruction.

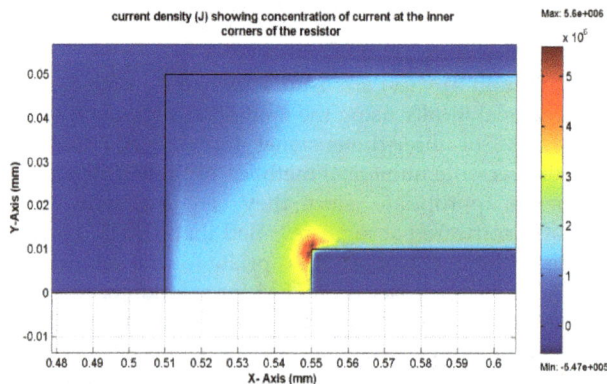

Fig. 5. Current density (J) at the inner corner of the resistive element

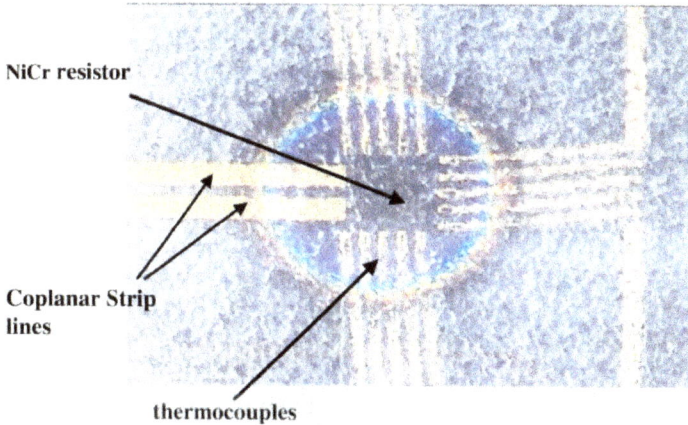

Fig. 6. Measured temperature distribution on the top surface of the sensor structure including thermocouple s

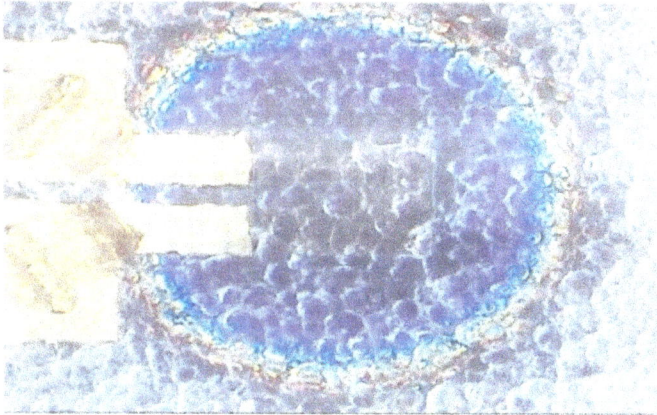

Fig. 7. Measured temperature distribution on the top surface of the sensor structure without thermocouples

Step 7: Using the Model

2-D simulation

Results obtained from 2-D simulation are shown in figure 8 through 11. Fig. 8 shows a 3D color plot of the normalized temperature distribution on the top surface of the sensor. It is shown here how temperature is concentrated in the Ni.Cr. resistor region with sharp decrease with distances from the resistor edges. In order to have a more quantitative picture, an enlarged view of the temperature contour around the resistor region is shown in Fig. 9. It can be seen that temperature decreases to about 67% of the peak value (at the resistor) at a distance of about 20 µm from the side arm of the resistor (y-direction). Further away from the resistor, temperature level reaches about only 33% of its peak value at a distance of about 180 µm. Along the x-axis away from the short arm of the resistor, the drop in temperature level is even more sharp reaching 67% at 10 µm, and 33% at 105 µm.

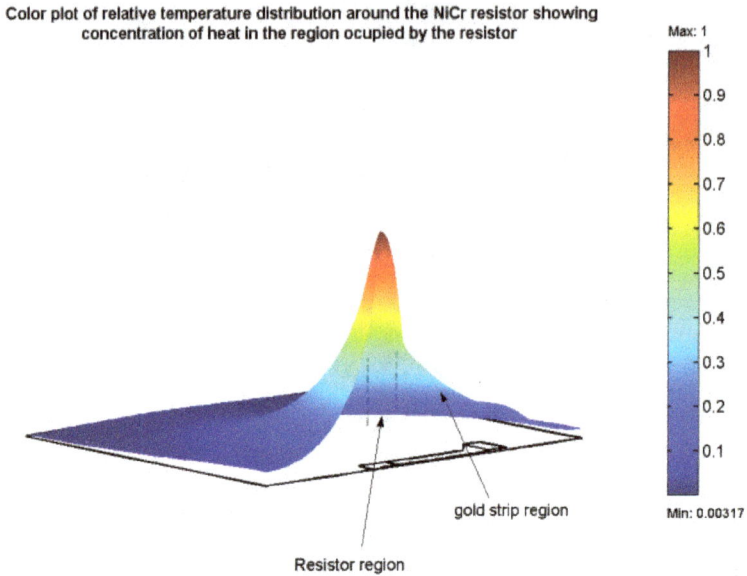

Fig. 8. Color plot of relative temperature distribution around the NiCr. Resistor.

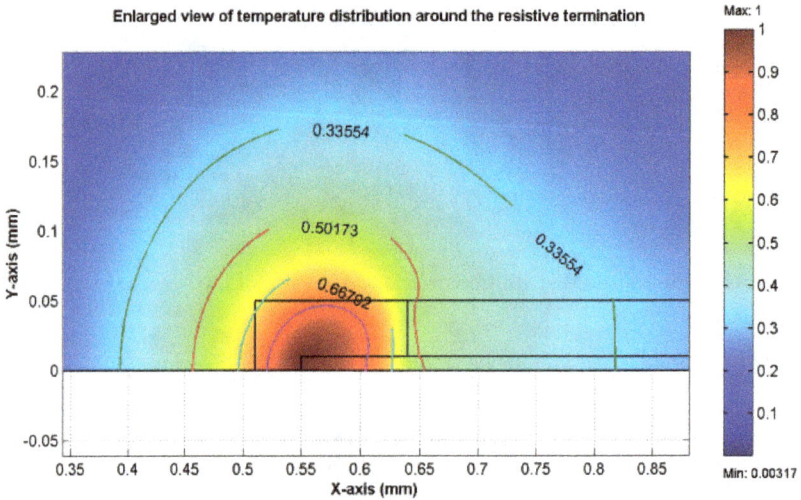

Fig. 9. Enlarged view of temperature distribution around the corner of the resistive termination

3-D simulation

The three-dimensional form of equations (5-8) were solved with the assumption of negligibly small resistor thickness. Fig. 10 shows a color plot of the temperature distribution on the top surface of a thin resistor on bulk substrate 150 μm thick. This figure compares very well with the experimentally obtained result of Fig. 9. 6,7, and clearly illustrates the elliptic form of surface temperature distribution. To see the effect of the third, z-dimension, temperature distribution on a plane cut along the line y-y on Fig. 10 is plotted in Fig.11. This figure illustrates the diffusion of heat through the GaAs substrate with peak values of temperature directly under the two long arms of the resistor. Thus it shows the effect of the bulk substrate that leads to the spreading of temperature away from the resistor region and down into substrate. This reinforces the idea behind using thin membrane technology, in which case the bulk substrate is removed, in the construction of thermoelectric power sensors.

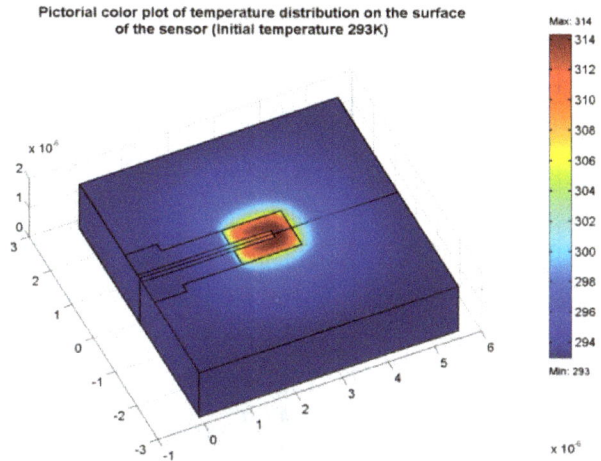

Fig. 10. Pictorial color plots of temperature distribution on the surface of the sensor (initial temperature 293K)

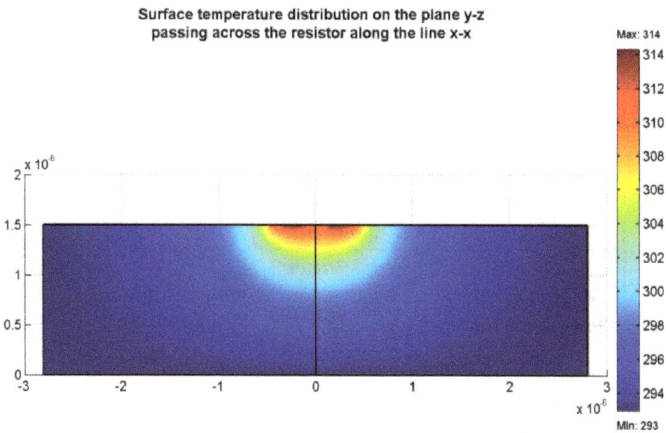

Fig. 11. Simulated temperature distribution on the top surface of the sensor structure along x-x

6.2 Other examples

Table 2a,b show respectively selected double and triple physics illustrative examples showing the type of MEMS, physical phenomena on which their operations are based, type of equations involved in their model and the technique and/or software tools used for their simulation. Information given in these tables are only highlights on the main principles and the interested reader is advised to consult cited references for more details.

Physical Phenomenon	MEMS type	Type of Equations	Simulation Software tool/ technique	Reference
Electromagnetism and Thermal	Thermal convertors for gas sensors	Electric current flow + heat equation	IntelliSuite TM	Ijaz et. al. (2005)
	Microwave power sensors	Maxwell's + heat equation	XFDTD+ FEMLAB (FDTD+ FEM method)	Ali, I.et. al. (2010)
	Fingerprint sensors	Maxwell's + heat equation	$\propto -Flow$	Ji-Song, et. al., (1999)
Electromagnetism and Mechanics	Parallel-plate capacitors	Maxwell's + Transport equation	FDTD method	Bushyager et. al. (2002)
	Stub tuners	Maxwell's + equation of motion	FDTD method	Bushyager et. al. (2001)
	Antennas with moving parts	Maxwell's + equation of motion	FDTD method	Yamagata, Michiko Kuroda & Manos M. Tentzeris (2005)
Magnetostaticts And Mechanics	Magneto- sensitive Elastometers	Stress tensor + magnetic field equation	COMSOL Multiphysics (F.E.M)	Bohdana Marvalova (2008)
	Magnetostrictive thin-film actuators	Stress tensor + magnetic field equation	Shell–Element Method	Heung-Shik Lee et. al. (2008)
Optics And Mechanics	Ring laser and fiber optic gyroscopes			Riccardo & Roberto (2008)

Table 2.a. Double-Physics Problems

Physical Phenomenon	MEMSMEMS type	Type of Equations	Simulation Software tool/ technique	Reference
Electromagnetism Thermal and Mechanical	Electrothermal Actuators (ETA)	Electric current flow (Jule heating)+ heat equation+ Mechanical deflection	COMSOL Multiphysics (F.E.M)	Fengyuan & Jason Clark (2009)
Electrical, Thermal and Mechanical	Self –Buckling of Micromechanical beams under resistive heating	Voltage equation + heat equation + Mechanical deformation	Analytical+ F.E.M	Chiao, Mu & David Lin, (2000)

Table 2.b. Triple-Physics Problems

7. Concluding remarks and future outlook

In this chapter we carefully looked at all MEMS features and nature which make their modeling and simulation a challenging task have been identified and summarized in the following:

1. Multidomain nature of MEMS calls for consideration of many interacting physical phenomena, thus leading to involvement of many types of equations that are coupled weakly or strongly depending on the type of MEMS.
2. Miniaturization: MEMS are by their nature tiny systems, sometimes with very large aspect ratios that make meshing a challenging task and demand considerable computer resources.
3. MEMS are very much affected by environmental conditions and need proper packaging, which in turn, complicates their modeling and simulation.

Different types of simulation techniques, as well as software tools based on these techniques, have been considered with advantages and limitations of each type. A detailed case study that illustrate proper modeling and simulation steps was made and some other successful modeling and simulation examples have been highlighted with proper reverence to their sources for interested reader.

The field of MEMS is very promising and much work is needed in the following areas:

1. The interdisciplinary nature of MEMS and the difficulties that face researchers and designers in this ever expanding field, calls for collaborative group work that comprises scientists and engineers with different background, such as electrical, mechanical, structural (civil) engineers and material scientists together with IT specialists in computer modeling and simulation.
2. Modified Finite difference time domain (FDTD) as well as the multiresolution time domain (MRTD) considered among the simulation techniques in this chapter, are promising due to their simplicity and efficiency compared to more mature finite element technique and more work is needed in development of software tools based on these techniques and specifically targeted to MEMS modeling and simulation.

8. References

Ali, I. Kabula M. & Hartnagel H. L.(2010). Electro-thermal Simulation of Micromachined Power Sensors at DC and Microwave Frequencies. Presented at The Seventh International Symposium on Machatronics and its Applications SIAM'10. American University of Sharjah, Sharjah, UAE, April 20-22-2010.
Available from: http://academic.research.microsoft.com/publication/4577492

Bohdana Marvalova (2008). Modeling of Magnetoresistive Elastometers. Modelling and Simulation, Giuseppe Petrone and Giuliano Cammarata (Ed.), ISBN: 978-3-902613-25-7, InTech, Available from:
http://www.intechopen.com/articles/show/title/multiphysics_modelling_and_si mulation_in_engineering

Bushyager, N. Intracell Modeling of Metal/Dielectric Interfaces for EBG/MEMS RF Structures Using the Multiresolution Time-Domain Method, (2003). Available from: http://academic.research.microsoft.com/publication/4575444.

Bushyager, N. Krista Lange, Manos Tentzeris, John Papapolymerou. (2002). Modeling and Optimization of RF-MEMS Reconfigurable Tuners with Computationally Efficient Time-Domain Techniques. Available from:
http://academic.research.microsoft.com/publication/4571811

Bushyager, N. Manos M.Tentzeris, Larry Gatewood, Jeff DeNatale, (2003). A Novel Adaptive Approach to Modeling MENS Tunable Capacitors Using MRTD and FDTD Techniques. Available from:
http://academic.research.microsoft.com/publication/4569517

Bushyager, N. Mc Garvey, B. & Tentzeris, M.M. (2001). Adaptive Numerical Modeling of RF Structures Requiring the Coupling of Maxwell's, Mechanical and Solid –State Equations. Submitted to The 17th Annual Review of Progress in Applied Computational Electromagnetics Session: "Wavelets in Electromagnetics" – Student Paper Competition.

Bushyager, N., Dalton, E., Papapolymerou, J. Tentzeris, M. (2002).Modeling of Large Scale RF-MEMS Circuits Using Efficient Time-Domairr Techniques. Available from: http://academic.research.microsoft.com/publication/4560023.

Bushyager, N.A. & Tentzeris, M.M. (2001). Modeling and Design of RF MEMS Structures Using Computationally Efficient Numerical Techniques. Electronic Components and Technology Conference.

Fengyuan Li & Jason Vaughn Clark, (2009). An Online Tool for Simulating Electro-Thermo-Mechanical Flexure Using Distributed and Lumped Analyses. Sensors & Transducers Journal, Vol.7, Special Issue, pp, 101-115, October 2009.

Heung-Shik Lee, Sung-Hoon Cho, Jeong-Bong Lee & Chongdu Cho, (2008).Numerical Modelling and Characterization of Micromachined Flexible Magnetostrictive Thin Film Actuator. IEEE Transactions on Magnetics, Vol.44, No. 11, November 2008.

Ijaz H. Jafri, Frank DiMeo Jr, Jeffrey, Neuner, Sue DiMascio, James Marchettl, (2005). Experimental investigation, modeling, and simulations for MEMS based gas sensor used for monitoring process chambers in semiconductor manufacturing.

Available from:
http://academic.research.microsoft.com/publication/11448778
Ji Song Han, Tadashi Kadowaki, Kazuo Sato, & Mitsuhiro Shikida (1999). Thermal Analysis of Fingerprint Sensor Having a Microheater Array. 1999 International Symposium on Micromechatronics and Human Science.
Khine Myint Mon A., Tin Thet New B. , Dr. Zaw Min Naing, C. & Yin Mon Myint,D. (2008). Analysis on Modeling and Simulation of Low Cost MEMS Accelerometer ADXL202., World Academy of Science, Engineering and Technology 42 2008.
Kohei Yamagata, Michiko Kuroda & Manos M. Tentzeris (2005). Numerical Modeling of Antennas with Mechanically/MEMS-Enabled Moving Parts. Available from: http://users.ece.gatech.edu/~etentze/APS05_MEMS_Kuro.
Lynn Khine & Moorthi Palaniapan, (2006). Behavioural modeling and system-level simulation of micromechanical beam resonators. Journal of Physics: Conference Series 34 (2006) 1053-1058, International MEMS Conference 2006.
Mu Chiao & Liwei Lin, Member, IEEE, (2000). Self-Buckling of Micromachined Beams Under Resistive Heating. Journal of Microelectromechanical Systems, Vol. 9, No. 1, March 2000.
Mutamba, K., Beilenhoff, K., Megej, A., Doerner, R., Genc, E.; Fleckenstein, A.; Heymann, P,; Dickmann, J. Woelk, C. (2001). Micromachined 60 GHz GaAs power sensor with integrated receiving antenna. 2001 IEEE MTT-S Digest, pp.2235-2238.
Peyrou David, Coccetti Fabio, Achkar Hikmat, Pennec Fabienne, Pons Patric & Plana Robert (2008). A new methodology for RF MEMS Simulation. Recent Advances in Modeling and Simulation.House.
Riccardo Antonello & Roberto Oboe (2011). MEMS Gyroscopes for Consumers and Industrial Applications, Microsensors, Igor Minin (Ed.), ISBN: 978-953-307-170-1, InTech, Available from: http://www.intechopen.com/articles/show/title/mems-gyroscopes-for-consumers-and-industrial-applications
Shanmugavalli, M. Uma, G. Vasuki, B. & Umapathy, M. (2006). Design and Simulation of MEMS using Interval Analysis. Journal of Physics: Conference Series 34 (2006) pp. 601-605, International MEMS Conference 2006.
Silvester, P. P. & Ferrari, R. L0 (1990). Finite Elements for Electrical Engineers, 2nd Ed. Cambridge University Press, C Cambridge, 1990.
Taflove, A. & Hagness, S., (2001). Computational Electrodynamics, the finite difference time domain approach. 2nd Edn. Boston, Artech.
Tentzeris,M., Bushyager,N. K.Lim, R.Li, Davis, M. Pinel, Laskar, J. Zheng, & Papapolymerou,J., (2002). Analysis of MEMS and Embedded Components in Multilayer Packages using FDTD/MRTD for system-on-Package Application. Available from: http://academic.research.microsoft.com/publication/4577492
Yong Zhu & Horacio Espinosa, D. (2003). Electromechanical Modeling and Simulation of RF MEMS Switches. Proceedings of the 4th International Symposium on MEMS and

Nanotechnology, the 2003 SEM Annual Conference and Exposition on Experimental and Applied Mechanics, June 2-4, Charlotte, North Carolina, Session 03, Paper 190, pp.8-11, 2003.

Permissions

The contributors of this book come from diverse backgrounds, making this book a truly international effort. This book will bring forth new frontiers with its revolutionizing research information and detailed analysis of the nascent developments around the world.

We would like to thank Mojtaba Kahrizi, Professor, for lending his expertise to make the book truly unique. He has played a crucial role in the development of this book. Without his invaluable contribution this book wouldn't have been possible. He has made vital efforts to compile up to date information on the varied aspects of this subject to make this book a valuable addition to the collection of many professionals and students.

This book was conceptualized with the vision of imparting up-to-date information and advanced data in this field. To ensure the same, a matchless editorial board was set up. Every individual on the board went through rigorous rounds of assessment to prove their worth. After which they invested a large part of their time researching and compiling the most relevant data for our readers. Conferences and sessions were held from time to time between the editorial board and the contributing authors to present the data in the most comprehensible form. The editorial team has worked tirelessly to provide valuable and valid information to help people across the globe.

Every chapter published in this book has been scrutinized by our experts. Their significance has been extensively debated. The topics covered herein carry significant findings which will fuel the growth of the discipline. They may even be implemented as practical applications or may be referred to as a beginning point for another development. Chapters in this book were first published by InTech; hereby published with permission under the Creative Commons Attribution License or equivalent.

The editorial board has been involved in producing this book since its inception. They have spent rigorous hours researching and exploring the diverse topics which have resulted in the successful publishing of this book. They have passed on their knowledge of decades through this book. To expedite this challenging task, the publisher supported the team at every step. A small team of assistant editors was also appointed to further simplify the editing procedure and attain best results for the readers.

Our editorial team has been hand-picked from every corner of the world. Their multi-ethnicity adds dynamic inputs to the discussions which result in innovative outcomes. These outcomes are then further discussed with the researchers and contributors who give their valuable feedback and opinion regarding the same. The feedback is then collaborated with the researches and they are edited in a comprehensive manner to aid the understanding of the subject.

Apart from the editorial board, the designing team has also invested a significant amount of their time in understanding the subject and creating the most relevant covers. They scrutinized every image to scout for the most suitable representation of the subject and create an appropriate cover for the book.

The publishing team has been involved in this book since its early stages. They were actively engaged in every process, be it collecting the data, connecting with the contributors or procuring relevant information. The team has been an ardent support to the editorial, designing and production team. Their endless efforts to recruit the best for this project, has resulted in the accomplishment of this book. They are a veteran in the field of academics and their pool of knowledge is as vast as their experience in printing. Their expertise and guidance has proved useful at every step. Their uncompromising quality standards have made this book an exceptional effort. Their encouragement from time to time has been an inspiration for everyone.

The publisher and the editorial board hope that this book will prove to be a valuable piece of knowledge for researchers, students, practitioners and scholars across the globe.

List of Contributors

Gunasekaran Venugopal
Jeju National University, Department of Mechanical Engineering, Jeju, South Korea
Karunya University, Department of Nanosciences and Technology, Tamil Nadu, India

Sang-Jae Kim
Jeju National University, Department of Mechatronics Engineering, Jeju, South Korea
Jeju National University, Department of Mechanical Engineering, Jeju, South Korea

Shrikant Saini
Jeju National University, Department of Mechanical Engineering, Jeju, South Korea

Tai-Chang Chen and Robert Bruce Darling
University of Washington, USA

Fei Xu, Jun-long Kou, Yan-qing Lu and Wei Hu
College of Engineering and Applied Sciences and National Laboratory of Solid State Microstructures,
Nanjing University, Nanjing, P. R. China

Hongmin Lee
Kyonggi University, Korea

Shefiu S. Zakariyah
Advanced Technovation Ltd., Loughborough Innovation Centre, Loughborough, UK

Xianghua Wang
Key Lab of Special Display Technology, Ministry of Education, National Engineering Lab of Special Display Technology, National Key Lab of Advanced Display Technology, Academy of Photoelectric Technology,
Hefei University of Technology, Hefei, China

Giuseppe Yickhong Mak and Hoi Wai Choi
Department of Electrical and Electronic Engineering, The University of Hong Kong, Hong Kong

T. Gietzelt and L. Eichhorn
Karlsruhe Institute of Technology, Campus Nord, Institute for Micro Process Engineering, Karlsruhe, Germany

Mohammad Yeakub Ali, Reyad Mehfuz and Ahsan Ali Khan and Ahmad Faris Ismail
International Islamic University, Malaysia

Salvador Mendoza-Acevedo, Mario Alfredo Reyes-Barranca, Edgar Norman Vázquez-Acosta and José Antonio Moreno-Cadenas
CINVESTAV-IPN, Electrical Engineering Department, México

José Luis González-Vidal
UAEH, Computing Academic Area, México

H.-T. Liu and E. Schubert
OMAX Corporation, USA

P.J. French and P.M. Sarro
Delft University of Technology, The Netherlands

Anjali Vishwas Kulkarni
Centre for Mechatronics, Indian Institute of Technology Kanpur, India

Idris Ahmed Ali
Dean, College of Engineering & Applied Sciences Al Ghurair University, Academic City, Dubai, UAE